AF356533

Remote Sensing and Geospatial Approaches for Studying the Environment Affected by Human Activities

Remote Sensing and Geospatial Approaches for Studying the Environment Affected by Human Activities

Editors

Jun Li
Xinyi Shen
Qiusheng Wu
Chengye Zhang

Basel • Beijing • Wuhan • Barcelona • Belgrade • Novi Sad • Cluj • Manchester

Editors

Jun Li
China University of Mining &
Technology
Beijing
China

Xinyi Shen
University of Connecticut
Storrs
USA

Qiusheng Wu
University of Tennessee
Knoxville
USA

Chengye Zhang
China University of Mining &
Technology
Beijing
China

Editorial Office
MDPI
St. Alban-Anlage 66
4052 Basel, Switzerland

This is a reprint of articles from the Special Issue published online in the open access journal *Remote Sensing* (ISSN 2072-4292) (available at: https://www.mdpi.com/journal/remotesensing/special_issues/RS_Human_Environment).

For citation purposes, cite each article independently as indicated on the article page online and as indicated below:

Lastname, A.A.; Lastname, B.B. Article Title. *Journal Name* **Year**, *Volume Number*, Page Range.

ISBN 978-3-0365-9328-9 (Hbk)
ISBN 978-3-0365-9329-6 (PDF)
doi.org/10.3390/books978-3-0365-9329-6

Contents

About the Editors

Jun Li

Jun Li, Professor, Ph.D., Vice Dean of Earth Science and Surveying and Mapping Engineering, China University of Mining and Technology, Beijing. His research interests include remote sensing of ecology and environment; geospatial analysis; ecology and environment in mining areas; machine learning; and spatiotemporal data mining. He has hosted more than 20 projects commissioned by the National Natural Science Foundation of China, the National Key R&D Programmed, the National Energy Group, etc. He has published more than 60 peer-reviewed papers.

Xinyi Shen

Xinyi Shen, Assistant Professor in the School of Freshwater Sciences at the University of Wisconsin-Milwaukee. His research interests include interactions of humans, disasters, and the built environment under climate change; flood-inundation modeling and observatory using hydrological modeling and remote sensing; AI applications in natural hazards and climate change; AI applications in desert biodiversity under climate change; and remote sensing, photogrammetry and GIS. He has led and participated in more than ten research projects with NASA, NOAA, and international cooperation projects.

Qiusheng Wu

Qiusheng Wu, Associate Professor in the Department of Geography & Sustainability at the University of Tennessee, Knoxville. His research interests include Geographic Information Science (GIS); remote sensing; and environmental modeling. More specifically, He is interested in applying geospatial big data, machine learning, and cloud computing (e.g., Google Earth Engine) to study environmental change, especially surface water and wetland inundation dynamics. He has published a total of 72 papers. He has reviewed 170+ times for 40+ journals since 2015.

Chengye Zhang

Chengye Zhang, Associate Professor in Earth Science and Surveying and Mapping Engineering at the China University of Mining and Technology, Beijing. His research interests include quantitative remote sensing; multi-spectral and hyper-spectral remote sensing; remote sensing of vegetation; machine learning; and radiative transfer models. He has presided over and participated in more than ten scientific research projects. He has published more than 30 peer-reviewed papers.

Preface

We are delighted to introduce this Reprint, showcasing a compilation of insightful scientific contributions spanning diverse fields. This collection centers on the significant realm of environments influenced by human activities, highlighting the essential role of remote sensing and geospatial approaches. Our goal with this compilation is to offer a comprehensive overview of state-of-the-art research and advancements in the covered subjects.

As Guest Editors, we are curating this Reprint with the aim of promoting knowledge sharing and fostering collaboration among experts and enthusiasts in these fields. We are confident that this collection will be a useful resource for our audience.

This compilation from the perspectives of new theories, datasets, methods, findings, and applications delves into essential questions: How can we monitor the environmental alteration by human activities at different spatiotemporal scales? How can we quantify, evaluate, and predict the impact of human activities on the environment? How can we decompose or evaluate the compound influence of various natural disturbances and human alteration on the environment? What is the corresponding pattern or mechanism of the environmental evolution to different human activities?

Through this Special Issue, we aim to foster knowledge exchange, encourage innovative research from diverse perspectives, and make scientific contributions to the environmental protection and governance affected by a variety of human activities.

We would like to express our sincere gratitude to all the authors who submitted their high-quality manuscripts to this Special Issue. We also thank all the reviewers who provided constructive comments and suggestions to improve the manuscripts. We appreciate the support and guidance from the editorial team of Remote Sensing.

We hope this compilation serves as a source of inspiration and prompts further exploration in the realms of "Remote Sensing and Geospatial Approaches for Studying the Environment Affected by Human Activities", making a significant impact on scientific advancement and environmental stewardship.

Jun Li, Xinyi Shen, Qiusheng Wu, and Chengye Zhang

Editors

Article

Monitoring the Characteristics of Ecological Cumulative Effect Due to Mining Disturbance Utilizing Remote Sensing

Quansheng Li, Junting Guo *, Fei Wang and Ziheng Song

State Key Laboratory of Water Resource Protection and Utilization in Coal Mining, Beijing 102209, China; quansheng.li@chnenergy.com.cn (Q.L.); 20047006@chnenergy.com.cn (F.W.); ziheng.song@chnenergy.com.cn (Z.S.)
* Correspondence: junting.guo.a@chnenergy.com.cn

Abstract: This study conducted land cover classification and inversion analysis to estimate land surface temperature, soil moisture, specific humidity, atmospheric water vapor density, and relative humidity using remote sensing and multi-source mining data. Using 1990–2020 data from the Shendong mining area in Inner Mongolia, China, the eco-environmental evolution and the ecological cumulative effects (ECE) of mining operations were characterized and analyzed at a long-term scale. The results show that while the eco-environment was generally stable, mining activities affected the eco-environment at the initial stage (1990–2000) to a certain degree. During the rapid development stage of coal mining, the eco-environment was severely damaged, and the ECE were significant at the temporal scale. The absolute value of the change rate of ecological parameters was increasing. Due to an increased focus on ecological restoration, starting in 2010, the environmental indicators gradually stabilized and the eco-environment improved considerably, ushering in a period of stability for coal mining activities. The absolute value of the change rate of ecological parameters became stable. Analysis of the change in eco-environmental indicators with distance and comparison to the contrast area showed the ECE characteristics from mining disturbance at the spatial scale. This study shows that remote sensing technology can be used to characterize the ECE from mining operations and analyze eco-environmental indicators, providing crucial information in support of ecological protection and restoration, particularly in coal mining areas.

Keywords: ecological cumulative effect; coal mining area; quantitative remote sensing; ecology and environment; land surface temperature; soil moisture

Citation: Li, Q.; Guo, J.; Wang, F.; Song, Z. Monitoring the Characteristics of Ecological Cumulative Effect Due to Mining Disturbance Utilizing Remote Sensing. *Remote Sens.* **2021**, *13*, 5034. https://doi.org/10.3390/rs13245034

Academic Editor: Parth Sarathi Roy

Received: 15 November 2021
Accepted: 6 December 2021
Published: 10 December 2021

Publisher's Note: MDPI stays neutral with regard to jurisdictional claims in published maps and institutional affiliations.

1. Introduction

Ecological cumulative effect (ECE) refers to the cumulative impact on the eco-environment caused by activities from the past to the future. This metric is particularly useful for understanding that, while the individual effects of some operations may be small or subtle, the combined effects could be significant [1]. Characterized by long-term lagging, accumulation, and interaction [2], the ECE consists of temporal and spatial aspects. The temporal cumulative effect refers to the accumulation generated when the time interval between two disturbances is less than that required by environmental restoration. The spatial cumulative effect refers to the accumulative phenomenon generated when the distance between two disturbances is less than that required by the attenuation [3].

Understanding the cumulative effects on the environment of policies, industries, and other economic activities is extremely important when evaluating choices and addressing the long-term effects caused by various anthropogenic endeavors. One such industry where the holistic comprehension of environmental effects is crucial is coal mining. Although coal mining has brought huge economic benefits, it also has resulted in the continuous decline in the quality of soil, water, vegetation, and atmosphere in the mining area, and has caused long-term adverse effects on the ecological environment [4–6]. Given its small object and long-duration features, the eco-environmental change caused by coal mining activities shows typical cumulative effect characteristics, which could be reflected by the superposition of the damage degree at the temporal scale and the attenuation with distance at the spatial scale. The cumulative effect from coal mining activities has accelerated the degradation of eco-environmental quality. Therefore, analyzing the cumulative effect of coal mining would be extremely useful for ecological restoration and sustainable development.

The impact of mining and restoration activities in coal mining areas is a continuous and intense process. Environmental effects vary considerably at the different stages of mining (i.e., initial stage, intense phase, stable period, reduction phase, and closure stage), as well as before, during, and after ecological restoration. Generally speaking, there is a coupling relationship between mining and restoration activities and various eco-environmental factors. This means that monitoring and evaluating the eco-environment in coal mining areas require specialized observational techniques involving large-scale, high-frequency, continuous, long-term, and comprehensive observations and quantitative analyses. However, traditional approaches, such as field surveys, ground monitoring stations, and sensor networks are unable to meet these requirements.

In recent years, remote sensing has been used to monitor and evaluate the eco-environment in mining areas, given its ability to provide rapid synchronous observations in large areas [7–11]. The combination of remotely sensed images and ground data can generate land cover information and physical-chemical parameters of various ecological elements, which are essential in estimating the ECE of mining activities [12]. Numerous studies have employed and developed remote sensing techniques to measure ecological parameters in mining areas. For example, in terms of land cover, Balaniuk et al. [13] used Sentinel-2 satellite images to conduct nationwide automatic identification and classification of open-pit mines and tailings dams in Brazil, with an accuracy of more than 95%. A systematic image analysis method based on geographical objects (GEOBIA) was proposed by Nascimento et al. [14], and land-use changes of open-pit mines were quantified using high-resolution satellite images from different sensors, with an accuracy of more than 90%. In terms of ecological parameters, Wu et al. [15] adopted the BFAST algorithm to detect abrupt changes in vegetation cover from open-pit mining using Landsat time series. This algorithm can automatically detect the time of initial mining and evaluate the spatial distribution of vegetation destroyed by mining. Fu et al. [16] monitored NDVI (Normalized Difference Vegetation Index) in Xilinhot using the rate of change in greenness (RCG) and coefficient of variation (CV) as indicators. They used correlation analysis and stepwise regression to analyze the driving factors of vegetation and the impact of mining activities. Using SPOT 5/6 and WorldView-2 data, Liu et al. [17] analyzed soil moisture (SM) in the Dariuta Coal Mine through the Scaled SM Monitoring Index (S-SMMI). Cao et al. [18] analyzed land use, vegetation cover, and land surface temperature (LST) of the Jixi mining area in Heilongjiang with the Landsat 8 Operational Land Imager (OLI) data and used correlation analysis to explore the influence of land-use types, vegetation cover and coal mining activities on LST. García Millán et al. [19] detected water distribution change through multi-temporal analysis of multispectral data and extracted mine-related flood zones using high-resolution hyperspectral data. Using Hyperion hyperspectral data and Landsat data, Kayet et al. [20] conducted indirect monitoring of dust in opencast iron ore mines using eight vegetation indices. When analyzed and compared with field measurements, the remote sensing estimates were found to be reliable. In terms of cumulative effect [21–24], a pixel-based Ecosystem Service Value (ESV) time series model was proposed that quantified the ECE of the Yanzhou Coalfield over the past 30 years. However, to the best of our knowledge, there has been no study estimating the ECE of coal mining operations and analyzing its temporal and spatial dimensions using remote sensing.

Using long-term remote sensing and multi-source spatial data from 1990 to 2019, this study performed land cover classification and change detection and quantitatively estimated land surface (LST and SM) and atmospheric parameters (specific humidity, atmospheric water vapor density, and relative humidity). The characteristics of ECE due to coal mining disturbance were also analyzed temporally and spatially.

2. Study Area and Datasets

2.1. Study Area

The study area is located in the Ejin Horo Banner, Inner Mongolia Autonomous Region, and the Shenmu, Shaanxi Province, China (Figure 1).

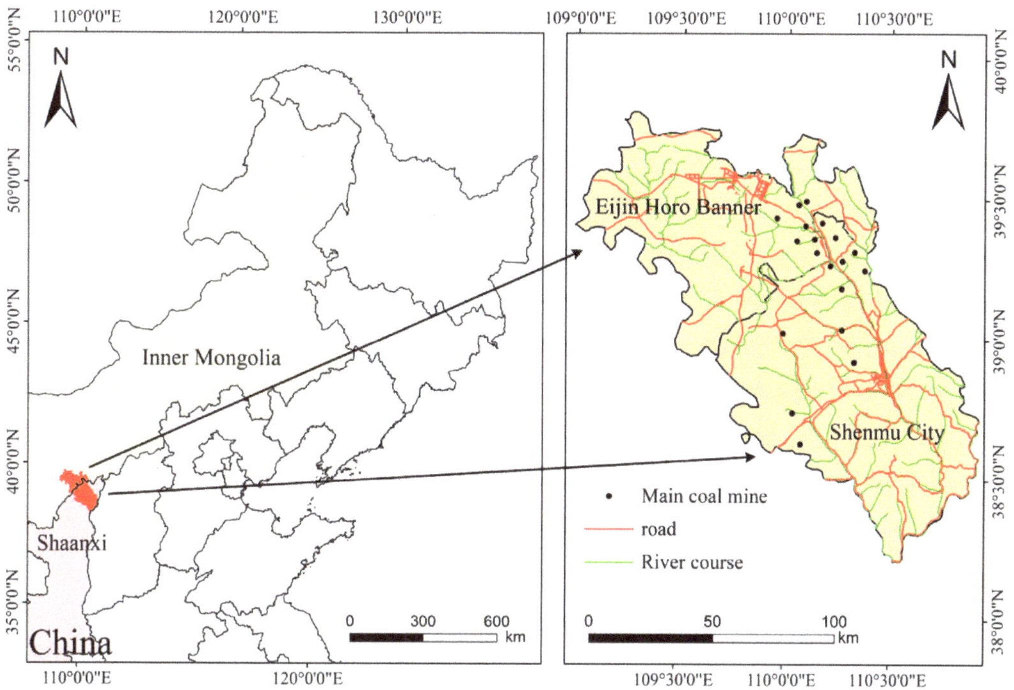

Figure 1. Map of the study area.

Ejin Horo Banner is located in the southeastern part of Ordos, Inner Mongolia, China. The southern part comprises the northern edge of the Mu Us Desert. The area has an average annual temperature of ~6.9 °C, and the terrain is high in the northwest and low in the southeast. Having a typical temperate continental climate, the region is dry, experiences little rainfall, and has annual precipitation less than evaporation. The vegetation is mostly herbaceous and small shrubs. Ejin Horo Banner is rich in mineral resources, the third-largest coal-producing county in China, and is considered a critical national energy strategic base [25], with coal reserves of 56 billion tons and an annual output of about 200 million tons. Shenmu, found in Yulin, Shaanxi, is a transitional zone from grassland, forest, and hill to steppe and desert. The area has a typical temperate continental climate and has an annual average temperature of ~8.9 °C. Shenmu is located in the southeastern part of the Mu Us Desert and has proven coal reserves of over 50 billion tons, with stable coal occurrence and excellent mining conditions [26,27].

The Shendong mining area across the Ejin Horo Banner and Shenmu is one of the most famous coal fields and mining bases with the largest proven reserves in China. According to the statistical data of raw coal output in Ordos and Yulin, from 1990 to 2019 [28], the study area underwent three stages of development based on the intensity of coal mining activities: initial stage (1990–2000), rapid development stage (2001–2010) and stable operation stage (2011–2019). Ecological restoration efforts in the mining site have increased considerably since 2005.

2.2. Datasets

The Landsat data used in this paper have been processed by cloud removing, radiometric calibration, atmospheric correction, mosaicking and clipping. The time range from 1 July to 1 September during 1990–2020 was selected as the time range of the image. Images with less than 5% cloud cover were utilized. All of the steps were completed on Google Earth Engine's remote sensing cloud computing platform. In this study, Landsat images were used for retrieving the ecological parameters (Figure 2).

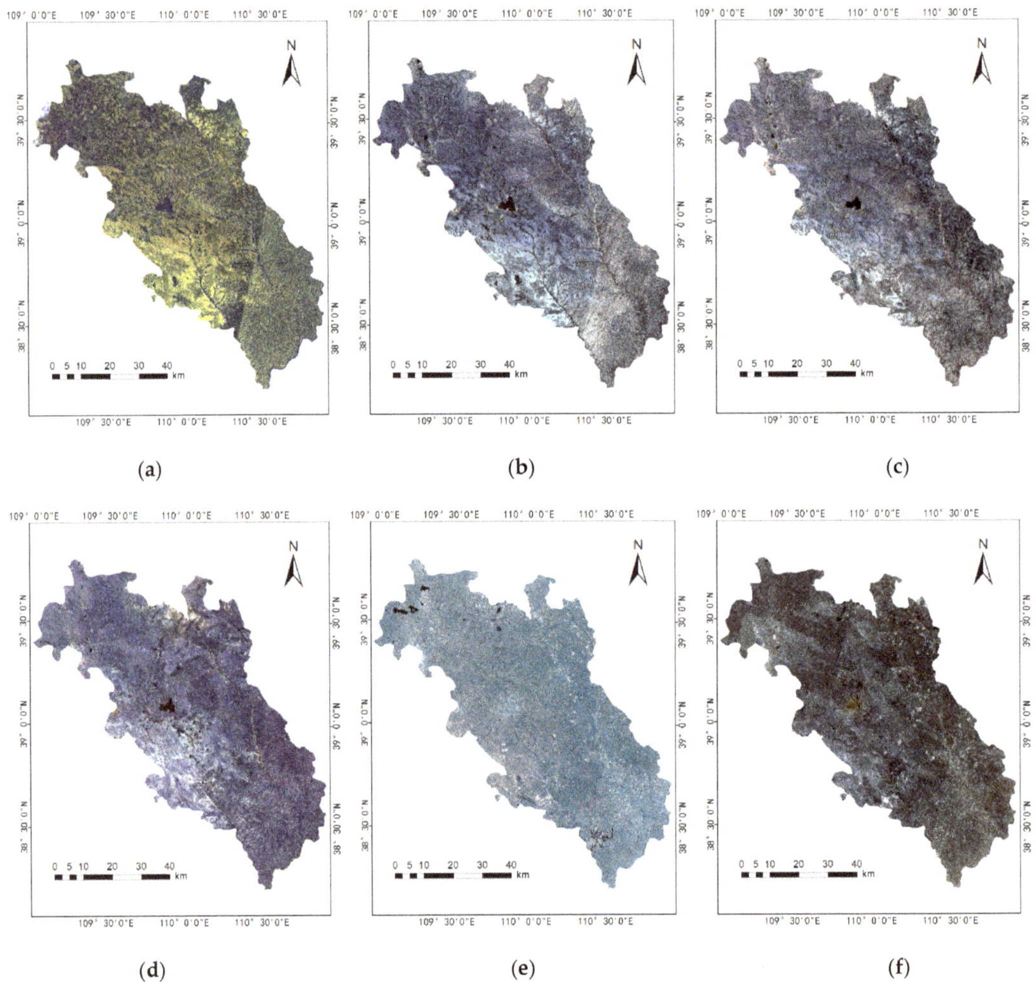

Figure 2. Landsat satellite images in the study area: (**a**) 1990, (**b**) 2000, (**c**)2005, (**d**) 2010, (**e**) 2015, (**f**) 2019.

In addition, ASTER-GDEM (Advanced Spaceborne Thermal Emission and Reflection Radiometer–Global Digital Elevation Model) images were used in land cover classification. FLDAS (Famine Early Warning Systems Network Land Data Assimilation System), Chinese farmland soil moisture ten-day dataset, and other soil moisture products were used for estimating soil moisture, while the Atmospheric reanalysis ERA5 dataset was used for determining the atmospheric parameters.

3. Methodology

The main methods used in the study are as follows: multi-source remote sensing data processing, remote sensing classification of land cover in mining areas, inversion of land surface ecological parameters, reanalysis of atmospheric parameters, and analysis of impacts on the ecological environment. The corresponding flowchart of the technology framework is presented in Figure 3.

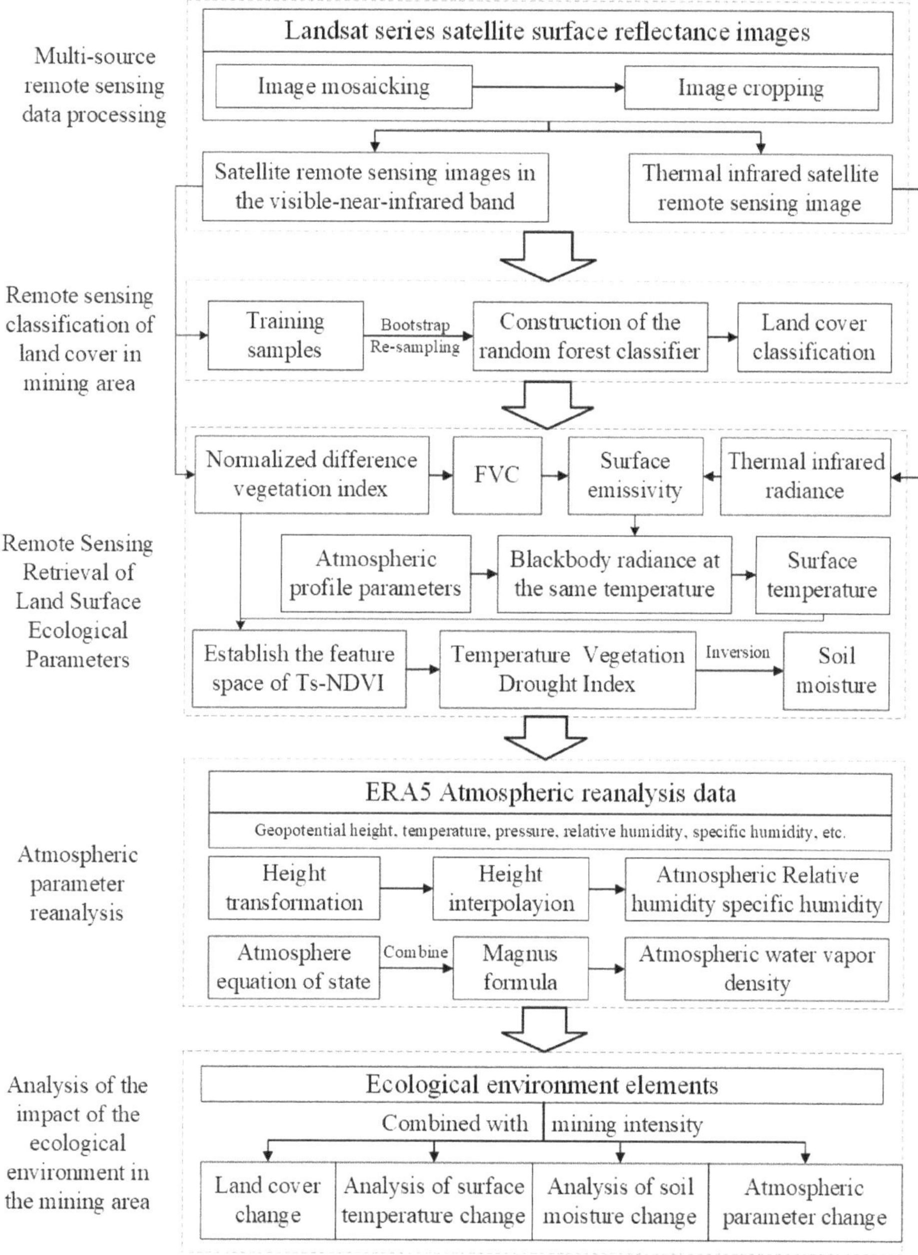

Figure 3. Flowchart of the technology frame.

3.1. Land Cover Classification

The study area was classified into six land cover types: vegetation, water area, cultivated land, bare land, mining land, and artificial land (i.e., urban and construction lands). Using high spatial resolution historical images, samples for each land cover type were selected using manual interpretation and were then used for model training and result verification. The Random Forest (RF) [29] classification algorithm, an integrated learning algorithm based on multiple decision trees, was adopted to generate land cover classification for the mining areas using remote sensing images. Figure 4 presents the schematic diagram of the RF algorithm.

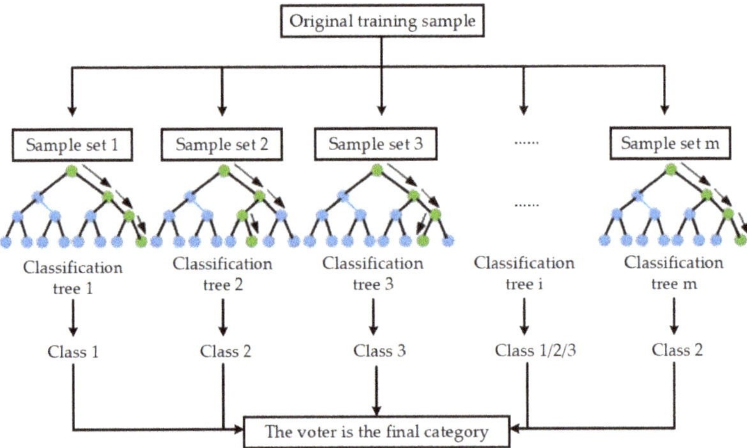

Figure 4. Schematic diagram of the random forest algorithm.

3.2. Inversion of Land Surface Ecological Parameters

3.2.1. Inversion of LST

The method proposed by Xu [30] was used in the inversion analysis of the land surface temperature (LST). After acquiring the vegetation coverage using the pixel dichotomy method [31–33] and empirically determining the surface emissivity of different ground objects [34], LST can be obtained using the formula:

$$T_S = T/[1 + (\lambda T/\rho)\ln\varepsilon]$$

(1)

where T is the temperature value at the sensor, which can be obtained by radiation calibration; λ is the center wavelength of the thermal infrared band of the sensor; $\rho = 1.438 \times 10^{-2}$ mK; and ε is the emissivity of the surface.

3.2.2. Inversion of SM

The Temperature Vegetation Dryness Index (TVDI) [35] was used for the inversion of soil volumetric moisture (unit: m^3/m^3). The Normalized Difference Vegetation Index (NDVI) is calculated by the formula:

$$NDVI = \frac{NIR - R}{NIR + R}$$

(2)

where NIR represents the reflectivity in the near-infrared waveband, and R represents the reflectivity in the red-light waveband.

The TVDI is determined by constructing the feature space of LST (Ts) and NDVI (Figure 5) and calculated using the following equations:

$$\text{TVDI} = \frac{T_S - T_{S_{min}}}{T_{S_{max}} - T_{S_{min}}} \tag{3}$$

$$T_{S_{min}} = a_1 + b_1 * \text{NDVI} \tag{4}$$

$$T_{S_{max}} = a_2 + b_2 * \text{NDVI} \tag{5}$$

where a_1, b_1, a_2, and b_2 are the undetermined coefficients of the dry-wet edge, obtained by fitting the dry-wet edge equations using the least square method [36]. Equation (6) is obtained by substituting Equations (4) and (5) into Equation (3).

$$\text{TVDI} = \frac{T_S - (a_1 + b_1 * \text{NDVI})}{(a_2 + b_2 * \text{NDVI}) - (a_1 + b_1 * \text{NDVI})} \tag{6}$$

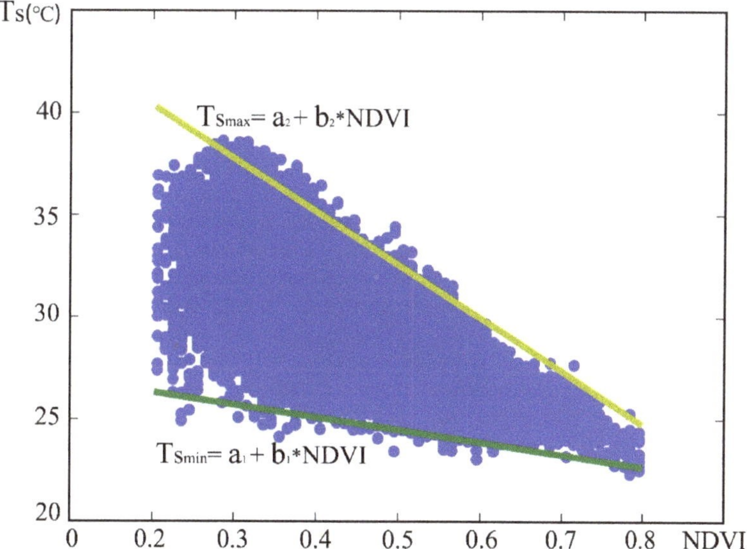

Figure 5. Schematic diagram of TVDI.

After scale conversion, the FLDAS dataset and the Chinese farmland soil moisture ten-day dataset were used to construct the remote sensing regression model of SM with the TVDI.

3.3. Atmospheric Parameter Reanalysis

The ERA5 reanalysis data provide temperature, pressure, relative humidity, and specific humidity values using 37 pressure levels. Note that the geopotential height should be converted into geometric height using the formula [37]):

$$h = \frac{R_e(\varphi) \cdot H}{g(\varphi)/g_0 \cdot R_e(\varphi) - H} \tag{7}$$

where h is the ellipsoidal height (km), H is the geopotential height (km), φ is the latitude, $g_0 = 9.80665 \text{ ms}^{-2}$, $R_e(\varphi)$ is the radius of curvature of the Earth at latitude φ, and $g(\varphi)$ represents the gravity on the geoid.

The converted meteorological parameters were then interpolated to obtain the atmospheric relative humidity and the atmospheric specific humidity at the Earth's surface. The atmospheric water vapor density ρ_v can be obtained using the atmospheric equation of state:

$$\rho_v = \frac{e}{R_v T} \tag{8}$$

where T is the temperature in Kelvin; R_v is the gas constant of water vapor equal to 461.5 J/kg; and e is the water vapor pressure, calculated from the relative humidity and temperature provided by ERA5 using the Magnus formula [38].

The water vapor humidity factors (i.e., density, relative humidity, and specific humidity) at the Earth's surface can then be obtained. Time series analysis for these factors was conducted for the different production periods, and the change rates in the different regions were fitted by the least square method.

4. Results and Discussion

4.1. Analysis of the Changes of Eco-Environmental Elements

4.1.1. Analysis of Land Cover Change

The land cover classification results using remote sensing are shown in Figure 6. To further explore the evolution of the land cover changes in the past 30 years, changes in each land cover type were calculated for the different time periods. The results are presented in Figure 7, and the land cover transfer matrix is presented in Table 1. The results show that the vegetation area remained stable in the first decade, showed rapid growth after 2000, and experienced steady growth since 2015. There were also three stages of change occurring on the water cover type, i.e., a sharp decline stage (1990–2005), a gentle decline stage starting (2005–2015), and a gradual growth stage (2015–2019). Mining land was almost zero in 1990 and increased slowly from 1990 to 2000. Rapid growth in mining lands occurred since 2000, gradually decelerating in recent years. Cultivated land exhibited a steady growth trend, while bare land declined continuously over 30 years. Growth in artificial land was observed from 1990 to 2019, increasing relatively more from 2005 to 2019. More lands were converted into vegetation land compared to vegetation lands converted into other cover types, increasing the extent of vegetation land in the study area. Many vegetation and bare lands were also transformed into mining and artificial lands over the years, resulting in a considerable increase in mining and artificial land extents.

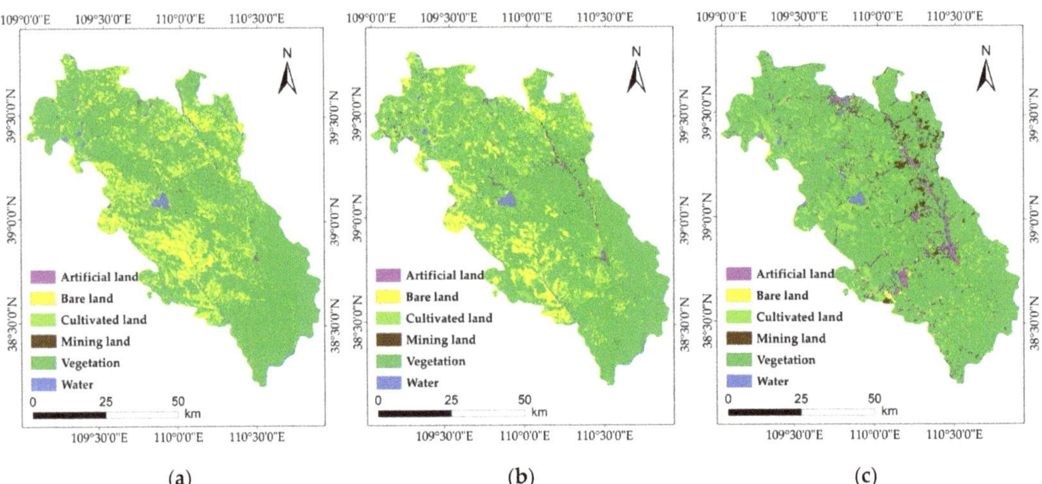

Figure 6. The results of the land cover classification in the study area: (**a**) 1990, (**b**) 2005, (**c**) 2019.

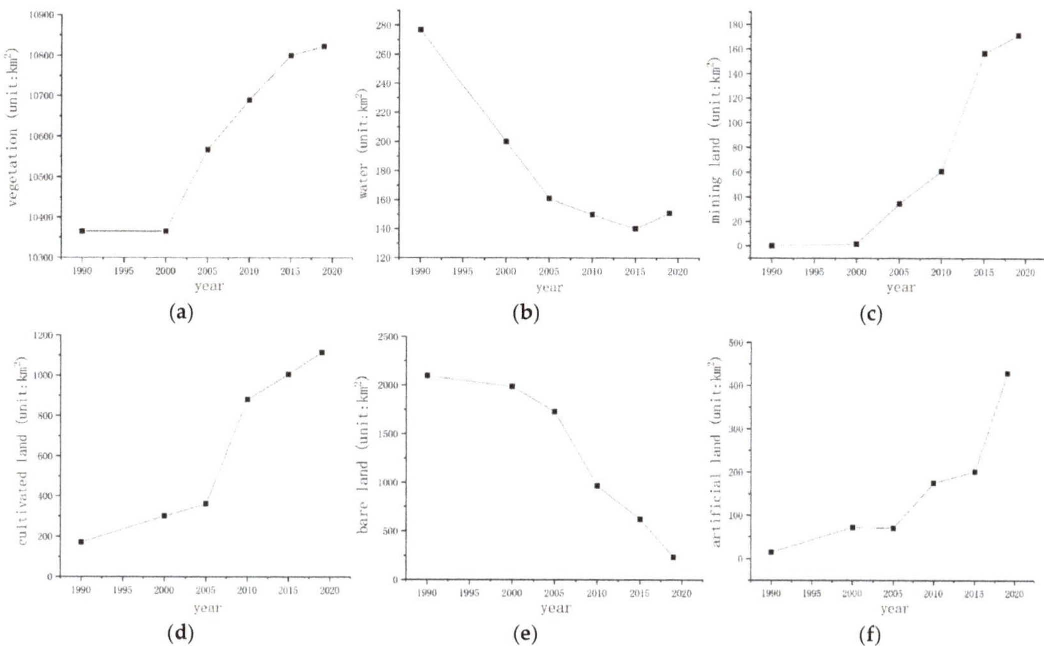

Figure 7. Changes of all kinds of land cover in the study area: (**a**) Vegetation, (**b**) Water, (**c**) Mining land, (**d**) Cultivated land, (**e**) Bare land, (**f**) Artificial land.

Table 1. Land cover transfer matrix (1990–2019) (km^2).

	Vegetation	Water	Mining Land	Cultivated Land	Bare Land	Artificial Land	Total Transfer out
vegetation	8811.29	42.52	130.82	886.88	178.79	311.28	1550.28
water	118.54	91.36	5.50	26.37	13.53	19.85	183.79
mining land	0.00	0.00	0.00	0.00	0.00	0.00	0.00
cultivated land	70.84	2.21	2.22	77.87	1.59	16.24	93.09
bare land	1811.39	14.56	32.32	122.15	39.61	74.57	2054.99
artificial land	5.45	0.15	0.23	2.33	0.12	6.54	8.28
total transfer in	2006.21	59.44	171.09	1037.74	194.03	421.93	

4.1.2. Analysis of the LST Inversion Results

The results of the LST inversion using remote sensing images (mean values from June to August in summer) are shown in Figure 8, while the mean LST values for summer at different years are presented in Figure 9. The results show the mean LST value increased annually from 1990, reaching its peak in 2005 with the value at over 40 °C. The LST decreased rapidly from 2005 to 2010, slightly increased in 2010–2015, and stabilized from 2015 to 2019. In particular, as shown in Figure 8, the LST has a more pronounced increase at the northeastern section of the central image from 1990 to 2005. This is because many coal mining areas are located in this region and have increased significantly over the years.

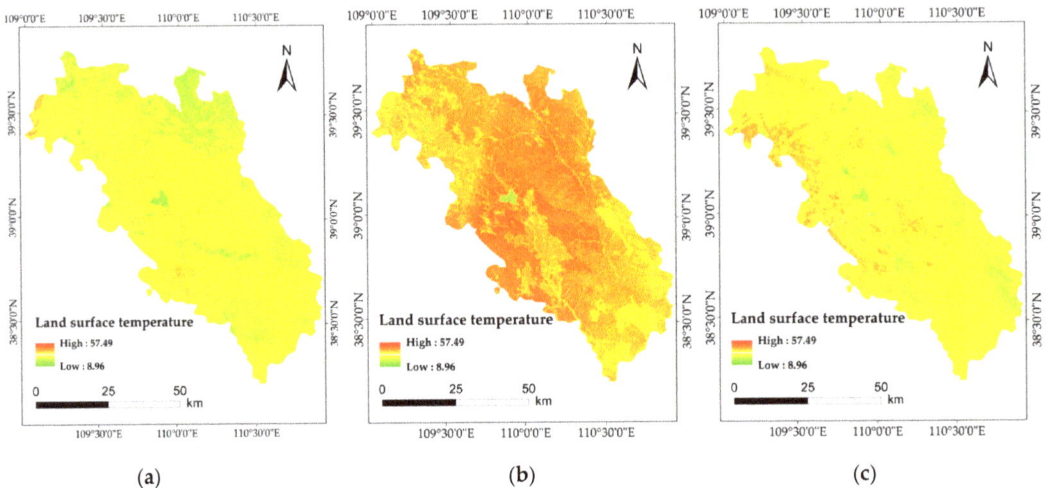

Figure 8. Inversion results of LST in the study area (°C): (**a**) 1990, (**b**) 2005, (**c**) 2019.

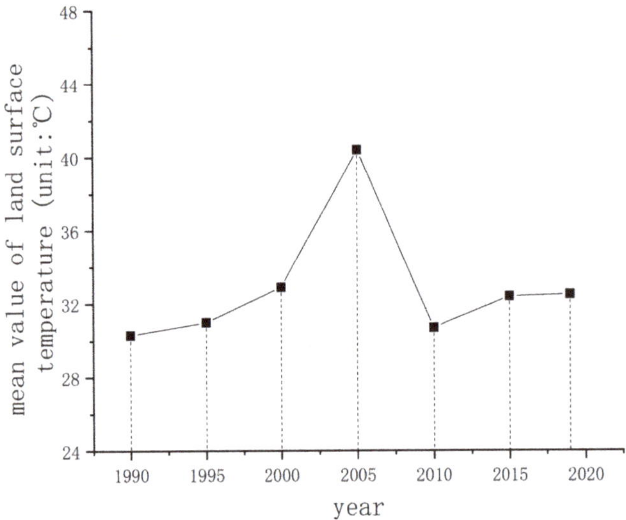

Figure 9. The mean value of LST in the study area.

The LST change rates varied considerably during the 30-year study period. The increase in LST was accelerating from 1990 to 2005, which was the initial and rapid development stage of coal mining activities. The increased influence of mining activities accelerated the upsurge in LST, showing a significant ECE at the temporal scale. This was because more attention was given to ecological restoration and green mine construction, and the impacts caused by mining activities were artificially improved. When coal mining activities became stable after 2010, the LST change rate decelerated, and the impact of mining activities gradually stabilized.

4.1.3. Analysis of the SM Inversion Results

The inversion results using remote sensing images (mean values from June to August in summer) are shown in Figure 10, while the mean SM values for summer (June–August) at different years are presented in Figure 11. As shown in the figures, SM experienced two different trends, i.e., a decreasing trend in 1990–2005 and the increasing trend in 2005–2019. The highest and lowest mean SM values were observed in 2005 and 2019, respectively.

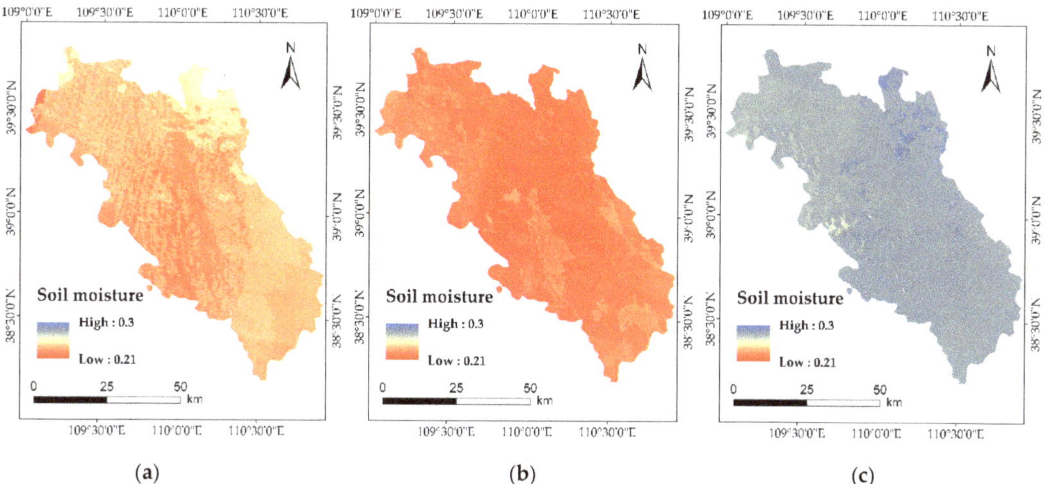

Figure 10. Inversion results of SM in the study area (m^3/m^3): (**a**) 1990, (**b**) 2005, (**c**) 2019.

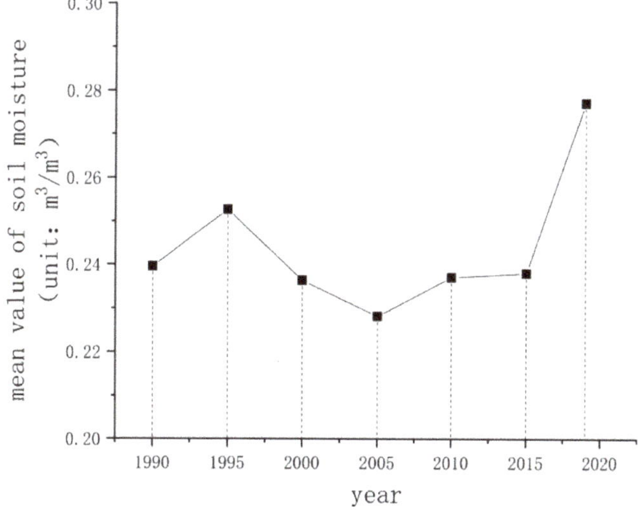

Figure 11. The mean value of SM in the study area.

The rate of increase in mean SM in 1990–1995 was comparable to its rate of decrease in 1995–2000. Note that the rate of decline in SM decelerated during the rapid development of coal mining activities (from 2000 to 2005). The reason may be that while there may have been increased mining activities during this period, people have started to pay greater attention to ecological restoration, environmental protection, and sustainability. Since 2005, the SM has gradually increased, particularly in 2015–2019, exhibiting apparent ecological cumulative effects related to progress in sustainable mining practices and focus on environmental protection.

Remote Sens. **2021**, _13_, 5034

4.1.4. Analysis of Atmospheric Parameters

ERA5 reanalysis data were used to obtain mean values for relative humidity, specific humidity, and atmospheric water vapor density (June to August). Figures 12–14 show the spatial distribution of these parameters for 1990, 2005, and 2019. Specific humidity and atmospheric water vapor density exhibited pronounced spatial variations in each interval. In terms of spatial distribution, the parameter values were higher in the southern region than in the north. The three atmospheric parameters were also found to have similar trends in the time scale, showing a downward trend in 1990–2005 and a rising trend from 2005 to 2019. The three parameters reached their peak in 1990 and their minimum in 2005.

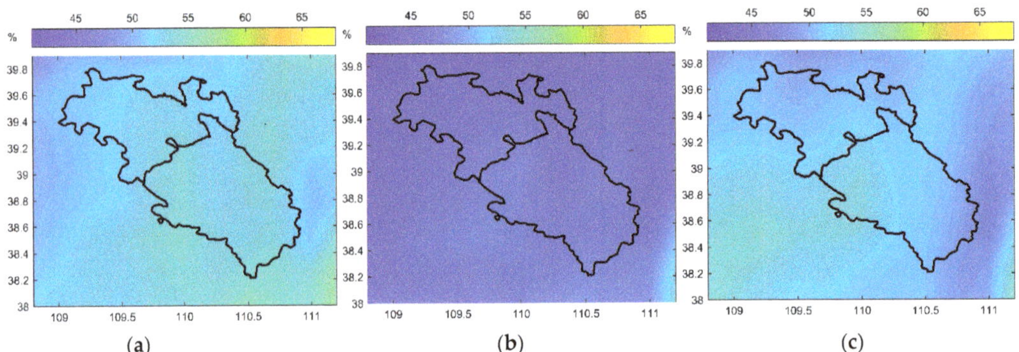

Figure 12. Inversion results of relative humidity in the study area (%): (**a**) 1990, (**b**) 2005, (**c**) 2019.

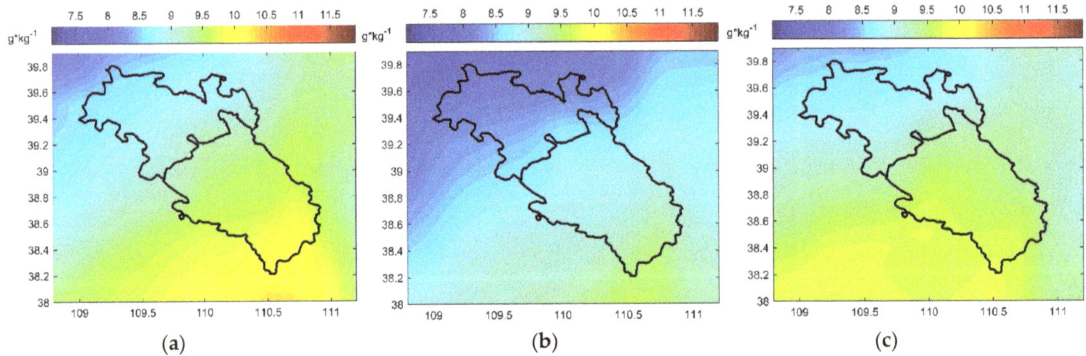

Figure 13. Inversion results of atmospheric specific humidity in the study area (g/kg): (**a**) 1990, (**b**) 2005, (**c**) 2019.

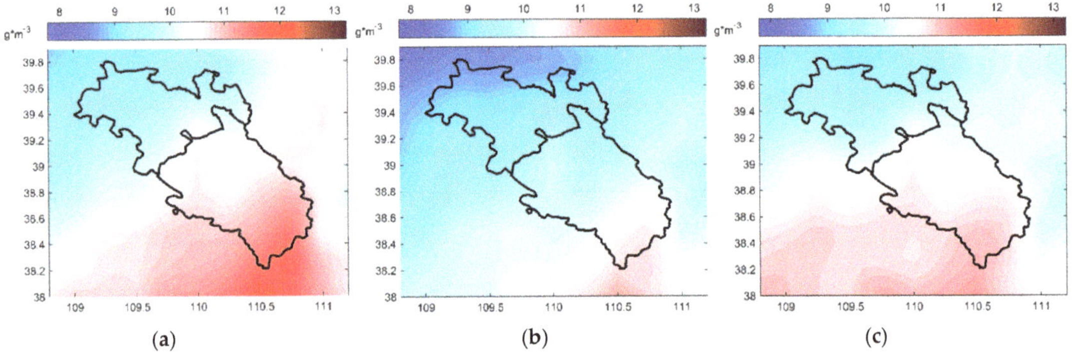

Figure 14. Inversion results of atmospheric water vapor density in the study area (g/cm^3): (**a**) 1990, (**b**) 2005, (**c**) 2019.

Figures 15 and 16 show the interannual mean values (from June to August) for temperature and precipitation from 1990 to 2019. The mean temperature has been relatively stable for the past 30 years. The lowest temperature was recorded in 2003 at 19.9 °C, while the highest temperature occurred in 1999 and 2010 at 22.5 °C. The mean precipitation (June to August) changed significantly over the past 30 years. It reached its maximum value in 2016 at 65.8 mm and was relatively low in 2005, 2009, and 2010.

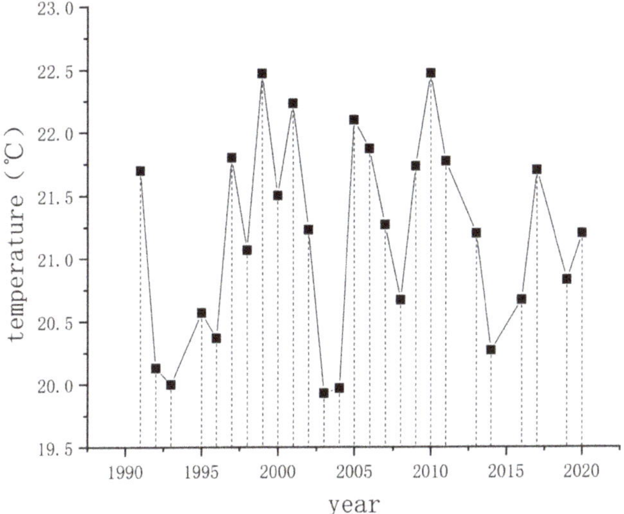

Figure 15. Average value of temperature in the study area.

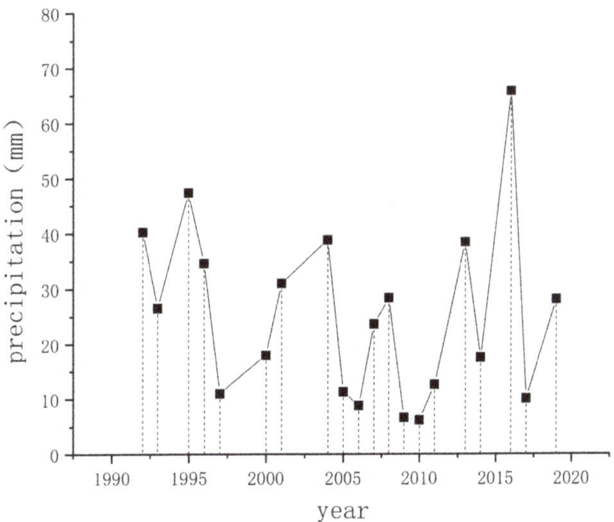

Figure 16. Average value of summer precipitation in the study area.

4.2. Monitoring of Key Mining Area—Shangwan Coal Mine

Shangwan coal mine (see Figure 17), located in Wulanmulun Town, Ejin Horo Banner, Ordos City, China, is one of the ultra-large modern backbone mines of Shenhua Shendong Coal Group Co., Ltd. [39]. The mine is a subterranean coal mine (pithead industry coal mine) with an area of 61.8 km^2 and a geological reserve of 1.23 billion tons [40]. It started its operation in 2000, with an approved production capacity of 14 million tons per year.

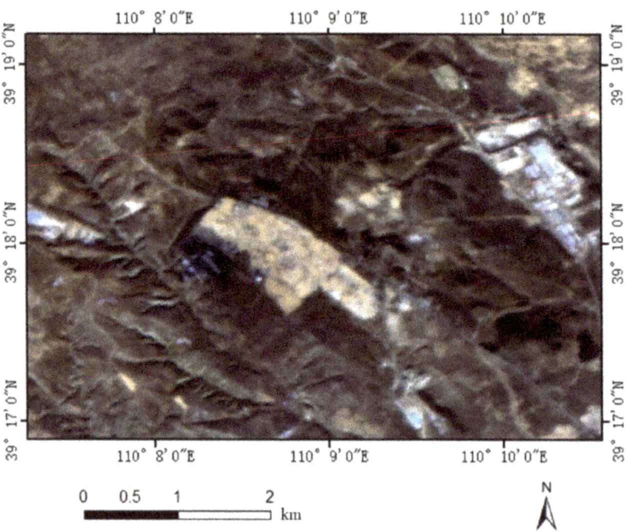

Figure 17. Research area of Shangwan Coal Mine.

4.2.1. Land Cover Classification of the Shangwan Coal Mine

The land cover classification results for the Shangwan coal mining area at different years are shown in Figure 18, and the tabulated results for each cover type are summarized in Table 2. In 1990, almost no mining land and artificial land was found in the Shangwan coal mining area; the main cover types were vegetation and bare lands. In 2005, the main cover types were still vegetation and bare land. Patches of cultivated land and mining land were observed, but there was still no artificial land. In 2019, most of the land had vegetation cover type; bare land decreased, while mining land and artificial land increased significantly. In the past 30 years, changes in the land cover types in the mining area have been pronounced. The primary land cover type was vegetation. Bare land and water had decreased considerably, while mining land and artificial land have increased year by year.

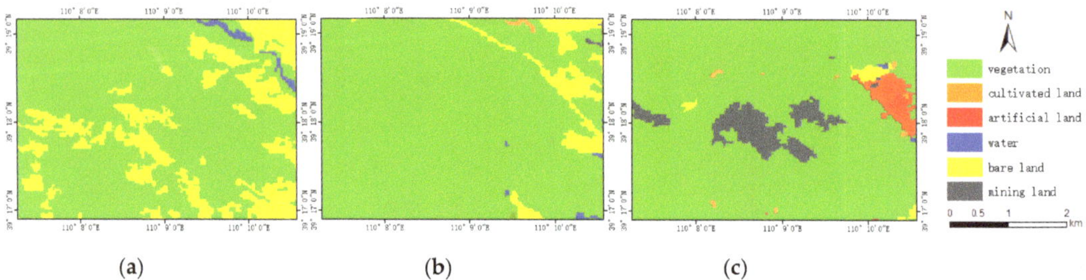

Figure 18. Land cover classification in the Shangwan coal mining area for (**a**) 1990, (**b**) 2005, and (**c**)2019.

Table 2. Statistics of various types of land coverage in different years in Shangwan coal mining area.

Coal Mine	Year	Class	Vegetation	Water	Mining Area	Cultivated Land	Bare Land	Artificial Land
Shangwan coal mine	1990	area (km^2)	15.45	0.20	0.00	0.00	4.43	0.00
		Percentage %	77.00	1.00	0.00	0.00	22.00	0.00
	2005	area (km^2)	18.52	0.08	0.04	0.06	1.38	0.00
		Percentage %	92.20	0.40	0.20	0.30	6.90	0.00
	2019	area (km^2)	17.70	0.02	1.41	0.08	0.24	0.63
		Percentage %	88.14	0.10	7.02	0.40	1.20	3.14

In Figure 19, the vegetation area in the Shangwan coal mining area showed an accelerated growth trend from 1990 to 2005. The effects of mining activities on the vegetation were minimal at this time, given that it was still the initial and early stages of development for mining activities. During the period of rapid development of coal mining activities (from 2005 to 2010), vegetation showed pronounced ECE, which gradually increased due to its cumulative effects year by year.

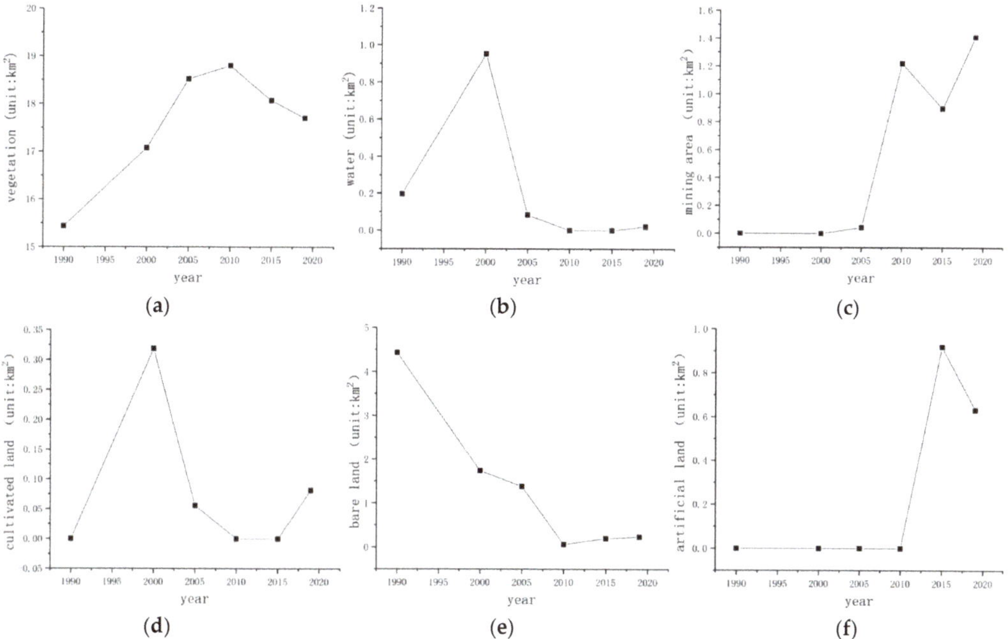

Figure 19. Changes for all land cover types in the Shangwan coal mining area: (**a**) Vegetation, (**b**) Water, (**c**) Mining land, (**d**) Cultivated land, (**e**) Bare land, (**f**) Artificial land.

There was a slow decline in vegetation from 2010 to 2019. Compared with the previous period, vegetation in the Shangwan coal mine had increased overall, while water and bare lands had downward trends. In 2000, lands classified as water were more extensive than usual, possibly due to the heavy precipitation occurring in that year. The expansion of artificial lands accelerated in 2010–2015; coal mining processing plants (screening fields) and other artificial lands were built in the eastern part of the study area, given increased mining activities in the region. Cultivated lands had increased rapidly in 1990–2000 and fluctuated in the following decades. These changes are closely related to changes in human activities.

4.2.2. Analysis of LST and SM

The LST inversion results for the Shangwan coal mining area are shown in Figure 20, and the mean LST summer values (June to August) at different years are presented in Figure 21. The LST exhibited an overall upward trend in 1990–2005, reaching its peak in 2005 before declining in subsequent years. As shown in Figures 11 and 21, the change trend in LST for the Shangwan coal mining area is comparable to that of the entire study area. The LST in the mining area gradually surged in 1990–2000 because the impact from mining on the surrounding environmental factors was weak at the initial stage of coal mining operations. From 2000 to 2010, the LST initially increased and then declined. To a certain extent, these changes were influenced by the cumulative effects of mining activities, showing particular ECE characteristics. The LST in the mining area increased gradually in 2010–2019, given the plateauing of coal mining activities, and the impact on the LST diminished.

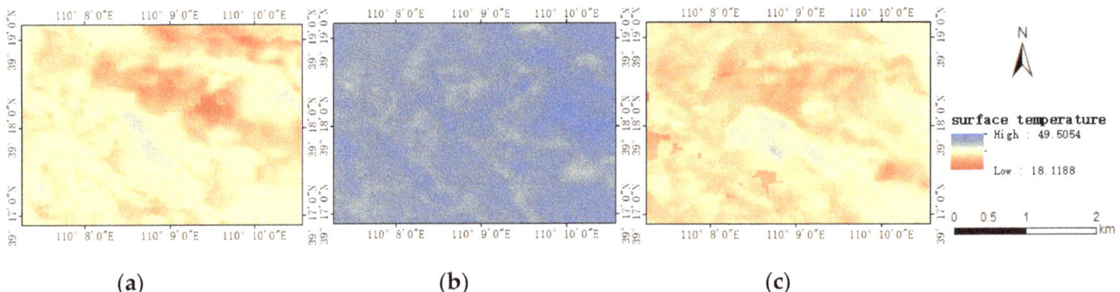

Figure 20. LST inversion results of the Shangwan coal mining area (°C): (**a**) 1990, (**b**) 2005, (**c**) 2019.

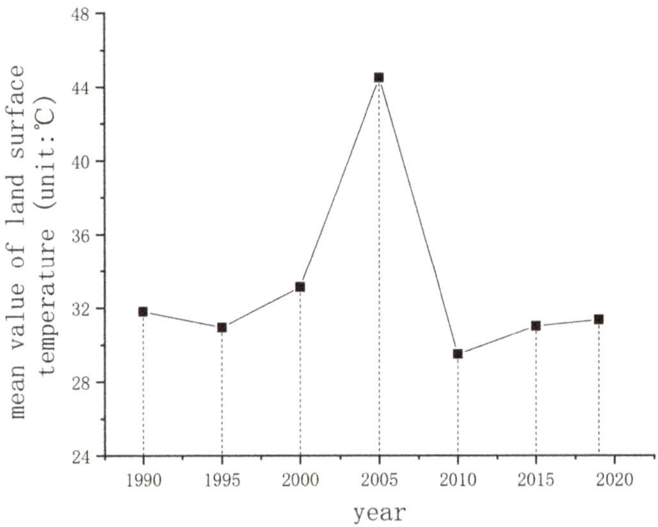

Figure 21. The mean LST value in the Shangwan coal mining area.

The SM inversion results of the Shangwan coal mining area are shown in Figure 22, and the mean SM values for summer (June to August) at different years are presented in

Figure 23. SM had an overall fluctuating trend, with the lowest value occurring in 2005 and the highest in 2019. SM was steady from 1990 to 2000. From 2000 to 2005, it dropped rapidly; this was caused mainly by long-term mining activities. From 2005 to 2020, SM was in an upward trajectory, surging particularly in 2015–2019. The pronounced SM changes in the Shangwan coal mining area are related, not only to the climate elements, but also to the ecological restoration measures being implemented at that time.

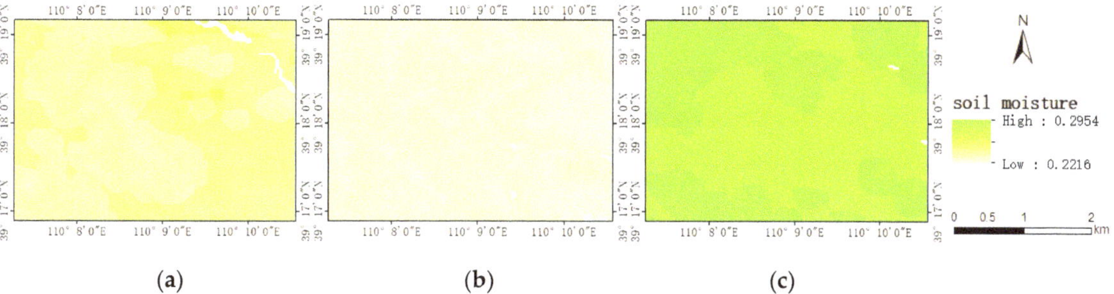

(a)　　　　(b)　　　　(c)

Figure 22. SM inversion results of the Shangwan coal mining area (m^3/m^3): (**a**) 1990, (**b**) 2005, (**c**) 2019.

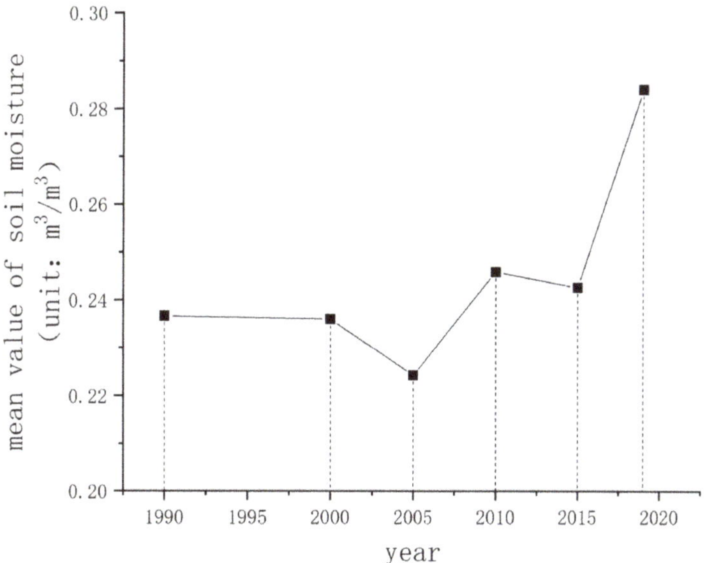

Figure 23. The mean value of SM in the Shangwan coal mining area.

4.3. Characteristics of Ecological Cumulative Effect of Mining Disturbance

4.3.1. Features on the Time-Scale

The spatial distribution of change rates in LST and SM are shown in Figures 24 and 25. The LST increased from 1990 to 2005, with the most pronounced surge occurring in 2000–2005. The LST in more than half of the study area decreased significantly in 2005–2010, especially in the eastern region. The LST increased from 2010 to 2019, comparable with the change in 1990–2000. A decrease in LST was observed in a small number of areas.

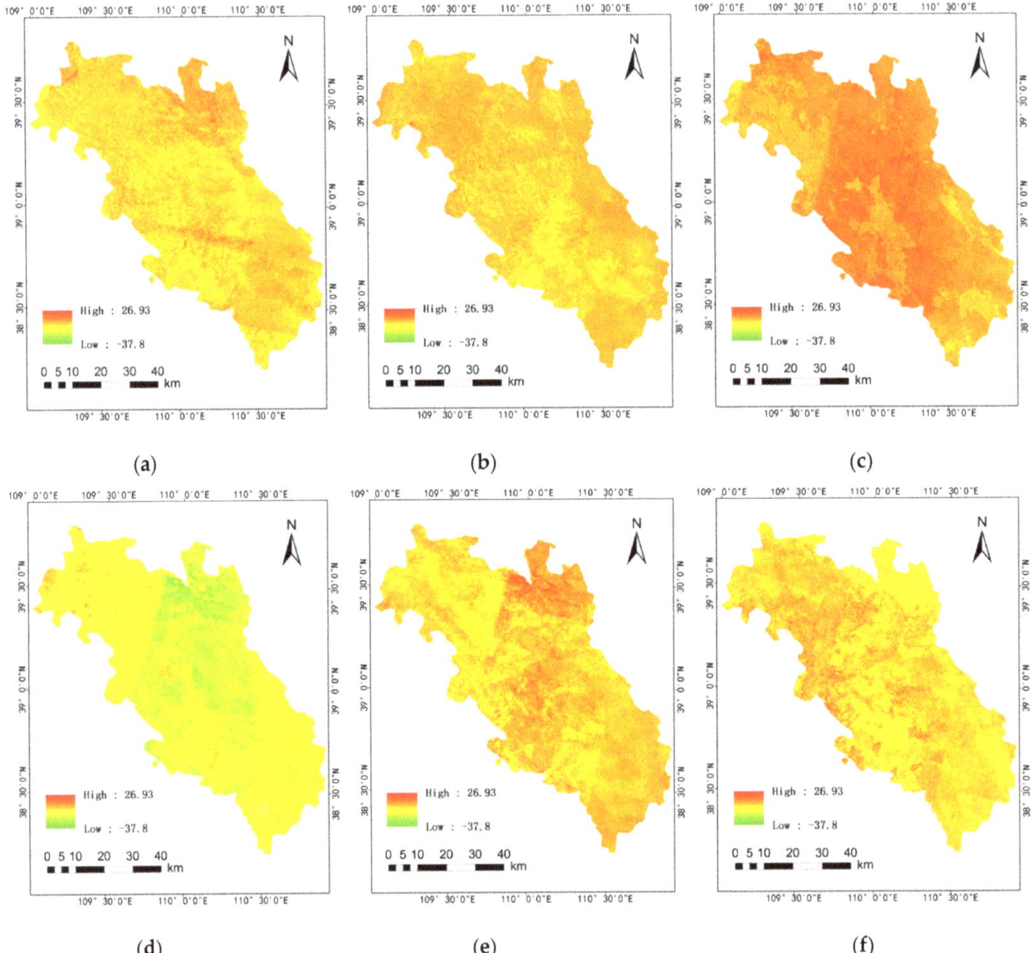

Figure 24. Change rate of LST in the study area: (**a**) 1990–1995, (**b**) 1995–2000, (**c**) 2000–2005, (**d**) 2005–2010, (**e**) 2010–2015, (**f**) 2015–2019.

For 1990–2005, the SM in the study area decreased overall. SM mainly increased in 1990–1995 and then declined in the following decade. From 2005 to 2010, the SM increased considerably in the eastern part of the study area and slightly increased in other areas. From 2010 to 2015, the SM was comparable to the 2005–2010 levels, with the main difference being that in the east-central area, the SM decreased slightly. From 2015 to 2019, the SM of the entire study area had drastically changed.

Figures 26–28 show the change rates in atmospheric specific humidity, atmospheric relative humidity, and atmospheric water vapor density. The atmospheric specific humidity increased from 1990 to 2000. However, the increase rate showed a decreasing trend from the northwest to the southeast, reflecting the variations in physical geography. The atmospheric specific humidity had a downward trend for 2000–2010, and the decrease rate near the mining area was significantly higher than that in other regions. From 2010 to 2020, the atmospheric specific humidity again surged, with a large increase rate and a significant ecological cumulative effect. The spatial and temporal distribution patterns of water vapor density and relative humidity change rate were consistent with that of atmospheric specific humidity.

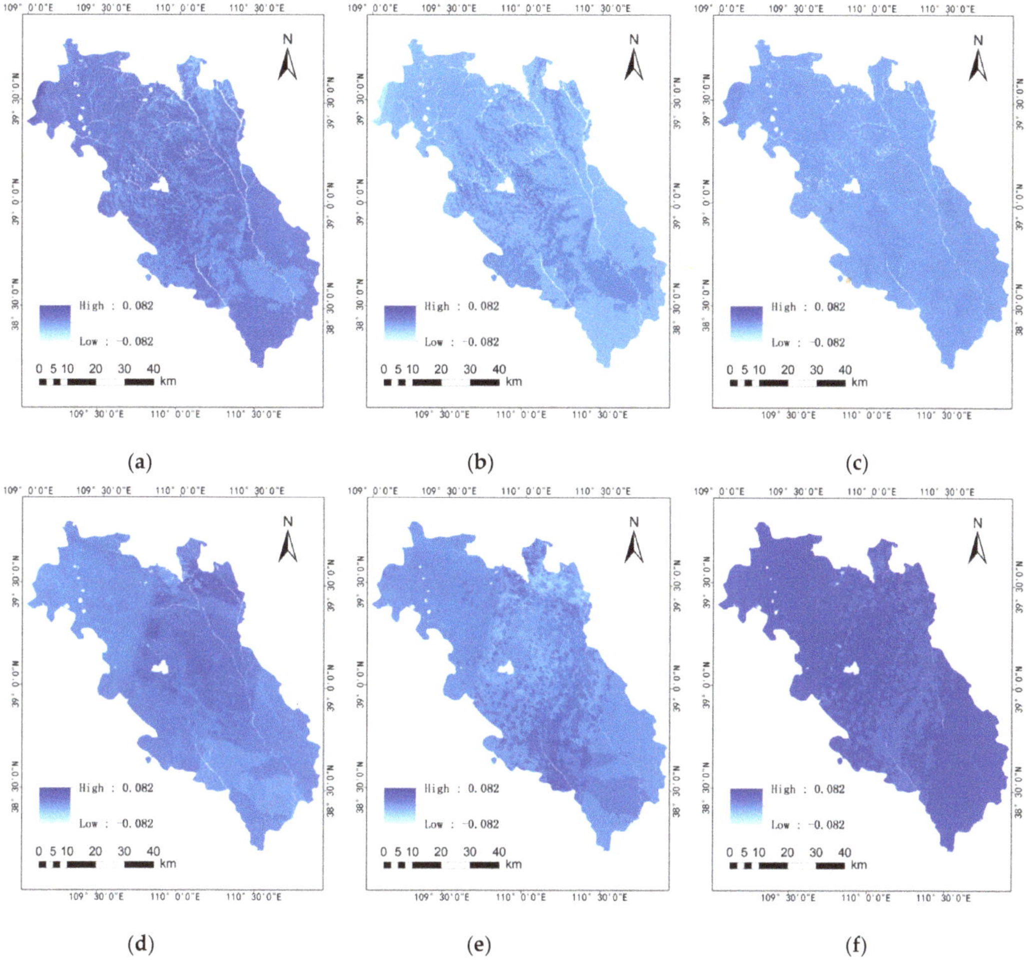

Figure 25. Change rate of SM in the study area: (**a**) 1990–1995, (**b**) 1995–2000, (**c**) 2000–2005, (**d**) 2005–2010, (**e**) 2010–2015, (**f**) 2015–2019.

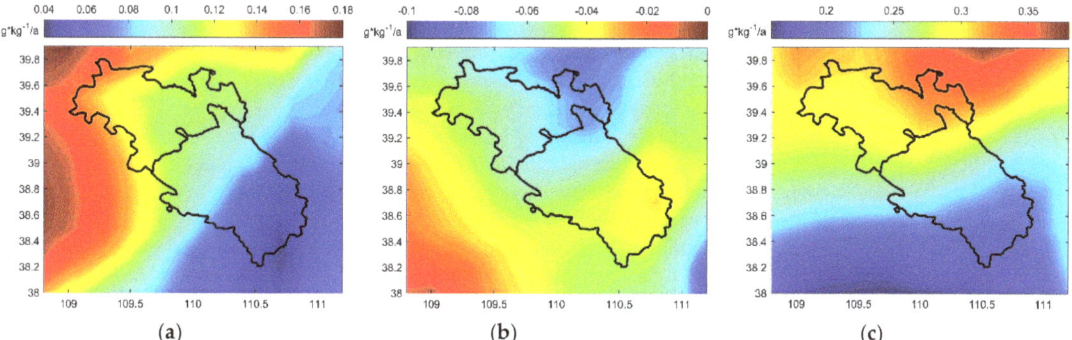

Figure 26. Change rate of atmospheric specific humidity in the study area: (**a**) 1990–2000, (**b**) 2000–2010, (**c**) 2010–2019.

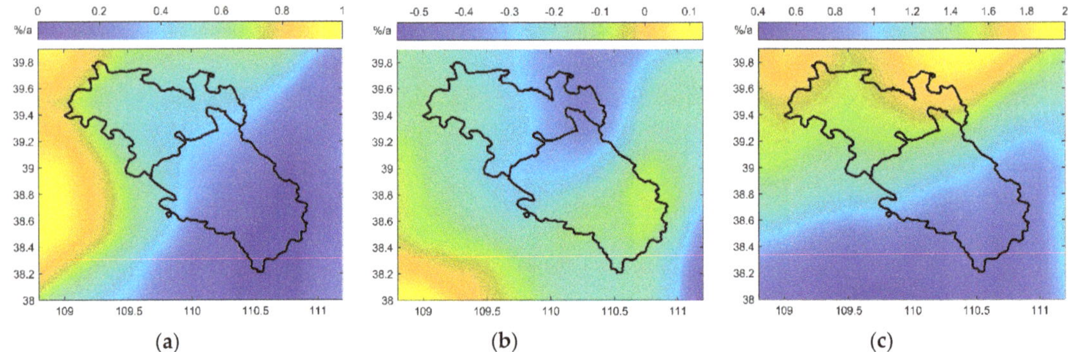

Figure 27. Change rate of atmospheric relative humidity in the study area: (**a**) 1990–2000, (**b**)2000–2010, (**c**)2010–2019.

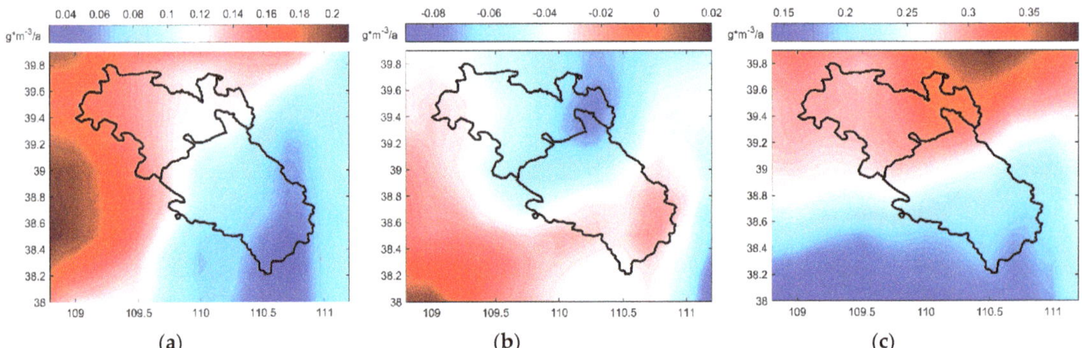

Figure 28. Change rate of atmospheric water vapor density in the study area: (**a**) 1990–2000, (**b**)2000–2010, (**c**)2010–2019.

The mean LST and SM change rates (in absolute value) at each time period are summarized in Figure 29. In the first four intervals (1990–1995, 1995–2000, 2000–2005, and 2005–2010), the mean LST change rate (in absolute value) showed an upward trend, indicating that the LST increase was accelerating. In 1990–2000, mining operations were still in the initial stage of development, and the succeeding decade (2000–2010) was the period for rapid development. During these years, with the gradual strengthening of mining activities, the influence of ecological factors also expanded, as indicated by the accelerated changes in LST. For 2010–2019, the values were in a downward trajectory, indicating a deceleration in LST change rates. At this period (2010–2019), coal mining activities were at the stage of steady development. Vigorous implementation of environmental protection and ecological restoration measures has gradually stabilized the ecological elements, as manifested by the decelerating changes in LST.

The change rate in mean SM (in absolute value) showed an upward trend in 1990–1995. From 1995 to 2010, the change rate of mean SM (in absolute value) decreased, indicating that it had weakened during this period. In 2015–2019, the slope showed an upward trend, indicating an accelerated change in mean SM.

Time accumulation is the cumulative phenomenon that occurs on the time scale when the time interval between two disturbances is less than the time required for environmental restoration. The external manifestation is an increase in the rate of change on the time scale. Therefore, the accelerated changes in LST and SM values show the ecological cumulative effect of mining disturbance on the time scale.

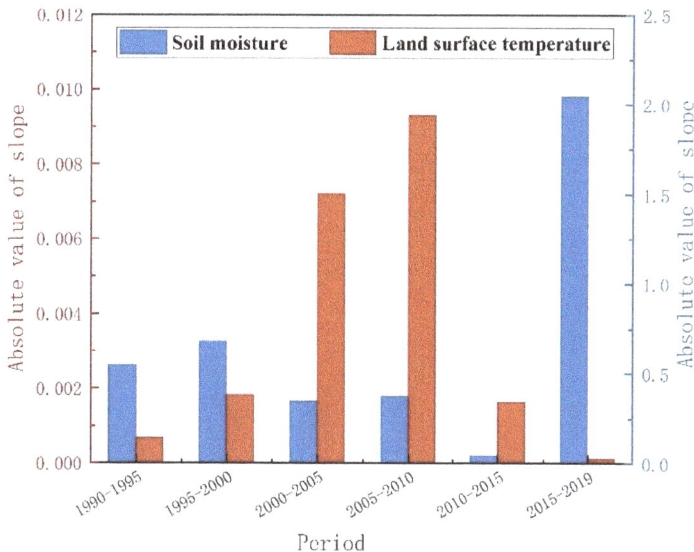

Figure 29. The absolute value of the change rate of mean LST and SM in the study area.

4.3.2. Features on Spatial Scale

As shown in Figure 30, seven buffer zones were set up, with the Shangwan Coal Mine as center and a 300 m interval between two adjacent buffers. The mean LST and SM values for each buffer zone in 2019 were then calculated, and the summary of results is presented in Figure 31.

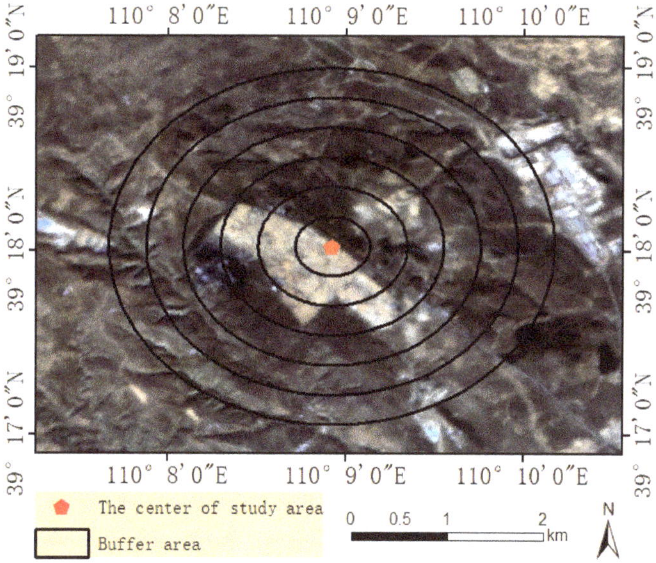

Figure 30. The buffer zones at the Shangwan coal mine.

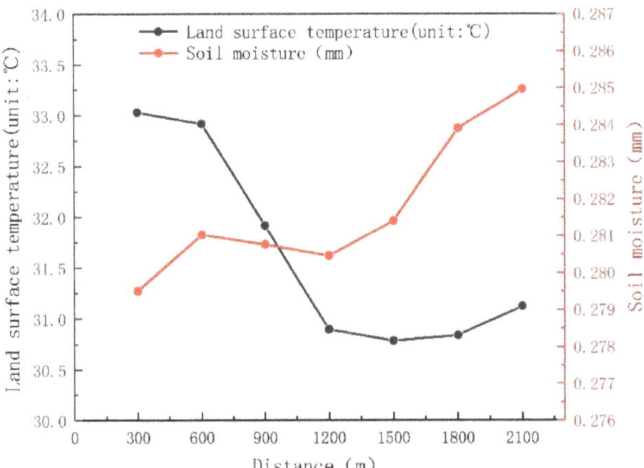

Figure 31. The relationship between the mean value of LST, SM, and the distance from the mine.

The first buffer zone (>300 m) had the highest mean LST and the lowest average SM values. The mean LST gradually declines as the distance increases, while the mean SM exhibits a gradual upward trend. Spatial accumulation refers to the accumulative phenomenon generated on the spatial scale when the spatial proximity between adjacent disturbance factors is less than the distance required to remove each disturbance. Externally, the influence of disturbance on the spatial scale attenuates with the distance. Therefore, the variations in LST and SM with regard to distance show pronounced ECE characteristics on the spatial scale. Based on the temporal and spatial cumulative effect characteristics (Wang et al., 2010), the source of the cumulative effect may be attributed to mining disturbance by way of mining subsidence and the land-use change.

To better analyze the environmental evolution characteristics, we chose an area unaffected by human activities to compare with the Shangwan coal mining area. The contrast area is rectangular, located two kilometers from the western boundary of the Shangwan coal mining area, and is similarly sized as the study site. The high-resolution image of the contrast area is shown in Figure 32.

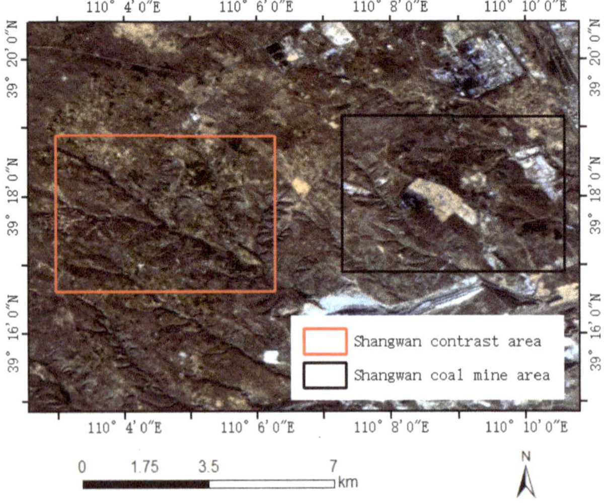

Figure 32. The contrast study area.

The land cover results of the contrast area at different years are shown in Figure 33, and the area statistics are summarized in Table 3. There was almost no mining or artificial land in the contrast area in 1990, and the main coverage types were vegetation (79%) and bare land (18%). In 2005, vegetation cover (90%) was still the primary coverage type, while barren land decreased to 8%. In 2019, vegetation (53%) decreased considerably, while bare land disappeared. Along with vegetation, cultivated land (47%) became a major land cover type, and new artificial lands (1%) were established. Results from the cover classification analysis reveal that the land cover in the contrast area has significantly changed; in particular, large portions of vegetation and bare lands were converted into cultivated land.

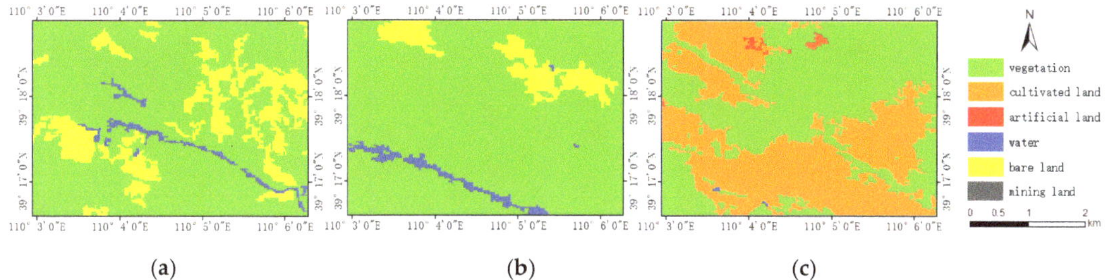

Figure 33. Results of land cover classification in the contrast area: (**a**) 1990, (**b**) 2005, (**c**) 2019.

Table 3. Statistics of various types of land coverage in the contrast area.

Coal Mine	Year	Class	Vegetation	Water	Mining Area	Cultivated Land	Bare Land	Artificial Land
Shangwan coal mine	1990	area (km^2)	15.92	0.48	0.00	0.00	3.68	0.00
		percentage %	79.28	2.39	0.00	0.00	18.33	0.00
	2005	area (km^2)	17.98	0.52	0.00	0.00	1.58	0.00
		percentage %	89.54	2.59	0.00	0.00	7.87	0.00
	2019	area (km^2)	10.55	0.01	0.00	9.38	0.00	0.14
		percentage %	52.54	0.05	0.00	46.71	0.00	0.70

Figure 34 shows the various land cover changes in the Shangwan coal mining area and the contrast area for the last 30 years. The vegetation cover in these two areas exhibited a growing trend in the first 20 years but has since trended downwards in the last decade. Vegetation in the contrast area decreased further and faster than in the mining area. The change trends for water, bare land, and cultivated lands in the two areas are comparable, but cultivated lands have expanded more rapidly in the contrast area in the past five years. Both mining and artificial lands had increased in the mining zones.

The LST inversion results for the contrast area are shown in Figure 35. The overall LST increased from 1990 until its peak in 2005 and then declined. The SM values initially decreased, reached the lowest value in 2005, and then increased to the highest point in 2019.

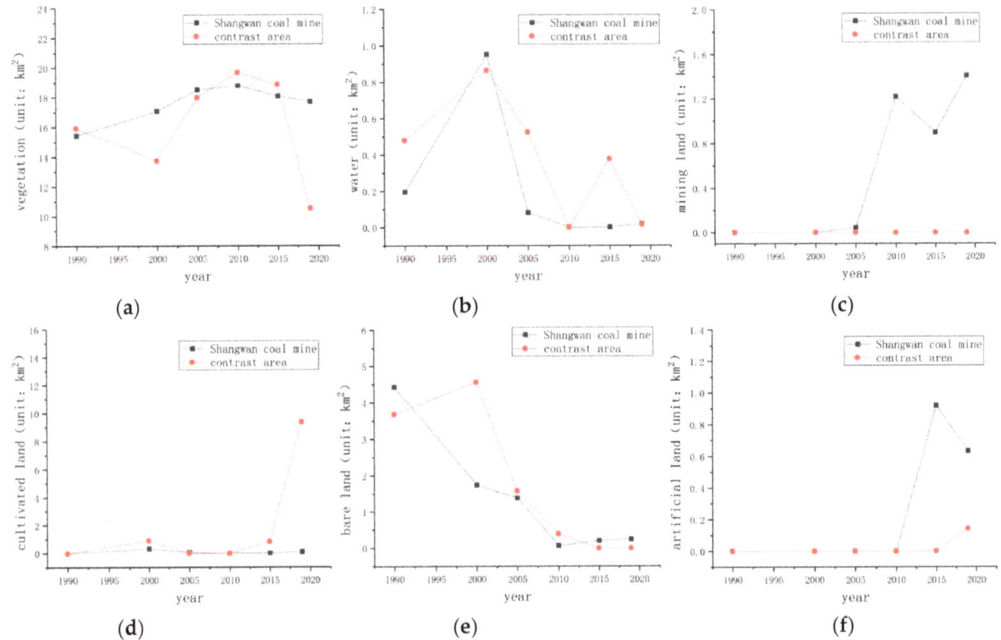

Figure 34. Changes of all kinds of land cover of the Shangwan coal mining area and contrast area: (**a**) Vegetation, (**b**) Water, (**c**) Mining land, (**d**) Cultivated land, (**e**) Bare land, (**f**) Artificial land.

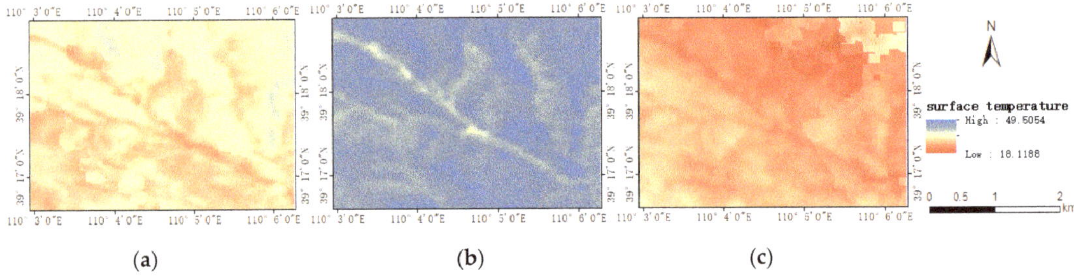

Figure 35. Inversion results of LST in the contrast area (°C): (**a**) 1990, (**b**) 2005, (**c**) 2019.

Figure 36 shows the LST box plot for the Shangwan coal mining area and the contrast area. Comparing Figures 20 and 35, the LST in the two regions showed a gradually increasing trend from 1990 to 2005. However, due to the emergence of mining land and artificial land in the Shangwan coal mining area in 2019, the coal mine processing plant and its surroundings yielded high LST values, while in other land cover types, the LST values were considerably lower. In the contrast area, the LST decreased overall. As shown in Figure 36, the Shangwan coal mining area had more concentrated LST values and a higher mean LST than the contrast area.

The SM inversion results for the contrast area are shown in Figure 37. The SM box plot between the Shangwan coal mining area and the contrast area is shown in Figure 38. Comparing Figures 22 and 37, the two regions had similar SM change trends, decreasing initially and then increasing. As shown in Figure 38, the distribution of the SM value for the two regions is relatively consistent and concentrated. However, the mining area has lower mean SM than the contrast area, which indicates that mining activities affect SM and that the impact is highly concentrated.

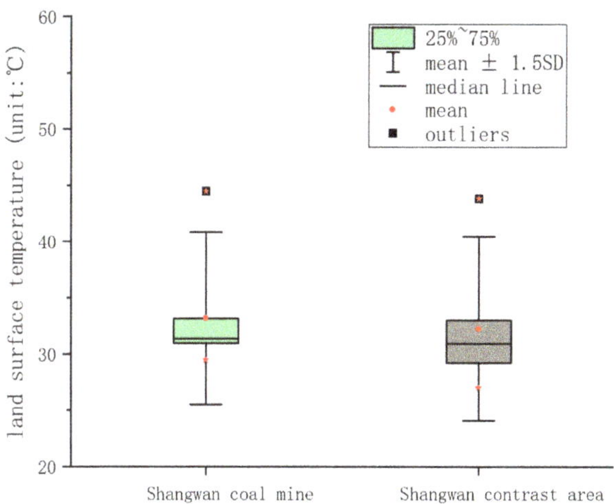

Figure 36. Comparison of LST of Shangwan coal mining area and contrast area.

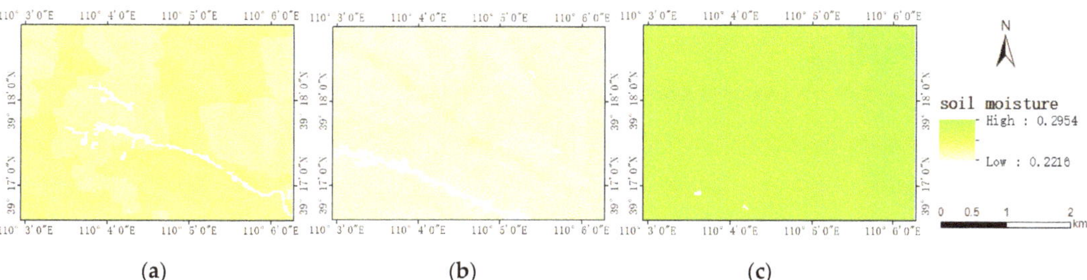

Figure 37. Inversion results of SM in the contrast area (m^3/m^3): (**a**) 1990, (**b**) 2005, (**c**) 2019.

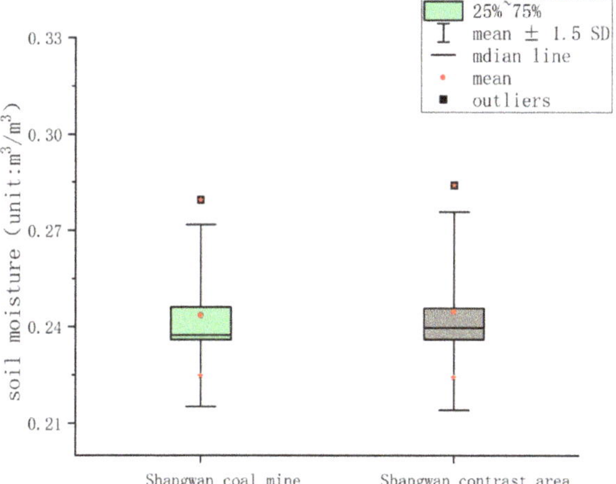

Figure 38. Comparison of SM of Shangwan coal mining area and contrast area.

5. Conclusions

Land cover classification and change analysis were carried out to retrieve land surface temperature, soil moisture, specific humidity, relative humidity, and atmospheric water vapor density using remote sensing and multi-source spatial data. The eco-environment change and the characteristics of ecological cumulative effect from coal mining operations were analyzed on a long-time scale. The Shangwan Coal mine was used as example in the comparative analysis between the mining zone and the contrast area. The main highlights of the study are as follows:

(1) In the initial stage of coal mining activities (1990–2000), the eco-environment was generally stable, but mining activities had impacted the eco-environment to a certain degree.

(2) During the rapid development stage of coal mining operations, the eco-environment was damaged severely, showing significant ECE, including temporal and spatial cumulative effects. The change rates for the eco-environmental parameters accelerated and showed pronounced ecological cumulative effects at the temporal scale. In 2010, coal mining activities entered a period of relative stability, and ecological restoration started to receive greater attention. The eco-environmental parameters gradually recovered, and the eco-environment generally improved. Results from the land surface temperature and soil moisture analyses and the spatial comparison with the contrast area show ECE characteristics due to mining disturbance at the spatial scale.

Future studies can comprehensively utilize remote sensing, ground investigation, and statistical data to conduct quantitative and high-frequency observations for more parameters at long-term scales in mining areas. The results would help support understanding the mechanisms and characteristics of the ecological cumulative effects caused by mining operations.

Author Contributions: Conceptualization, Q.L.; methodology, Q.L., J.G. and F.W.; software, J.G. and Z.S.; validation, F.W. and Z.S.; formal analysis, Q.L. and J.G.; investigation, Q.L. and J.G.; resources, Q.L. and J.G.; data curation, J.G. and F.W.; writing—original draft preparation, Q.L., J.G. and F.W.; writing—review and editing, Z.S.; visualization, F.W. and Z.S.; supervision, Q.L. and J.G.; project administration, Q.L. and J.G.; funding acquisition, Q.L. and J.G. All authors have read and agreed to the published version of the manuscript.

Funding: This research was funded by [Open Fund of State Key Laboratory of Water Resource Protection and Utilization in Coal Mining] grant number [GJNY-20-113-14], [State Key Laboratory of Coal Resources and Safe Mining] grant number [SKLCRSM19KFA04], [National Natural Science Foundation of China] grant number [41901291], and [Fundamental Research Funds for the Central Universities] grant number [2021YQDC02]. And The APC was funded by [GJNY-20-113-14].

Institutional Review Board Statement: Not applicable.

Informed Consent Statement: Not applicable.

Acknowledgments: The authors would like to thank the National Aeronautics and Space Administration for the open access to Landsat data and thank the European Centre for Medium-Range Weather Forecasts for the open access to ERA5 reanalysis data.

Conflicts of Interest: The authors declare no conflict of interest.

References

1. Zhou, Q. A discussion on accumulation effect in environment effect appraisal. *Energy Environ. Prot.* **2013**, *27*, 60–62.
2. Tollefson, C.; Wipond, K. Cumulative environmental impacts and aboriginal rights. *Environ. Impact Assess. Rev.* **1998**, *18*, 371–390. [CrossRef]
3. Sun, J.; Bu, Q.; Wu, D.; Liu, Y.; Jiang, W. A review on evaluation of ecological cumulative effect in coal mining areas. *Asian J. Ecotoxicol.* **2019**, *14*, 74–82.
4. Zhang, D.; Wen, P. The impact of coal mining on the ecological environment and its countermeasures. *Energy Environ.* **2019**, *2*, 74–75.
5. Wang, B.; Zhao, N.; Gao, H. Environment problems and ecological restoration in coal mining areas. *Shanxi For. Sci. Technol.* **2020**, *48*, 104–106, 113.
6. Hu, Z.; Xiao, W.; Zhao, Y. Re-discussion on coal mine eco-environment concurrent mining and reclamation. *J. China Coal Soc.* **2020**, *45*, 351–359.
7. Li, J.; Pei, Y.; Zhao, S.; Xiao, R.; Sang, X.; Zhang, C. A review of remote sensing for environmental monitoring in China. *Remote Sens.* **2020**, *12*, 1130. [CrossRef]

8. Liu, H.; Jiang, Y.; Xia, M.; Gao, J.; Jiang, Y. Ecological environment changes of mining area with 30 years' remote sensing monitoring: A case study around Nansihu Lake, Shandong Province. *Met. Mine* **2021**, *4*, 197–206.

9. Zhang, C.; Li, J.; Lei, S.; Yang, J.; Yang, N. Progress and prospect of the quantitative remote sensing for monitoring the eco-environment in mining areas. *Met. Mine.* in press.

10. Werner, T.; Mudd, G.; Schipper, A.; Huijbregts, M.; Taneja, L. Global-scale remote sensing of mine areas and analysis of factors explaining their extent. *Glob. Environ. Chang.* **2020**, *60*, 102007. [CrossRef]

11. Mezned, N.; Dkhala, B.; Abdeljaouad, S. Multitemporal and multisensory Landsat ETM+ and OLI 8 data for mine waste change detection in Northern Tunisia. *J. Spat. Sci.* **2018**, *63*, 135–153. [CrossRef]

12. Zhang, H.; Zhang, K.; Liu, P.; Yu, Z.; Zhao, J. Ecological indexes extraction and safety assessment of coal mining area based on RS and GIS: Taking Jiaozuo Coal Mining Area as an example. *Coal Sci. Technol.* **2020**, *48*, 80–88.

13. Balaniuk, R.; Isupova, O.; Reece, S. Mining and tailings dam detection in satellite imagery using deep learning. *Sensors* **2020**, *20*, 6936. [CrossRef]

14. Nascimento, F.; Gastauer, M.; Souza-Filho, P.; Nascimento, W.; Santos, D.; Costa, M. Land cover changes in open-cast mining complexes based on high-resolution remote sensing data. *Remote Sens.* **2020**, *12*, 611. [CrossRef]

15. Wu, Q.; Liu, K.; Song, C.; Wang, J.; Ke, L.; Ma, R.; Zhang, W.; Pan, H.; Deng, X. Remote sensing detection of vegetation and landform damages by coal mining on the Tibetan Plateau. *Sustainability* **2018**, *10*, 3851. [CrossRef]

16. Fu, X.; Ma, M.; Jiang, P.; Quan, Y. Spatiotemporal vegetation dynamics and their influence factors at a large coal-fired power plant in Xilinhot, Inner Mongolia. *Int. J. Sustain. Dev. World Ecol.* **2016**, *24*, 433–438. [CrossRef]

17. Liu, Y.; Yue, H. Remote sensing monitoring of soil moisture in the Daliuta Coal Mine based on SPOT 5/6 and Worldview-2. *Open Geosci.* **2019**, *11*, 866–876. [CrossRef]

18. Cao, J.; Deng, Z.; Li, W.; Hu, Y. Remote sensing inversion and spatial variation of land surface temperature over mining areas of Jixi, Heilongjiang, China. *Peerj* **2020**, *8*, e10257. [CrossRef] [PubMed]

19. García Millán, V.; Faude, U.; Bicsan, A.; Klink, A.; Teuwsen, S.; Pakzad, K.; Müterthies, A. Monitoring flooding damages in vegetation caused by mining activities using optical remote sensing. *PFG—J. Photogramm. Remote Sens. Geoinf. Sci.* **2018**, *86*, 1–13. [CrossRef]

20. Kayet, N.; Pathak, K.; Chakrabarty, A.; Kumar, S.; Chowdary, V.; Singh, C.; Sahoo, S.; Basumatary, S. Assessment of foliar dust using Hyperion and Landsat satellite imagery for mine environmental monitoring in an open cast iron ore mining areas. *J. Clean. Prod.* **2019**, *218*, 993–1006. [CrossRef]

21. Li, J.; Liang, J.; Wu, Y.; Yin, S.; Yang, Z.; Hu, Z. Quantitative evaluation of ecological cumulative effect in mining area using a pixel-based time series model of ecosystem service value. *Ecol. Indic.* **2021**, *120*, 106873. [CrossRef]

22. Wan, X. *Study on Eco-Environmental Cumulative Effects in Coal Mining Area—Case Studies in Lu'an Mining Area*; China University of Mining Technology: Beijing, China, 2010.

23. Hodgson, E.E.; Halpern, B.S. Investigating cumulative effects across ecological scales. *Conserv. Biol.* **2019**, *33*, 22–32. [CrossRef] [PubMed]

24. Dong, J.; Ji, L.; Fang, A. Ecological cumulative effect of mining area in typical arid and semi-arid grassland. *J. China Coal Soc.* **2021**, *46*, 1945–1956.

25. Shen, Q.; Zhuang, S.; Jia, R. The Pearl of the Grassland, Ejin Horo Banner, Ordos. *Reg. Gov.* **2019**, 79–80.

26. Peng, P. *Analysis of LUCC and Driving Forces in Shenmu City*; Gansu Agricultural University: Lanzhou, China, 2017; pp. 1–6.

27. Zhou, Q.; Zhang, X.; Meng, Q.; Zhang, H.; Ma, B. Thoughts on the development of coal chemical industry in Shenmu. *Coal Process. Compr. Util.* **2017**, *10*, 10, 19–24.

28. Statistical Yearbook Sharing Platform. Available online: https://www.yearbookchina.com/index.aspx (accessed on 17 January 2021).

29. Breiman, L. Random forests. *Mach. Learn.* **2001**, *45*, 5–32. [CrossRef]

30. Xu, H. A remote sensing urban ecological index and its application. *Acta Ecol. Sin.* **2013**, *33*, 7853–7862.

31. Zhang, X.; Liao, C.; Li, J.; Sun, Q. Fractional vegetation cover estimation in arid and semi-arid environments using HJ-1 satellite hyperspectral data. *Int. J. Appl. Earth Obs. Geoinf.* **2013**, *21*, 506–512. [CrossRef]

32. Li, M. *The Method of Vegetation Fraction Estimation by Remote Sensing*; Institute of Remote Sensing and Digital Earth, Chinese Academy of Sciences: Beijing, China, 2003; pp. 41–46.

33. Jiapaer, G.; Chen, X.; Bao, A. A comparison of methods for estimating fractional vegetation cover in arid regions. *Agric. For. Meteorol.* **2011**, *151*, 1698–1710. [CrossRef]

34. Sobrino, J.; Jiménez-Muñoz, J.; Paolini, L. Land surface temperature retrieval from LANDSAT TM 5. *Remote Sens. Environ.* **2004**, *90*, 434–440. [CrossRef]

35. Sandholt, I.; Rasmussen, K.; Andersen, J. A simple interpretation of the surface temperature/vegetation index space for assessment of surface moisture status. *Remote Sens. Environ.* **2002**, *79*, 213–224. [CrossRef]

36. Qin, Q.; You, L.; Zhao, Y.; Zhao, Y.; Zhao, S.; Yao, Y. Soil line automatic identification algorithm based on two-dimensional feature space. *Trans. Chin. Soc. Agric. Eng.* **2012**, *28*, 167–171.

37. He, C.; Wu, S.; Wang, X.; Hu, A.; Wang, Q.; Zhang, K. A new voxel-based model for the determination of atmospheric weighted mean temperature in GPS atmospheric sounding. *Atmos. Meas. Tech.* **2017**, *10*, 2045–2060. [CrossRef]

38. Kraus, H. *Die Atmosphäre der Erde: Eine Einführung in die Meteorologie*; Springer: Berlin/Heidelberg, Germany, 2004; pp. 34–39.

39. Ji, Z.; Wang, X.; Hao, J.; Cui, D.; Wang, S. Study on the remarkable characteristics of mining in super high mining height working face of Shangwan coal mine. *China Min. Mag.* **2020**, *29*, 140–146.

40. Shenhua Shendong Coal Group Shangwan Coal Mine; China Coal Industry Association. *China's Large-Scale Modern Coal Mine Construction-National Large-Scale Coal Mine Construction Site Meeting and Promotion of Coal Production Scale Modernization Development forum Data Compilation*; China Coal Industry Association: Beijing, China, 2009; p. 2.

Article

A New Method for Quantitative Analysis of Driving Factors for Vegetation Coverage Change in Mining Areas: GWDF-ANN

Jun Li [1], Tingting Qin [1], Chengye Zhang [1,*], Huiyu Zheng [1], Junting Guo [2], Huizhen Xie [1], Caiyue Zhang [1] and Yicong Zhang [1]

[1] College of Geoscience and Surveying Engineering, China University of Mining and Technology-Beijing, Beijing 100083, China; junli@cumtb.edu.cn (J.L.); qting@student.cumtb.edu.cn (T.Q.); zhenghy@student.cumtb.edu.cn (H.Z.); huizhenxc@student.cumtb.edu.cn (H.X.); zcy_cumtb@student.cumtb.edu.cn (C.Z.); zhangyc@student.cumtb.edu.cn (Y.Z.)
[2] State Key Laboratory of Water Resource Protection and Utilization in Coal Mining, Beijing 102209, China; junting.guo.a@chnenergy.com.cn
* Correspondence: czhang@cumtb.edu.cn

Citation: Li, J.; Qin, T.; Zhang, C.; Zheng, H.; Guo, J.; Xie, H.; Zhang, C.; Zhang, Y. A New Method for Quantitative Analysis of Driving Factors for Vegetation Coverage Change in Mining Areas: GWDF-ANN. *Remote Sens.* **2022**, *14*, 1579. https://doi.org/10.3390/rs14071579

Academic Editors: Parth Sarathi Roy and Izaya Numata

Received: 30 November 2021
Accepted: 21 March 2022
Published: 24 March 2022

Abstract: Mining has caused considerable damage to vegetation coverage, especially in grasslands. It is of great significance to investigate the specific contributions of various factors to vegetation cover change. In this study, fractional vegetation coverage (FVC) is used as a proxy indicator for vegetation coverage. We constructed 50 sets of geographically weighted artificial neural network models for FVC and its driving factors in the Shengli Coalfield. Based on the idea of differentiation, we proposed the geographically weighted differential factors-artificial neural network (GWDF-ANN) to quantify the contributions of different driving factors on FVC changes in mining areas. The highlights of the study are as follows: (1) For the 50 models, the average RMSE was 0.052. The lowest RMSE was 0.007, and the highest was 0.112. For the MRE, the average value was 0.007, the lowest was 0.001, and the highest was 0.023. The GWDF-ANN model is suitable for quantifying FVC changes in mining areas. (2) Precipitation and temperature were the main driving factors for FVC change. The contributions were 32.45% for precipitation, 24.80% for temperature, 22.44% for mining, 14.44% for urban expansion, and 5.87% for topography. (3) Over time, the contributions of precipitation and temperature exhibited downward trends, while mining and urban expansion showed positive trajectories. For topography, its contribution remains generally unchanged. (4) As the distance from the mining area increases, the contribution of mining gradually decreases. At 200 m away, the contribution of mining was 26.69%; at 2000 m away, the value drops to 17.8%. (5) Mining has a cumulative effect on vegetation coverage both interannually and spatially. This study provides important support for understanding the mechanism of vegetation coverage change in mining areas.

Keywords: quantify; GWDF-ANN; FVC; vegetation cover; mining area

1. Introduction

Coal is an important source of energy in the world. For example, China's primary energy source is mainly coal [1]. Mining has caused land subsidence, occupation, excavation, and pollution [2,3], considerably reducing vegetation areas. Mining also affects the inherent growth of vegetation. For example, when dust from open-pit mines falls on the plant leaf surface, the stomatal conductance of the leaf decreases, and the amount of carbon dioxide exchange is also reduced. Large-scale mining has severely damaged the grassland ecosystem [4] and accelerated changes in vegetation coverage [5]. In addition to mining, many researchers have shown that precipitation, temperature [6–9], topography [10], and urban expansion [11] are important driving factors altering vegetation coverage in mining areas. However, the contribution of each factor to the change of vegetation coverage in mining areas remains unclear. Therefore, it is crucial to accurately quantify the contributions of the different driving factors in altering the fractional vegetation coverage (FVC) in mining areas. The quantitative analysis of these factors may also help reveal the internal mechanisms of vegetation coverage change in mining areas.

There are two main approaches based on remote sensing used to investigate the impact of mining on vegetation. One is to use vegetation parameters obtained from satellite monitoring as proxy indicators. Direct changes in proxy indicators are then used to analyze the impact of mining on vegetation. Li and Wang used the temporal and spatial changes in the FVC to reveal the changes in vegetation coverage in the Baorixile mining area, the Shengli mining area, and the Yellow River basin mining area [12–14]. Li used NDVI to analyze the impact of mining on vegetation change in Jungar Banner, China, from 2000 to 2017 [15]. Yang used NDVI and LandTrendr algorithms to monitor vegetation disturbance and restoration in the Kurah coal mining area [16]. However, the proxy parameter for vegetation is direct change, which is affected by the complex coupling of multiple factors and cannot be attributed solely to mining.

The other uses statistical methods to construct a regression model based on ordinary linearity for vegetation parameters and their driving factors. For example, Li used multiple linear regression, spatial correlation, and partial least square regression to analyze the impact of mining on the NDVI changes in the grasslands of Chenbarhu Banner, Inner Mongolia, China [17]. Fu Xiao used the rate of change in greenness (RCV) and the coefficient of variation (CV) to conduct correlation and stepwise regression analysis in studying the driving factors for the vegetation change in Xilinhot, particularly mining activities [18]. These studies mainly use a global regression model, in which the relationship between vegetation parameters and driving factors is assumed to be spatially constant. However, such a relationship often varies significantly in space because the vegetation coverage in mining areas may be affected by spatial factors such as climate, topography, and mining. Each driving factor may also have different effects on FVC at various locations. The phenomenon of how spatial relationships vary at different geographic locations is called spatial nonstationarity [19]. Simple linear regression modeling cannot accurately describe the complex driving processes of these different factors in changing the vegetation coverage in mining areas.

Therefore, the local modeling method, geographically weighted regression (GWR), was developed to deal with spatial nonstationarity. GWR has been widely applied in various fields, such as real estate economics [20,21], land-use science [22,23], ecology [24], and criminology [25,26]. However, GWR is essentially linear modeling. The geographically weighted artificial neural network (GWANN) was proposed to address the limitations of the GWR in describing complex nonlinear processes. GWANN uses the distance attenuation kernel function and bandwidth to calculate the geographic weight of the observed value results based on the GWR, solving the problem in spatial nonstationarity modeling. The model uses ANN to construct a nonlinear function model without making any assumption in completing complex spatial prediction tasks. It has been used to solve problems such as real estate valuation [27], agricultural output estimation [28], and wildfire analysis [29]. GWANN has been shown to be significantly better than the GWR when modeling synthetic data and actual data with nonlinearity and high spatial variance, respectively [27].

Vegetation coverage can be evaluated using the FVC determined using satellite imagery [30]. FVC is the ratio of the vertical projection of vegetation (e.g., leaves, stems, branches) on the ground in each pixel, which directly reflects the amount of vegetation in the target area [31,32]. FVC describes and measures vegetation growth and is the most important and sensitive indicator for vegetation coverage change [33,34].

In this study, we developed the geographically weighted differential factors-artificial neural network (GWDF-ANN). It is a mixed methodology based on the GWR and ANN, which constructs a nonlinear model while considering the nonstationarity of variables. Most importantly, this approach can quantify the contributions of different driving factors to the FVC. We applied the above methods to the various driving factors (i.e., temperature, precipitation, topography, mining, and urban expansion) and quantified each driving factor's contribution on FVC in the Shengli Coalfield of Inner Mongolia for 2004–2020.

2. Methods

2.1. Geographically Weighted Artificial Neural Network

The artificial neural network is a computing model consisting of many neurons and one-way connections [35]. Each neuron i represents a specific output function called an activation function. The connection between the two neurons i and j are controlled by the weight w_{ij}, which is the weighted value of the signal through the connection. These neurons are usually organized in layers, and each neuron is directionally connected to the neuron in the next layer. GWANN is different from the traditional ANN. It consists of an input layer, a hidden layer, and an output layer. A network with a single hidden layer can better approximate any continuous function on the closed subset and bounded subset of n-dimensional Euclidean space, given enough hidden neurons.

An input neuron is a grid cell that contains location information, y-value, and x-values. In this paper, y-value refers to FVC, and x-values refer to temperature, precipitation, topography, mining, and urban expansion. The modeling process of GWANN is shown in Figure 1. Many input neurons from the input layer pass into the hidden layer and are then aggregated and converted in the hidden layer using Equation (1). The output of neuron i is calculated using Equation (2). In neurons, people often use a nonlinear hyperbolic tangent activation function, which has the necessary conditions for calculating the network's continuous and differentiable error gradient [36], as shown in Equation (3). The output for each neuron is then passed to the output layer. Each output neuron of the output layer is assigned to a location in the geographic space and predicts a global observed value. However, the observed value is still not accurate enough. To simulate an exact nonlinear relationship model, the connection weight of the ANN must be adjusted.

$$Layer_j = \sum_{i \in S_j} w_{ij} o_i \tag{1}$$

$$o_i = \Phi(Layer_j) \tag{2}$$

$$f(x) = \frac{e^x - e^{-x}}{e^x + e^{-x}} \tag{3}$$

where $Layer_j$ is the network input of neuron j, w_{ij} is the connection weight between neuron i and j, o_i is the output of neuron i, S_j is a group of neurons that have output connection with neuron j, Φ is the activation function, $f(x)$ is the transfer value after activation of the neuron, and x is the parameter value before neuron activation.

Two main steps are usually required to adjust the ANN. First, backpropagation is used to calculate the error signal for the observed value of each neuron [37]. Unlike traditional ANN, GWANN uses a geographically weighted error function instead of a quadratic error function to calculate the error signal. GWANN weighs the difference between the output neuron and the target value (y-value) according to the spatial distance between the location of the output neuron and the observed value. These weights can be interpreted as the GWR model. When the output neuron's position is closer to the observed value, it is given a higher weight than those farther from the observed value. The geographically weighted error function is defined as Equation (4), and the backpropagation error signal is determined using Equation (5).

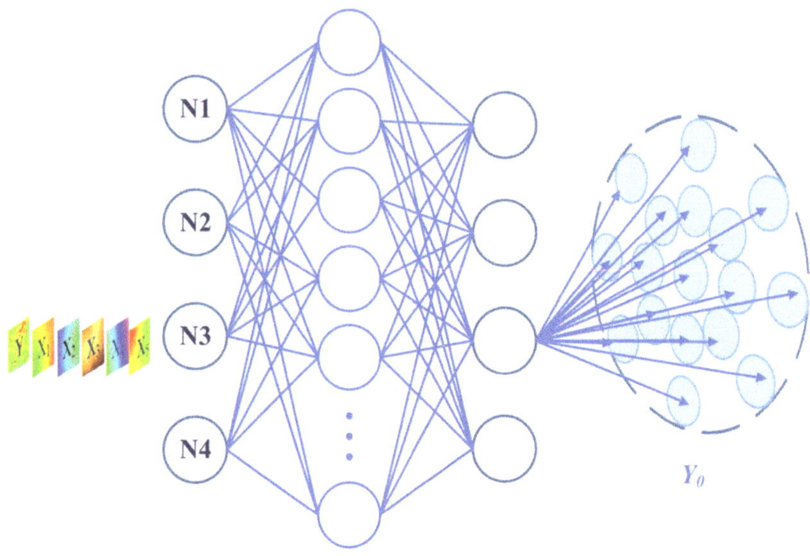

Figure 1. The modeling process of GWANN. N1, N2, N3, and N4 are the multiple input neurons composing the input layer. The rectangles on the left represent the attributes of each input neuron (*y*-value and *x*-values), and Y_0 is the prediction.

Gradient descent is used to adjust the connection weight in the second step, as in Equation (6). We use Nesterov's acceleration gradient [38], which significantly improves training performance [39] and makes training more robust. These two steps are repeated until the termination condition is reached (e.g., the error rate is lower than a predetermined threshold) and the GWANN model construction is completed. Finally, the modeling is completed, and we obtain the prediction(Y_0). The GWANN can be implemented on RStudio software. An R package providing an implementation of GWANN can be obtained from https://github.com/jhagenauer/gwann (accessed on 1 July 2021) [27], the purpose of which was to predict Y_0 in this study.

$$E = \frac{1}{2}\sum_{i=1}^{n} v_i(t_i - o_i)^2 \tag{4}$$

$$\delta_j = \begin{cases} \Phi'(Layer_j)v_i(o_j - t_j) & \text{if } j \text{ is an output neuron} \\ \Phi'(Layer_j)\sum_k \delta_k w_{jk} & \text{otherwise} \end{cases} \tag{5}$$

$$\Delta w_{ij} = -\eta\frac{\partial E}{\partial w_{ij}} = -\eta\delta_j o_i \tag{6}$$

where o_i is the output of neuron i, o_j is the output of neuron j, t_i is the target value of neuron i, t_j is the target value of neuron j, v_i is the geographically weighted distance between the observed value and the position of output neuron i, v_j is the geographically weighted distance between the observed value and the location of output neuron j, n is the number of output neurons, $Layer_j$ is the network input of neuron j, Φ' is the derivative of the activation function, δ is the value of neuron j error signal, w_{jk} is the connection weight between neuron j and k, w_{ij} is the connection weight between neuron i and j, E is the error function, and η is the learning rate.

2.2. Geographically Weighted Differential Factors-Artificial Neural Network

The GWANN can be used for value prediction but not in exploring the contributions of driving factors. Hence, we developed the geographically weighted differential factors-artificial neural network (GWDF-ANN), which can be obtained from https://figshare.com/s/3d06bb3dc4660396d539 (accessed on 15 November 2021). In GWDF-ANN, we adapted the existing architecture (GWANN) and we added a bias for each driving factor before prediction. The idea of differentiation is then utilized to quantify the contribution of each factor. The steps are as follows.

First, we use the FVC and five driving factors (temperature, precipitation, topography, mining, and urban expansion) (x_i) to build the GWANN model and predict Y_0. Second, a bias (Δx_i) is added to the x_i, which is calculated using Equation (7). The bias is added to a specific driving factor for all input neurons, while the other driving factors for each input neuron are kept constant. The bias does not affect the learning in the hidden layer of other inputs in computing the contribution of one factor. The setting of the bias is carried out in a series of experiments. If the bias is greater than 0.001, the contribution of the driving factor always changes. But if the bias is less than 0.001, the contribution of the driving factor hardly varies. From experiments, we found that 0.001 is the threshold value at which the contribution of the driving factor tends to stabilize. The bias is added separately for each driving factor and then input into the model to predict Y (Y_{xi}). The partial derivative for each driving factor is then calculated separately, using Equation (8). Finally, the partial derivative results are normalized using Equation (9), and the contribution of each driving factor to Y is obtained. The modeling process of GWDF-ANN is shown in Figure 2.

$$\triangle x_i = x_i * 0.001 \tag{7}$$

$$G_{x_i} = \frac{Y_{x_i} - Y_0}{\Delta x_i} \tag{8}$$

$$W_{x_i} = \frac{G_{x_i}}{\sum\limits_{i=1}^{n} G_{x_i}} \tag{9}$$

where x_i is the value of the driving factor i, Δx_i is the bias of x_i, G_{x_i} is the partial derivative of x_i, Y_{x_i} is the prediction after adding a bias to x_i, Y_0 is the prediction of original data, n is the number of driving factors, and W_{x_i} is the contribution of the x_i factor.

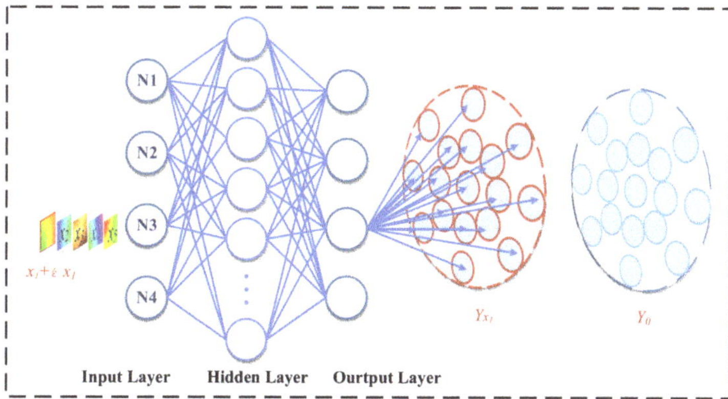

Figure 2. *Cont.*

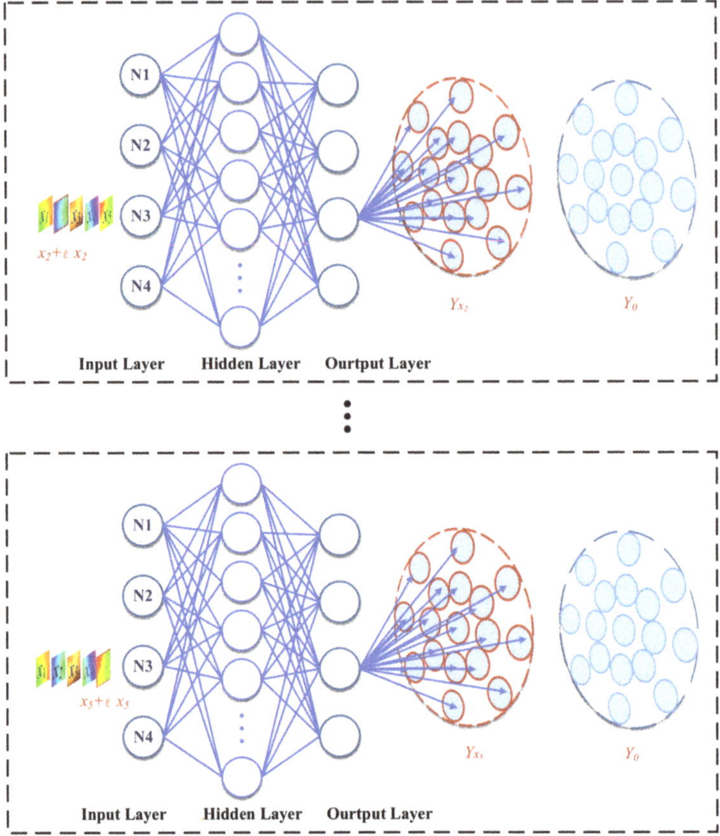

Figure 2. The modeling process of GWDF-ANN. N1, N2, N3, and N4 are the multiple input neurons composing the input layer. The first rectangular box shows the model after the first driving factor has been added with a bias to x_1 ($x_1 + \Delta x_1$). The Yx_1 represents the predicted result after adding a bias. The second rectangular box shows the model after the first driving factor has been added with a bias to x_2 ($x_2 + \Delta x_2$). The Yx_2 represents the predicted result after adding a bias. We do not show all the driving factors. The last rectangular box shows the model after the first driving factor has been added with a bias to x_5 ($x_5 + \Delta x_5$). The Yx_5 represents the predicted result after adding a bias.

3. A Case Study

3.1. Study Area

Shengli Coalfield is located in Xilinhot, Inner Mongolia, China ($115°18'\sim117°06'$ E, $43°02'\sim44°52'$ N), as shown in Figure 3. It belongs to the mid-temperate semiarid continental monsoon climate. From 2004 to 2020, the monthly 24 h average temperature ranges from $-18.4\ °C$ in January to $22.4\ °C$ in July, with an annual mean temperature of $2.76\ °C$. Most of the 250 mm (9.84 inches) annual rainfall occur in July and August. The average precipitation and temperature from January to December 2004–2020 are shown in Figure 4a. In winter, northwest winds predominate, while south and southeast winds are more prevalent in summer. Xilinhot's usable pasture area is 1,369,000 hectares and is a significant processing and production base for green livestock products in China. The grasslands are composed of meadow grassland, typical grassland, and dune grassland. Xilinhot is also rich in mineral resources, with more than 30 billion tons of proven coal reserves.

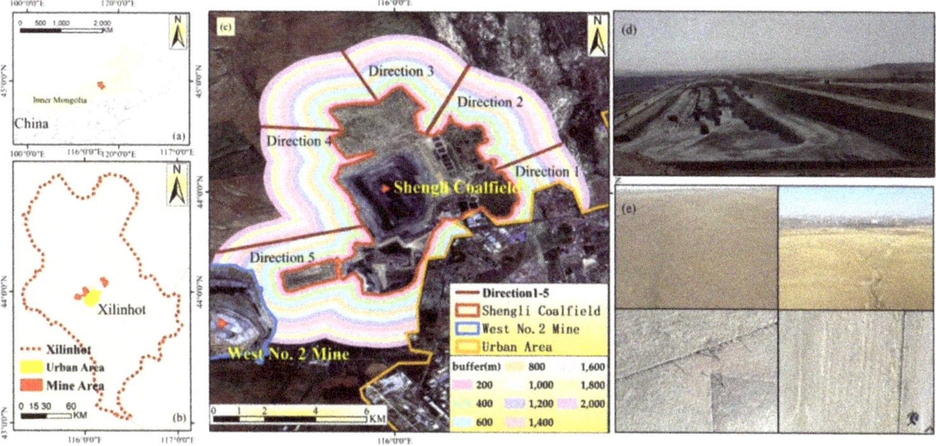

Figure 3. The geographical location of Shengli Coalfield in Xilinhot, Inner Mongolia, China. (**a**) Location of Xilinhot in China; (**b**) boundaries of the mine and urban areas in Xilinhot; (**c**) extent of the study area; (**d**) drone image of the mining area; (**e**) drone images of the grassland.

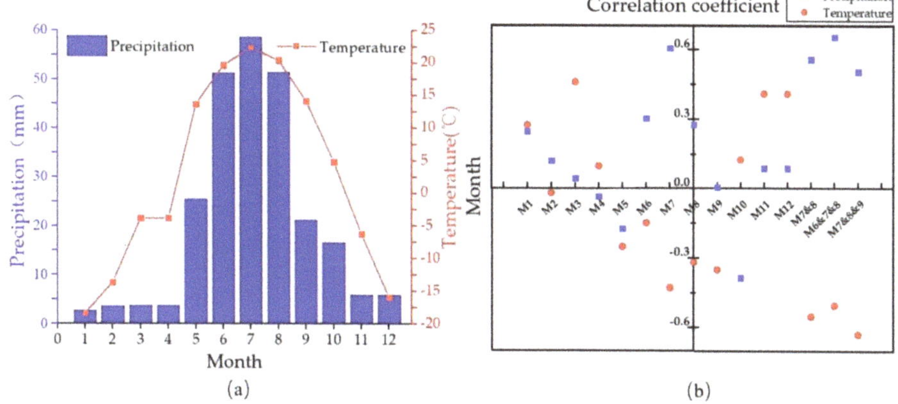

Figure 4. (**a**) The average precipitation and temperature from January to December during 2004−2020; (**b**) the correlation coefficient between FVC and precipitation and temperature.

The Shengli Coalfield is a lignite coalfield with the thickest coal seam, and it holds the largest reserves (estimated at 22.7 billion tons) in China. Its thick coal seam, shallow burial, and simple geological structure are suitable for centralized development. The coal quality is well suited for power generation, liquefaction, and chemical industries. The coalfield is located three kilometers north of Xilinhot. In 2004, Shengli Coalfield started the mining project and entered the large-scale construction phase in 2006.

A field survey was carried out in April 2020, in which drone images of the study area were taken and discussed with the residents. We found that the impact of mining activities on the vegetation two kilometers away from the coalfield boundary was minimal. While a larger area would have been more favorable, we were hampered by data acquisition constraints. In this paper, we selected a portion of the Shengli Coalfield to investigate the effects of mining and other factors on the vegetation using our proposed approach. We set a two-kilometer buffer and selected five directions (directions 1–5, Figure 3c) with uniform and random distribution to acquire more representative results. Given that the south of the Shengli Coalfield is mainly urban communities with little vegetation coverage, we did not include the south direction in this study. The placement of direction 5 was determined by considering the land cover, which was incompatible with the other four. Each direction was set up with ten analysis units placed at 200 m intervals. GWDF-ANN models were constructed separately for each analysis unit.

3.2. Fractional Vegetation Coverage

In this study, FVC was used as a proxy for vegetation coverage change. FVC can be calculated by the Normalized Difference Vegetation Index (NDVI) using the pixel dichotomy model. First, Landsat Collection 2 surface reflectance products were accessed on Google Earth Engine (GEE) from 2004 to 2020. All Collection 2 surface reflectance products were created with a single-channel algorithm jointly produced by the Rochester Institute of Technology (RIT) and National Aeronautics and Space Administration (NASA) Jet Propulsion Laboratory (JPL). This 30 m spatial resolution product is derived from Landsat 5 TM and Landsat 8 OLI sensors and has undergone radiation calibration and atmospheric correction.

We selected satellite imageries from July 1 to September 30, when vegetation grows vigorously. The masking and the removal of clouds on the images were implemented on the GEE. Cloud removal was achieved using the QA quality band of Landsat and filtering the pixel values by bit manipulation. Masking was then used to remove clouds, cloud shadows, and snow pixels. The processed satellite images were synthesized using the maximum function in GEE. The maximum algorithm reduces image collection by calculating the maximum values at each pixel across the stack of all matching bands from July to September for each year. In other words, we used this reduction function for the images of every year. Years 2008, 2012, and 2018 were excluded due to excessive cloudiness. The information of satellite imageries is shown in Table 1.

We then calculated the NDVI for each pixel in each year using Equation (10). We corrected the NDVI for Landsat 5 TM to Landsat 8 OLI to resolve sensor differences between satellites. Previous studies have proposed methods to calibrate Landsat 5 TM, Landsat 7 ETM+, and Landsat 8 OLI [40–42]. We followed their procedure and selected some sample plots in Xilinhot to calibrate these sensors. We extracted the NDVI from TM and ETM+ for the same dates and performed regression analysis on the two NDVI groups to obtain the correction equation for TM to ETM+. The same method was used to calibrate ETM+ to OLI. Finally, the TM was corrected to OLI.

We then calculated the FVC according to the pixel dichotomy model using Equation (11). Pixel dichotomy assumes that ground pixels exist in a linear mixture of vegetation and soil. The pixel's NDVI value is only affected by vegetation coverage. We selected plots that were completely bare. The NDVI pixel values from 2013 to 2020 were sorted in descending order, and the 95% percentile was used as NDVIsoil. The NDVIveg and NDVIsoil are stable from year to year with the long time series. We then selected some with complete vegetation cover during the peak growing season. The NDVI values in these plots were sorted in ascending order, and the 95% percentile was selected as NDVIveg. The long time series of pixel values can ensure the interannual stability of NDVIveg and NDVIsoil. The NDVIveg and NDVIsoil values are suitable since the different satellite sensors were corrected for consistency. The use of the 95% percentile pixel values reduces anomalous noisy pixels. The NDVIsoil was calculated to be 0.08, while NDVIveg was 0.7.

$$NDVI = \frac{\rho_{nir} - \rho_{red}}{\rho_{nir} + \rho_{red}} \tag{10}$$

$$FVC = \frac{NDVI - NDVI_{soil}}{NDVI_{veg} - NDVI_{soil}} \tag{11}$$

where ρ_{nir} is the surface reflectance in the near-infrared band, ρ_{red} is the surface reflectance in the red band, $NDVI_{soil}$ is the NDVI for pure bare soil pixel, and $NDVI_{veg}$ is the NDVI for pure vegetation pixel.

Table 1. Information of satellite imageries.

Sensor	Year	Path/Row	Date						Number
Landsat 8 OLI	2020	124/29	9 July	25 July	10 August	26 August	11 September	27 September	16
		124/30	9 July	25 July	10 August	26 August	11 September	27 September	
		125/29	16 July	1 August	2 September	18 September			
	2019	124/29	7 July	23 July	8 August	24 August	9 September	25 September	16
		124/30	7 July	23 July	8 August	24 August	9 September	25 September	
		125/29	14 July	31 July	31 August	16 September			
	2017	124/29	1 July	17 July	18 August	3 September	19 September		16
		124/30	1 July	17 July	18 August	3 September	19 September		
		125/29	8 July	24 July	9 August	25 August	10 September	26 September	
	2016	124/29	14 July	30 July	21 August	16 September			14
		124/30	14 July	30 July	21 August	16 September			
		125/29	5 July	21 July	6 August	22 August	7 September	23 September	
	2015	124/29	12 July	28 July	13 August	29 August	14 September	30 September	18
		124/30	12 July	28 July	13 August	29 August	14 September	30 September	
		125/29	3 July	19 July	4 August	20 August	5 September	21 September	
	2014	124/29	9 July	25 July	10 August	26 August	11 September	27 September	16
		124/30	9 July	25 July	10 August	26 August	11 September	27 September	
		125/29	16 July	1 August	2 September	18 September			
	2013	124/29	6 July	22 July	7 August	23 August	8 September	24 September	16
		124/30	6 July	22 July	7 August	23 August	8 September	24 September	
		125/29	13 July	30 July	30 August	15 September			
Landsat 5 TM	2011	124/29	1 July	17 July	18 August	3 September	19 September		16
		124/30	1 July	17 July	18 August	3 September	19 September		
		125/29	8 July	24 July	9 August	25 August	10 September	26 September	
	2010	124/29	14 July	30 July	21 August	16 September			14
		124/30	14 July	30 July	21 August	16 September			
		125/29	5 July	21 July	6 August	22 August	7 September	23 September	
	2009	124/29	11 July	27 July	12 August	28 August	13 September	29 September	18
		124/30	11 July	27 July	12 August	28 August	13 September	29 September	
		125/29	2 July	18 July	3 August	19 August	4 September	20 September	

Table 1. *Cont.*

Sensor	Year	Path/Row	Date						Number
	2007	124/29	6 July	22 July	7 August	23 August	8 September	24 September	
		124/30	6 July	22 July	7 August	23 August	8 September	24 September	16
		125/29	13 July	30 July	30 August	15 September			
	2006	124/29	3 July	29 July	4 August	20 August	5 September	21 September	
		124/30	3 July	29 July	4 August	20 August	5 September	21 September	17
		125/29	10 July	27 July	27 August	12 September	28 September		
	2005	124/29	16 July	17 August	2 September	18 September			
		124/30	16 July	17 August	2 September	18 September			14
		125/29	7 July	23 August	7 August	24 August	9 September	25 September	
	2004	124/29	13 July	29 July	14 August	30 August	15 September		
		124/30	13 July	29 July	14 August	30 August	15 September		16
		125/29	4 July	20 July	5 August	21 August	6 September	22 September	

3.3. Driving Factors

For this study, temperature, precipitation, topography, mining, and urban expansion were selected as driving factors. Precipitation and temperature data were obtained from the China Meteorological Data Network (http://data.cma.cn (accessed on 1 January 2021)) (station number 54,102). We collected the monthly cumulative precipitation (unit: mm) and monthly average temperature (unit: °C) for Xilinhot in 2004–2020 (except 2008, 2012, and 2018) from station number 54,102. Numerous studies have shown that vegetation response to climate and precipitation may exhibit varying time lag effects for different regions [43,44]. It remains unclear how precipitation and temperature correlate with the FVC in the study area at different months of the year. Therefore, we performed a Pearson correlation analysis on temperature, precipitation, and FVC for each month. The correlation coefficients between FVC and precipitation and temperature are presented in Figure 4b. The results show that the cumulative precipitation from July to September and the average temperature from July to September have the highest correlations with the FVC at 0.660 and −0.616, respectively. Hence, they were selected as driving factors.

The topography data were derived from the digital elevation model (DEM) of the Shuttle Radar Topography Mission (SRTM), released by NASA in 2014. The spatial resolution is 30 m. Since topography does not significantly change year to year and annual topography data is unavailable, the same topography data was used for each year (see Figure 5). The mining factor was determined using the annual coal production, and the shortest Euclidean distance between the grid cell and the mining boundary and is calculated by Equation (12). Aside from the Shengli Coalfield, we also accounted for the effect of the West No. 2 Mine. This Euclidean distance refers to the shortest distance from each grid cell to each mine boundary. Coal production refers to the total output of the Shengli Coalfield and the West No. 2 Mine. The urban expansion was quantified using the annual urban population of Xilinhot and the shortest Euclidean distance between the grid cell and the mining boundary, as shown in Equation (13).

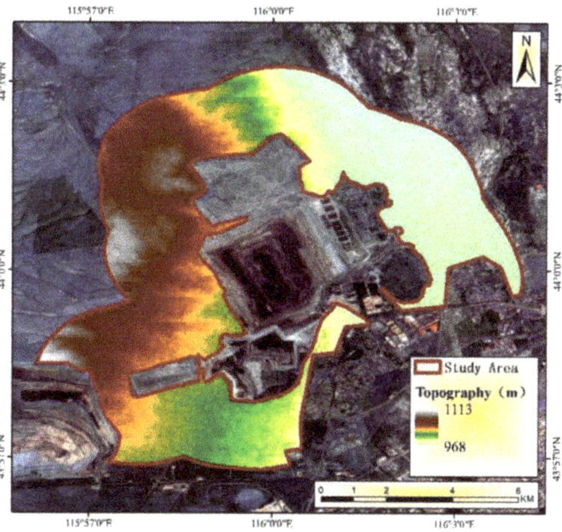

Figure 5. The spatial distribution of topography.

Annual coal production was provided by the National Energy Investment Group Co., Ltd. (Beijing, China). Coal production did not increase annually. The annual coal production from 2004 to 2020 is shown in Figure 6a. The urban population was obtained from the Xilinhot Statistical Yearbook, while the mining and urban boundaries were generated using Landsat imageries. In this study, we quantified mining and urban expansion from 2004 to 2020 (except 2008, 2012, and 2018). The quantitative results for the mining factor in 2020 are presented in Figure 6b. The results for the urban expansion factor in 2020 are shown in Figure 7.

$$X_{mine} = \frac{MA}{ED_{mine} + 1} \tag{12}$$

$$X_{urban} = \frac{NP}{ED_{urban} + 1} \tag{13}$$

where MA is the mined amount (unit: 10^4 m^3), ED_{mine} is the shortest Euclidean distance between the grid cell and the boundary of the mining area (unit: km), NP is the urban population of Xilinhot, and ED_{urban} is the shortest Euclidean distance between the grid cell and the urban boundary (unit: km).

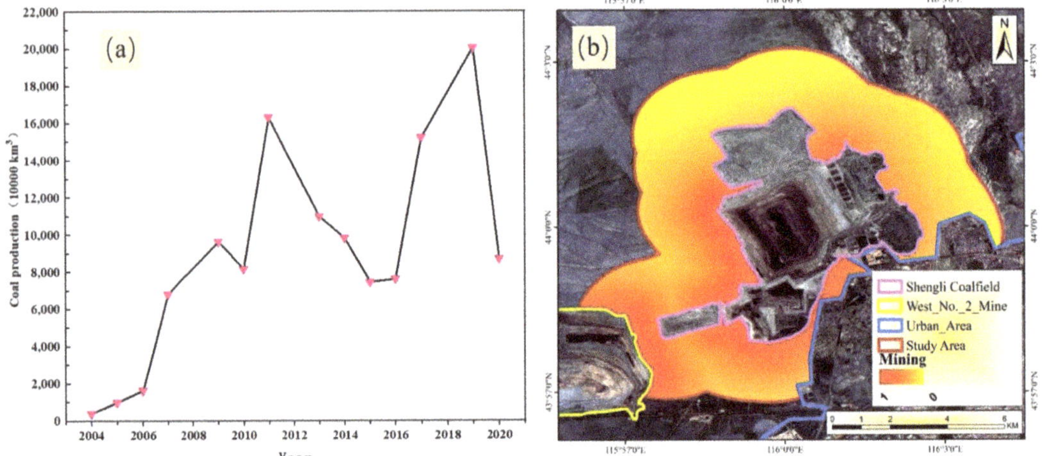

Figure 6. (**a**) Coal production from 2004 to 2020; (**b**) quantitative results of the mining factor in 2020 (normalized).

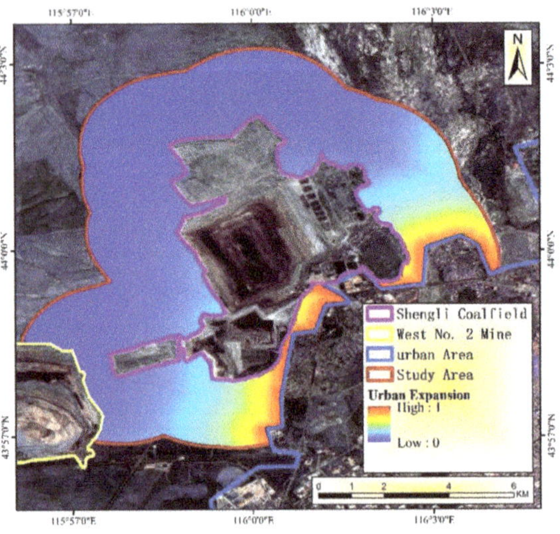

Figure 7. Quantitative results of the urban expansion factor in 2020 (normalized).

3.4. Model Building

We set a two-kilometer buffer in the study area and delineated the buffer zone uniformly and randomly in five directions (directions 1–5, Figure 3c). Each direction had ten analysis units placed at 200 m intervals. The analysis unit was composed of 81 grid cells, separated by 30 m. Each unit was a circular area with a 90 m radius and was used to train a model; 50 driving models were established.

Before model training, we normalized the FVC, temperature, precipitation, topography, mining, and urban expansion factors for 2004–2020 (except 2008, 2012, and 2018) using Equation (14). The normalized data were then extracted to the corresponding grid cells (*y*-value and *x*-values). In this case, the input values were mapped on a 0–1 range with no dimension.

$$Z_i = \frac{x_i - \max_{1 \le i \le n}(x_i)}{\max_{1 \le i \le n}(x_i) - \min_{1 \le i \le n}(x_i)} \tag{14}$$

where Z_i is the normalized value of the Z factor in grid cell *i*, x_i is the original value of the grid cell *i*, and *n* is the number of grid cells.

The 15-year grid cells (*y*-value and *x*-values) for each analysis unit were input into the model as the input layer. The FVC value (Y_0) was predicted for each analysis unit. The input data were then modified. Biases were added to the five driving factors ($x_i + \Delta x_i$) and were input back into the model to predict the FVC value (Y_{xi}) for each driving factor. Finally, the GWDF-ANN method was used to obtain the contribution of each grid cell. The average value of all the grid cells in each analysis unit was calculated to determine the contribution of the analysis unit. The procedure was repeated for each analysis unit to obtain the contribution (*W*) of the different analysis units. Each analysis unit uses an independent GWDF-ANN model.

4. Results

4.1. Modeling Results and Accuracy

Models were generated for the different analysis units. We selected the predicted FVC results (Y_0) of the first analysis unit in each direction and compared them with the actual FVC results (*y*-value), as shown in Figure 8. The predicted FVC values are similar to the actual results. RMSE and MRE were used as the model evaluation indicators, and the calculation methods are shown in Equations (15) and (16). The RMSE and MRE are typical accuracy evaluation indicators widely used in evaluating modeling accuracy in geospatial modeling [45,46]. The results of the RMSE and MRE are shown in Table 2. The average RMSE for the 50 groups of models is 0.052. The minimum RMSE for analysis unit 16 is 0.007, and the highest RMSE for analysis unit 10 is 0.112. The average MRE value is 0.007. The lowest MRE for analysis unit 6 is 0.001, while the highest MRE for analysis unit 46 is 0.023. The results suggest that the GWDF-ANN model is suitable for quantifying the change of FVC in mining areas.

$$\text{RMSE} = \sqrt{\frac{1}{n}\sum_{i=1}^{n}\left(\text{FVC}_{\text{pred}_i} - \text{FVC}_{\text{true}_i}\right)^2} \tag{15}$$

$$\text{MRE} = \frac{1}{n}\sum_{i=1}^{n}\frac{\left|\text{FVC}_{\text{pred}_i} - \text{FVC}_{\text{true}_i}\right|}{\text{FVC}_{\text{true}_i}} \tag{16}$$

where *n* is the total number of grid cells in training data, $\text{FVC}_{\text{pred}_i}$ is the FVC value of grid cell *i* predicted by the model, and $\text{FVC}_{\text{true}_i}$ is the actual FVC of grid cell *i*.

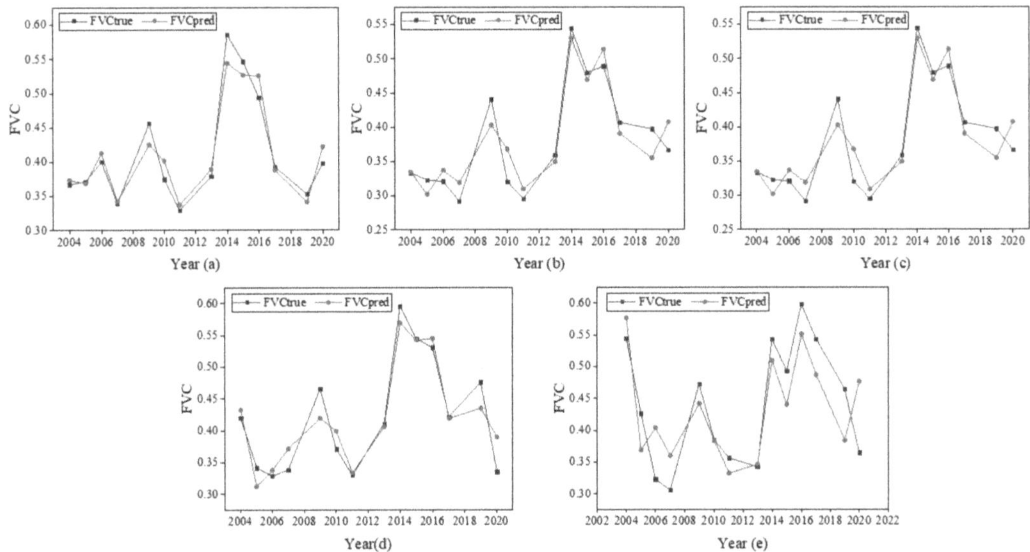

Figure 8. The comparison of actual FVC and predicted FVC results. (**a–e**): Directions 1–5.

Table 2. Model accuracy.

	Model 1	Model 2	Model 3	Model 4	Model 5	Model 6	Model 7	Model 8	Model 9	Model 10
RMSE	0.0577	0.0567	0.0654	0.0642	0.0506	0.0493	0.0548	0.0554	0.0853	0.1117
MRE	0.0064	0.0084	0.0061	0.0035	0.0019	0.0014	0.0040	0.0006	0.0023	0.0162
	Model 11	**Model 12**	**Model 13**	**Model 14**	**Model 15**	**Model 16**	**Model 17**	**Model 18**	**Model 19**	**Model 20**
RMSE	0.0537	0.0621	0.0190	0.0173	0.0262	0.0066	0.0334	0.0148	0.0935	0.0141
MRE	0.0095	0.0056	0.0076	0.0032	0.0085	0.0044	0.0061	0.0078	0.0080	0.0015
	Model 21	**Model 22**	**Model 23**	**Model 24**	**Model 25**	**Model 26**	**Model 27**	**Model 28**	**Model 29**	**Model 30**
RMSE	0.0468	0.0930	0.0139	0.0954	0.0142	0.0630	0.0597	0.0709	0.0805	0.0963
MRE	0.0218	0.0042	0.0058	0.0025	0.0076	0.0083	0.0039	0.0087	0.0086	0.0020
	Model 31	**Model 32**	**Model 33**	**Model 34**	**Model 35**	**Model 36**	**Model 37**	**Model 38**	**Model 39**	**Model 40**
RMSE	0.0706	0.0688	0.0178	0.0230	0.0167	0.0522	0.0256	0.0352	0.0931	0.0888
MRE	0.0038	0.0068	0.0042	0.0073	0.0043	0.0096	0.0078	0.0085	0.0057	0.0072
	Model 41	**Model 42**	**Model 43**	**Model 44**	**Model 45**	**Model 46**	**Model 47**	**Model 48**	**Model 49**	**Model 50**
RMSE	0.0494	0.0598	0.0671	0.0223	0.0246	0.0660	0.0399	0.0883	0.0143	0.0569
MRE	0.0074	0.0078	0.0034	0.0120	0.0093	0.0234	0.0056	0.0126	0.0056	0.0130

4.2. FVC Spatial Changes and Quantitative Results of Driving Factors

Figure 9 shows the spatial distribution of FVC values within two kilometers from the Shengli Coalfield. The FVC values are in the 0–1 range; the higher the value, the greater the vegetation coverage. From 2004 to 2010 (except 2008), FVC decreased annually. The average FVC was 0.48, 0.40, 0.37, 0.34, 0.34, and 0.30, decreasing by 37.02% in six years. The FVC had a short-term increase from 2011 to 2013 and then decreased from 2013 to 2019. The lowest FVC value was 0.30, occurring in 2010, and the highest was 0.63 in 2020.

We calculated the average contributions of all the analysis units in the five directions. We found that the most prominent driving factor for FVC is precipitation (32.45%), followed by temperature (24.80%), mining (22.44%), and urban expansion (14.44%). The topography had the lowest contributions, with 5.87%. When combined, precipitation and temperature drive more than half (57.25%) of the FVC changes.

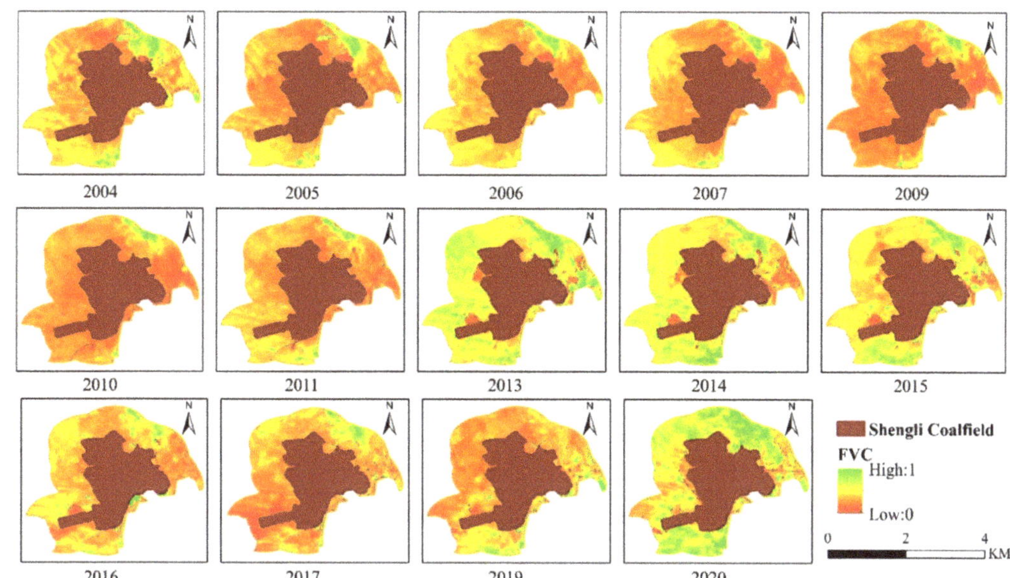

Figure 9. The spatial distribution of FVC from 2004 to 2020 (except 2008, 2012, and 2018).

We then computed the average factor contributions of the ten analysis units for a given year at varying distances from the boundary of the mining area, and the results are displayed in Figure 10. The contribution of each driving factor shows similar characteristics. Over the years, the contributions of the different driving factors constantly changed. Precipitation and temperature have a downward trend, while mining and urban expansion have an increasing trajectory. The topography contributions remained unchanged.

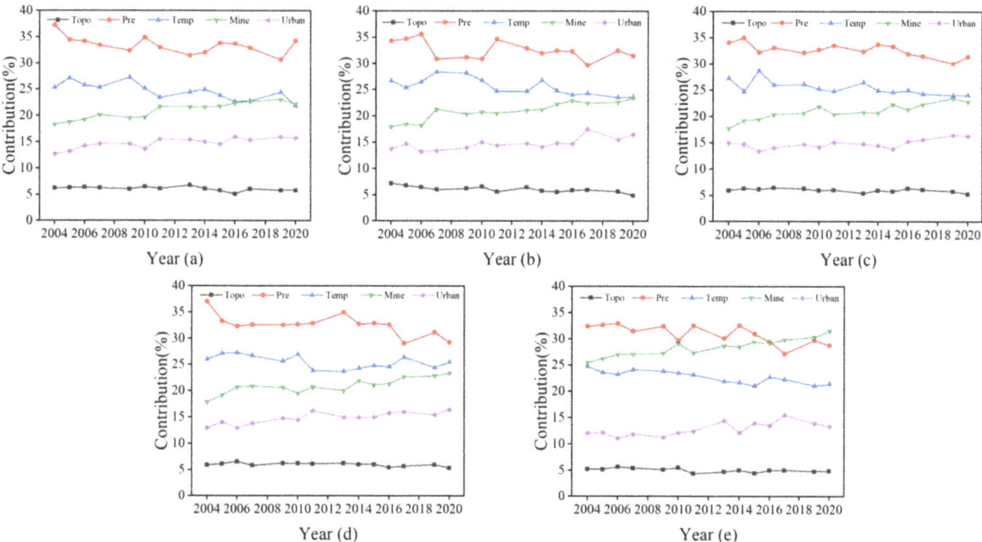

Figure 10. The contribution of driving factors from 2004 to 2020. (**a–e**): Directions 1–5. Topo represents the factor of topography; Pre represents the factor of precipitation; Temp represents the temperature factor; Mine represents the factor of mining, and Urban represents the factor of urban expansion.

To compare the contribution of a particular driving factor in different years, take direction 1 as an example. Precipitation had the highest contribution at 37.28% in 2004 and the lowest contribution in 2019 at 30.68%. Temperature's contribution was at 25.37% in 2004 and it contributed the least in 2020 at 21.88%. The mining effect increased year by year, closely following temperature. Temperature's contributions in 2016, 2017, and 2020 were comparable. The trend suggests that temperature has become the second major driving factor. The contribution of mining from 2004 to 2020 increased from 18.38% to 22.23%, in which the fastest growth occurred in 2010–2011, rising by 10.04 percentage points. Urban expansion had a relatively flat growth trend and had declined in specific years. Still, urban expansion's overall contributions have grown over the years, increasing from 12.73% in 2004 to 15.77% in 2020. For topography, the interannual contributions had very little change, only fluctuating within a 1% variation range.

The interannual contributions of each driving factor in the five directions had similar fluctuation characteristics, with the fifth direction being slightly different. The contribution of mining in direction 5 increased significantly interannually. In 2004, the major contributing factors were precipitation (32.42%), mining (25.47%), and temperature (24.75%). For 2020, mining (31.52%), precipitation (28.75%), and temperature (21.43%) were the main contributors. During the early stages of mining in 2004, its contribution was the second leading driving factor, surpassing that of temperature. However, over the years, temperature's role gradually diminished while mining's influence increased. From 2004 to 2020, contributions from mining increased from 25.47% to 31.52%, growing by 23.77 percentage points. After 2016, mining surpassed precipitation and became the leading driving factor.

Interannually, the contributions from precipitation and temperature show a fluctuating trend. The increase in the contribution of the precipitation is accompanied by the decrease in the influence of temperature. Likewise, the reduction in the contribution of precipitation is accompanied by the increase in the contribution of temperature. In direction 4, the contribution of precipitation showed a downward trend in 2004–2006, 2014–2017, and 2019–2020, and an upward trend in other years. Temperature showed an upward trajectory in 2004–2006, 2014–2017, and 2019–2020, and had a downward trend in other years.

We then took the average contribution of each factor in the ten analysis units at varying distances from the boundary of the mining area in the same direction. We calculated the change in the contribution of each factor in different directions, and the results are shown in Figure 11. The contribution of driving factors in direction 5 is noticeably different from other directions. In direction 5, the contribution of mining increased significantly, while the contributions of precipitation, temperature, urban expansion, and topography decreased.

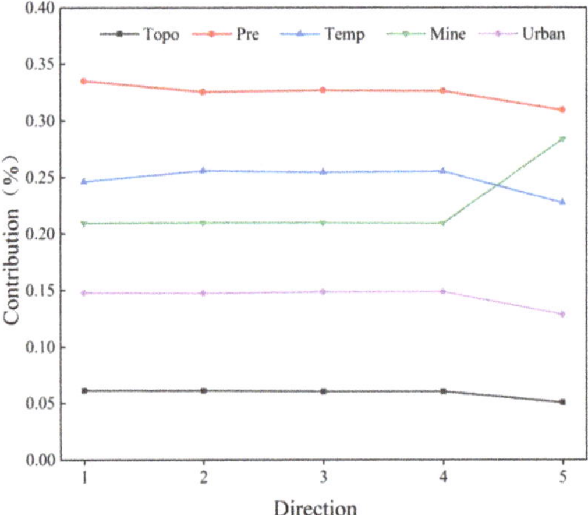

Figure 11. The contribution of driving factors in different directions.

We computed the mean contribution of each factor in the ten analysis units at different distances from the boundary of the mining area in the same direction for 2004–2020 (except 2012 and 2018). We drew pie charts detailing the factor contribution of different directions (see Figure 12). Precipitation had the maximum contribution of 33.50% in direction 1 and a minimum of 30.94% in direction 5. For temperature, the maximum contribution was in direction 2 at 25.61%, and the minimum in direction 5 at 22.76%. The maximum contribution of mining in direction 5 is 28.36%, and the minimum in direction 1 is 20.93%. Urban expansion had the largest share of contribution in direction 4 at 14.89% and the smallest in direction 5 at 12.88%. For topography, the maximum contribution was 6.13% in direction 1 and the minimum at 5.06% in direction 5.

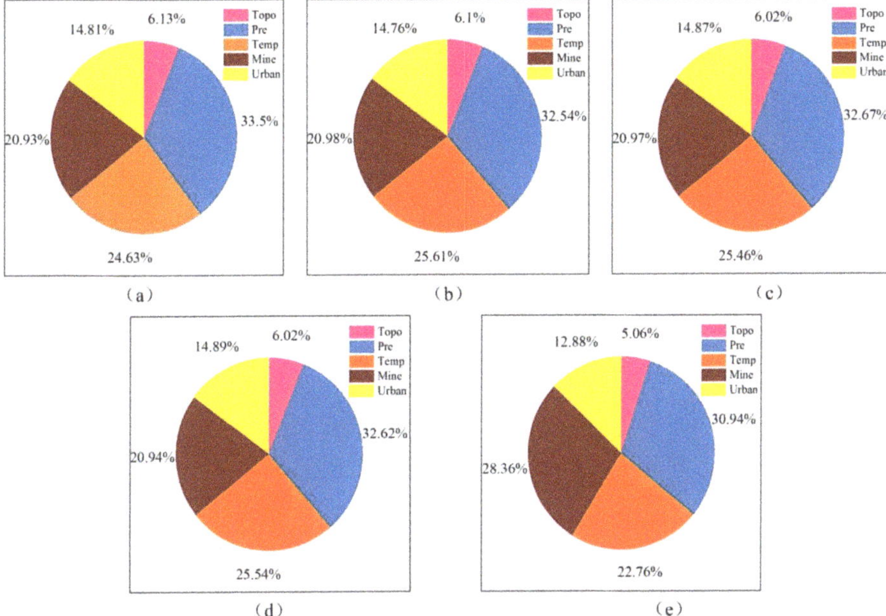

Figure 12. The pie chart of driving factor contribution in different directions. (**a–e**): Directions 1–5. Topo represents the factor of topography; Pre represents the factor of precipitation; Temp represents the factor of temperature; Mine represents the factor of mining, and Urban represents the factor of urban expansion.

The contributions of the various driving factors varied for the different directions. In directions 1–3, the order of driving factors from highest to lowest contribution was precipitation, temperature, mining, urban expansion, and topography. In direction 4, the ranking from highest to lowest was precipitation, temperature, mining, topography, and urban expansion. In this direction, the topography influence exceeded that of urban expansion. In direction 5, the order from highest to lowest contribution was precipitation, mining, temperature, urban expansion, and topography. Mining surpassed temperature to become the number two leading driving factor, second only to precipitation.

4.3. Contribution Analysis of the Mining Factor

We took the average value of mining contribution for each analysis unit from 2004 to 2020 (except 2008, 2012, and 2018) and analyzed the mining contributions at varying distances from the Shengli Coalfield boundary. We plotted the mean values for the different directions, and the results are presented in Figure 13. In all five directions, the contribution of mining showed apparent distance attenuation. As the distance from the boundary of the Shengli Coalfield increased, the contribution from mining exhibited a downward trend. This means that the farther away a point is from the mining area, the lower the contribution of mining. The maximum contribution of mining in the five directions is 26.69% at 200 m from the mining area boundaries, which drops to 17.8% at 2000 m, decreasing by 33.31%. As the distance from the mining area boundary increases from 200 m to 2000 m, the contribution of mining in directions 1–5 decreases by 0.09, 0.10, 0.09, 0.08, and 0.08, respectively.

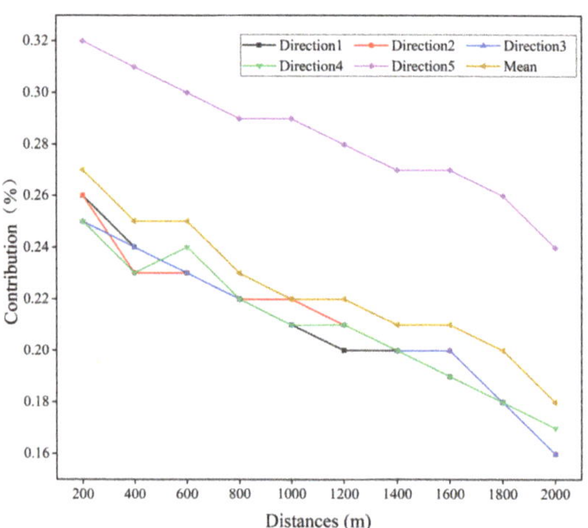

Figure 13. The contribution of the mining factor at different distances from the boundary of Shengli Coalfield.

5. Discussion

The GWDF-ANN method proposed in this study quantifies the contribution of driving factors to changes in the FVC. In particular, this study explores the impact of mining on FVC. Some results need further explanation and discussion.

The main driving factor of FVC around the mining area is not mining but precipitation, followed by temperature. In this study, the average contributions of precipitation and temperature from 2004 to 2020 in all analysis units were 32.45% and 24.80%, respectively. Since the study area was set as a buffer zone, two kilometers away from the mine boundary, the spatial heterogeneity of temperature and precipitation was almost absent. Therefore, the temperature and precipitation driving factors for each analysis unit were the same station data. The differences in temperature and precipitation were mainly between years. For example, the cumulative precipitation from July to September was high in 2004 and 2020 (274.3 mm and 268.4 mm) and low in 2014 and 2017 (93.9 mm and 93.1 mm). These values are consistent with our results, as in direction 1, the contribution of precipitation factor was high in 2004 and 2020 (37.28% and 33.27%) and low in 2014 and 2017 (32.07% and 32.93%). We found that more than 50% of the vegetation changes had been driven by precipitation and temperature and that the sum of their contributions reached 57.25%. Many studies have shown that vegetation change is very responsive to precipitation and temperature [47–50], consistent with the results of this study. In arid and semiarid grasslands, the contribution of precipitation on FVC is significantly greater than that of temperature.

In addition to climate factors, this study quantified the impact of mining and urban expansion. Hui and Aman Fang found that mining significantly degrades vegetation coverage [51,52], but its specific contribution to vegetation coverage change has not been thoroughly explored. From 2004 to 2020, the contribution of mining within two kilometers around the boundary of the Shengli Coalfield was 22.44%. Mining causes severe degradation in groundwater levels and noticeable soil desertification, resulting in harsh environments for vegetation growth. Coal mine dust falling on plant leaf surface may also affect plant growth and cause other serious problems. Around the mining area, FVC degradation caused by mining is apparent.

The city's rapid expansion and industrialization have resulted in decreased FVC. From 2004 to 2020, the contribution rate of urban expansion within two kilometers from the Shengli Coalfield is 14.81%. From 2004 to 2020, the urban area of Xilinhot expanded by 129.184 km^2, and the urban population increased by 42,697 people. Urban expansion reduces the dependence of plant growth on climate factors and has a particular impact on FCV, consistent with Song's and Noa's findings [53,54].

As for the topography, its influence on FVC is relatively small, and its interannual contribution remains unchanged. The topography showed little change from year to year, and its differentiation is mainly spatial. There is a 45 m elevation difference in the study area, high in the west and low in the east. However, the impact of topographic factors does not significantly change in different directions. This may be because the elevation of the mining area does not have a particular contribution to the vegetation.

The effects of mining have noticeable spatial and temporal variations. Temporally, the value of the mining factor varies with the coal production. The findings also show that the contribution of mining to FVC continued to increase annually, even as coal production decreased in certain years. In 2009, 2011–2015, and 2019, while coal production in the Shengli Coalfield declined from the previous year, the contribution of mining did not decrease. Hence, the results suggest that the impact of mining on FVC may have a cumulative effect in time. Years of continuous mining have exacerbated the adverse effects of mining operations on FVC.

Spatially, the further away from the mine boundary, the lower the value of the mining driving factor. The contribution of mining shows distance attenuation around the mining area. Within 200 m from the periphery of the mining area, mining's contribution was 26.69%; at 2000 m, it dropped to 17.84%. As the distance from the mining area increases, the contribution of mining decreases until it reaches some point where the value is zero. This suggests that the impact of mining on FVC may be limited. For example, Aman Fang studied the influence area of Bao's coal in eastern Inner Mongolia [51]. The scope of mining influence on FVC needs to be further explored.

The contribution of mining in direction 5 was significantly higher than in other directions. This may be because the Shengli Coalfield is close to the West No. 2 Mine (about three kilometers away), affecting the FVC. The West No. 2 Mine started operations in 2006, and by 2020, the cumulative coal produced was 825,110,800 m^3, about 0.91 times the cumulative coal production of the Shengli Coalfield. The high mining impact in direction 5 was likely caused by the cumulative effect of the Shengli Coalfield and the West No. 2 Mine.

Changes in the FVC are caused by the coupling of multiple factors. We only considered the contribution of precipitation, temperature, mining, urban expansion, and topography. Aside from these factors, overgrazing may have a considerable impact on the FVC [55]. Ecological environment restoration policies such as vegetation reclamation have also been found to significantly affect the FVC in mining areas [56,57]. Subsequent studies can analyze other parameters affecting vegetation coverage in mining areas, including grazing activities, environmental protection policies, evaporation, and soil types. In addition, there are currently few quantitative studies on mining and urban expansion. How to scientifically quantify the impact of mining and urban expansion requires further exploration.

6. Conclusions

This study proposed the GWDF-ANN to quantify the contributions of different driving factors on FVC changes in mining areas. Based on the results, some conclusions were reached, as follows.

(1)   For the 50 models, the average RMSE was 0.052 and the average MRE was 0.007. The GWDF-ANN model is suitable for quantifying FVC changes in mining areas.

(2)   Precipitation and temperature were the main driving factors for FVC change. The contributions were 32.45% for precipitation, 24.80% for temperature, 22.44% for mining, 14.44% for urban expansion, and 5.87% for topography.

(3)   The contributions of precipitation and temperature on vegetation cover exhibited downward trends, while mining and urban expansion showed positive trajectories. For topography, its contribution remains generally unchanged.

(4)   The contribution of mining showed apparent distance attenuation. At 200 m away, the contribution of mining was 26.69%; at 2000 m away, the value drops to 17.8%.

(5)   Mining has a cumulative effect on vegetation coverage both interannually and spatially.

In the future, more driving factors, such as grazing and soil quality, can be considered to improve model accuracy.

Author Contributions: Conceptualization, J.L. and C.Z. (Chengye Zhang); methodology, J.L., C.Z. (Chengye Zhang) and T.Q.; validation, T.Q., H.Z. and J.G.; formal analysis, J.L. and T.Q.; investigation, J.G., H.X., C.Z. (Caiyue Zhang) and Y.Z.; writing—original draft preparation, J.L. and T.Q.; writing—review and editing, T.Q. and C.Z. (Chengye Zhang); visualization, T.Q. and H.Z.; supervision, T.Q. and H.Z.; project administration, C.Z. (Chengye Zhang); funding acquisition, C.Z. (Chengye Zhang). All authors have read and agreed to the published version of the manuscript.

Funding: This research was funded by the State Key Laboratory of Water Resource Protection and Utilization in Coal Mining (Grant number GJNY-20-113-14), National Natural Science Foundation of China (Grant number 41901291), Fundamental Research Funds for the Central Universities (Grant number 2021YQDC02).

Institutional Review Board Statement: Not applicable.

Informed Consent Statement: Not applicable.

Data Availability Statement: The Landsat data and topography data can be downloaded from GEE (https://developers.google.com/earth-engine/datasets/, accessed on 15 November 2021). And the precipitation and temperature data can be downloaded from the China Meteorological Data Network (http://data.cma.cn, accessed on 15 November 2021).

Acknowledgments: The authors express gratitude to the National Energy Investment Group Co., Ltd. (Beijing, China) for providing the necessary information and data.

Conflicts of Interest: The authors declare no conflict of interest.

References

1.   Bai, X.; Ding, H.; Lian, J.; Ma, D.; Yang, X.; Sun, N.; Xue, W.; Chang, Y. Coal production in China: Past, present, and future projections. *Int. Geol. Rev.* **2017**, *60*, 535–547. [CrossRef]

2.   Wang, Y.J. Rearch Progress and Prospect on Ecological Disturbance Monitoring in Mining Area. *Acta Geod. Cartogr. Sin.* **2017**, *45*, 1705–1716.

3.   Liu, S.J.; Wang, Z.; Mao, Y.C.; Xu, B.S.; Wu, L.X. Multi-source Remote Sensing Technology for Monitoring Safety and Environment in Mine. *Geomat. Spat. Inf. Technol.* **2015**, *38*, 98–100.

4.   Wu, Z.; Lei, S.; Lu, Q.; Bian, Z. Impacts of Large-Scale Open-Pit Coal Base on the Landscape Ecological Health of Semi-Arid Grasslands. *Remote Sens.* **2019**, *11*, 1820. [CrossRef]

5.   Qian, T.; Bagan, H.; Kinoshita, T.; Yamagata, Y. Spatial–Temporal Analyses of Surface Coal Mining Dominated Land Degradation in Holingol, Inner Mongolia. *IEEE J. Sel. Top. Appl. Earth Obs. Remote Sens.* **2014**, *7*, 1675–1687. [CrossRef]

6.   Parmesan, C.; Yohe, G. A globally coherent fingerprint of climate change impacts across natural systems. *Nature.* **2003**, *421*, 37–42. [CrossRef] [PubMed]

7.   Chen, L.F.; Zhang, H.; Zhang, X.Y.; Liu, P.H.; Zhang, W.C.; Ma, X.Y. Vegetation changes in coal mining areas: Naturally or anthropogenically Driven? *Catena* **2022**, *208*, 105712. [CrossRef]

8.   Hao, L.; Sun, G.; Liu, Y.; Gao, Z.; He, J.; Shi, T.; Wu, B. Effects of precipitation on grassland ecosystem restoration under grazing exclusion in Inner Mongolia, China. *Landsc. Ecol.* **2014**, *29*, 1657–1673. [CrossRef]

9.   Chuai, X.W.; Huang, X.J.; Wang, W.J.; Bao, G. NDVI, temperature and precipitation changes and their relationships with different vegetation types during 1998–2007 in Inner Mongolia, China. *Int. J. Climatol.* **2013**, *33*, 1696–1706. [CrossRef]

10.   Pang, J.; Du, Z.Q.; Zhang, X.Y. Understanding of the relationship between vegetation change and physical geographic factors based on Geographical Detector. In Proceedings of the International Conference on Intelligent Earth Observing and Applications (IEOAs), Guilin, China, 9 December 2015; pp. 23–24.

11. Mu, S.J.; Chen, Y.Z.; Li, J.L.; Ju, W.M.; Odeh, I.O.A.; Zou, X.L. Grassland dynamics in response to climate change and human activities in Inner Mongolia, China between 1985 and 2009. *Rangel. J.* **2013**, *35*, 315–329. [CrossRef]

12. Li, J.; Deng, X.J.; Yang, Z.; Liu, Q.L.; Wang, Y.; Cui, L.Y. A Method of Extracting Mining Disturbance in Arid Grassland Based on Time Series Multispectral Images. *Spectrosc. Spectr. Anal.* **2019**, *39*, 3788–37893.

13. Wang, K.W.; Li, J.; Wang, R.G.; Fu, X. Spatial heterogeneity monitoring of temporal variation of vegetation coverage in Shengli mining area. *Bull. Surv. Mapp.* **2020**, *11*, 1–6. [CrossRef]

14. Li, J.; Yan, X.G.; Yan, X.X.; Guo, W.; Wang, K.W.; Qiao, J. Temporal and spatial variation characteristic of vegetation coverage in the Yellow River Basin based on GEE cloud platform. *J. China Coal Soc.* **2021**, *46*, 1439–1450. [CrossRef]

15. Li, X.H.; Lei, S.G.; Cheng, W.; Liu, F.; Wang, W.Z. Spatio-temporal dynamics of vegetation in Jungar Banner of China during 2000–2017. *J. Arid. Land* **2019**, *11*, 837–854. [CrossRef]

16. Yang, Y.; Erskine, P.D.; Lechner, A.M.; Mulligan, D.; Zhang, S.; Wang, Z. Detecting the dynamics of vegetation disturbance and recovery in surface mining area via Landsat imagery and LandTrendr algorithm. *J. Clean. Prod.* **2018**, *178*, 353–362. [CrossRef]

17. Yan, X.; Li, J.; Shao, Y.; Hu, Z.; Yang, Z.; Yin, S.; Cui, L. Driving forces of grassland vegetation changes in Chen Barag Banner, Inner Mongolia. *GISci. Remote Sens.* **2020**, *57*, 753–769. [CrossRef]

18. Fu, X.; Ma, M.; Jiang, P.; Quan, Y. Spatiotemporal vegetation dynamics and their influence factors at a large coal-fired power plant in Xilinhot, Inner Mongolia. *Int. J. Sustain. Dev. World Ecol.* **2016**, *24*, 433–438. [CrossRef]

19. Stewart Fotheringham, A.; Charlton, M.; Brunsdon, C. The geography of parameter space: An investigation of spatial non-stationarity. *Int. J. Geogr. Inf. Syst.* **1996**, *10*, 605–627. [CrossRef]

20. Bitter, C.; Mulligan, G.F.; Dall'erba, S. Incorporating spatial variation in housing attribute prices: A comparison of geographically weighted regression and the spatial expansion method. *J. Geogr. Syst.* **2006**, *9*, 7–27. [CrossRef]

21. Huang, B.; Wu, B.; Barry, M. Geographically and temporally weighted regression for modeling spatio-temporal variation in house prices. *Int. J. Geogr. Inf. Sci.* **2010**, *24*, 383–401. [CrossRef]

22. Yu, W.; Zang, S.; Wu, C.; Liu, W.; Na, X. Analyzing and modeling land use land cover change (LUCC) in the Daqing City, China. *Appl. Geogr.* **2011**, *31*, 600–608. [CrossRef]

23. Hagenauer, J.; Helbich, M. Local modelling of land consumption in Germany with RegioClust. *Int. J. Appl. Earth Obs. Geoinf.* **2018**, *65*, 46–56. [CrossRef]

24. Nelson, A.; Oberthür, T.; Cook, S. Multi-scale correlations between topography and vegetation in a hillside catchment of Honduras. *Int. J. Geogr. Inf. Sci.* **2007**, *21*, 145–174. [CrossRef]

25. Waller, L.A.; Zhu, L.; Gotway, C.A.; Gorman, D.M.; Gruenewald, P.J. Quantifying geographic variations in associations between alcohol distribution and violence: A comparison of geographically weighted regression and spatially varying coefficient models. *Stoch. Environ. Res. Risk Assess.* **2007**, *21*, 573–588. [CrossRef]

26. Wang, L.; Lee, G.; Williams, I. The Spatial and Social Patterning of Property and Violent Crime in Toronto Neighbourhoods: A Spatial-Quantitative Approach. *Isprs Int. J. Geo-Inf.* **2019**, *8*, 51. [CrossRef]

27. Hagenauer, J.; Helbich, M. A geographically weighted artificial neural network. *Int. J. Geogr. Inf. Sci.* **2021**, *36*, 215–235. [CrossRef]

28. Feng, L.; Wang, Y.; Zhang, Z.; Du, Q. Geographically and temporally weighted neural network for winter wheat yield prediction. *Remote Sens. Environ.* **2021**, *262*, 112514. [CrossRef]

29. Masrur, A.; Yu, M.; Mitra, P.; Peuquet, D.; Taylor, A. Interpretable Machine Learning for Analysing Heterogeneous Drivers of Geographic Events in Space-Time. Available online: https://pennstate.pure.elsevier.com/en/publications/interpretable-machine-learning-for-analysing-heterogeneous-driver (accessed on 15 September 2021).

30. Gitelson, A.A.; Kaufman, Y.J.; Stark, R.; Rundquist, D. Novel algorithms for remote estimation of vegetation fraction. *Remote Sens. Environ.* **2002**, *80*, 76–87. [CrossRef]

31. Purevdorj, T.S.; Tateishi, R.; Ishiyama, T.; Honda, Y. Relationships between percent vegetation cover and vegetation indices. *Int. J. Remote Sens.* **2010**, *19*, 3519–3535. [CrossRef]

32. Zhang, C.Y.; Li, J.; Lei, S.G.; Yang, J.Z.; Yang, N. Progress and prospect of the Quantitative Remote Sensing for Monitoring the Eco-Environment in Mining Areas. Available online: http://kns.cnki.net/kcms/detail/34.1055.TD.20210915.1042.002.html (accessed on 4 August 2021).

33. Bao, Y.H.; Bao, S.Y.; Shan, Y. Analysis on Temporal and Spatial Changes of Landscape Pattern in Dalinor Lake Wetland. In Proceedings of the 3rd International Conference on Environmental Science and Information Application Technology (ESIAT), Xi'an, China, 20–21 August 2011; pp. 2367–2375.

34. Li, F.; Lawrence, D.M. Role of Fire in the Global Land Water Budget during the Twentieth Century due to Changing Ecosystems. *J. Clim.* **2017**, *30*, 1893–1908. [CrossRef]

35. Lecun, Y.; Bengio, Y.; Hinton, G. Deep learning. *Nature* **2015**, *521*, 436–444. [CrossRef] [PubMed]

36. Rojas, R. *Neural Networks: A Systematic Introduction*; Springer Science & Business Media: Berlin/Heidelberg, Germany, 2013.

37. Rumelhart, D.E.; Hinton, G.E.; Williams, R.J.J.N. Learning representations by back-propagating errors. *Nature* **1986**, *323*, 533–536. [CrossRef]

38. Nesterov, Y. A Method of Solving a Convex Programming Problem with Convergence Rate. *Dokl. Chem.* **1983**, *27*, 372–376.

39. Sutskever, I.; Martens, J.; Dahl, G.; Hinton, G. On the importance of initialization and momentum in deep learning. In Proceedings of the International Conference on Machine Learning, Atlanta, GA, USA, 16 June 2013; pp. 1139–1147.

40. Flood, N. Continuity of Reflectance Data between Landsat-7 ETM+ and Landsat-8 OLI, for Both Top-of-Atmosphere and Surface Reflectance: A Study in the Australian Landscape. *Remote Sens.* **2014**, *6*, 7952–7970. [CrossRef]
41. Li, P.; Jiang, L.; Feng, Z. Cross-Comparison of Vegetation Indices Derived from Landsat-7 Enhanced Thematic Mapper Plus (ETM+) and Landsat-8 Operational Land Imager (OLI) Sensors. *Remote Sens.* **2013**, *6*, 310–329. [CrossRef]
42. She, X.; Zhang, L.; Cen, Y.; Wu, T.; Huang, C.; Baig, M. Comparison of the Continuity of Vegetation Indices Derived from Landsat 8 OLI and Landsat 7 ETM+ Data among Different Vegetation Types. *Remote Sens.* **2015**, *7*, 13485–13506. [CrossRef]
43. Bai, S.Y.; Wang, L.; Shi, J.Q. Time lag effect of NDVI response to climatic change in Yangtze River basin. *Chin. J. Agrometeorol.* **2012**, *33*, 579–586.
44. Wang, J.B.; Yang, X.L.; Zhang, Y.H.; Wang, X.W. Correlation Between NDVI and Meteorological Factors in Zhangye. *Chin. Agric. Sci. Bull.* **2019**, *35*, 85–90.
45. Liu, Y.L.; Guo, L.; Jiang, Q.H.; Zhang, H.T.; Chen, Y.Y. Comparing geospatial techniques to predict SOC stocks. *Soil Tillage Res.* **2015**, *148*, 46–58. [CrossRef]
46. Arora, A.; Pandey, M.; Mishra, V.N.; Kumar, R.; Rai, P.K.; Costache, R.; Punia, M.; Di, L. Comparative evaluation of geospatial scenario-based land change simulation models using landscape metrics. *Ecol. Indic.* **2021**, *128*, 107810. [CrossRef]
47. Chen, Z.F.; Wang, W.G.; Fu, J.Y. Vegetation response to precipitation anomalies under different climatic and biogeographical conditions in China. *Sci. Rep.* **2020**, *10*, 1–16. [CrossRef]
48. Shen, X.J.; Liu, B.H.; Li, G.D.; Yu, P.J.; Zhou, D.W. Impacts of grassland types and vegetation cover changes on surface air temperature in the regions of temperate grassland of China. *Theor. Appl. Climatol.* **2016**, *126*, 141–150. [CrossRef]
49. Liu, Y.; Shu, H.; Li, Y. Correlation Analysis of Spatio-temporal NDVI, Air Temperature, Precipitation, and Ground Temperature in the Bayinbuluk Grassland. In Proceedings of the IEEE International Geoscience and Remote Sensing Symposium (IGARSS), Denver, CO, USA, 1 July 2006; p. 3368.
50. Zhang, Y.; Liang, W.; Liao, Z.; Han, Z.; Xu, X.; Jiao, R.; Liu, H. Effects of climate change on lake area and vegetation cover over the past 55 years in Northeast Inner Mongolia grassland, China. *Theor. Appl. Climatol.* **2019**, *138*, 13–25. [CrossRef]
51. Fang, A.; Bao, M.; Chen, W.; Dong, J. Assessment of Surface Ecological Quality of Grassland Mining Area and Identification of Its Impact Range. *Nat. Resour. Res.* **2021**, *30*, 3819–3837. [CrossRef]
52. Hui, J.; Bai, Z.; Ye, B.; Wang, Z. Remote Sensing Monitoring and Evaluation of Vegetation Restoration in Grassland Mining Areas—A Case Study of the Shengli Mining Area in Xilinhot City, China. *Land* **2021**, *10*, 743. [CrossRef]
53. Ohana-Levi, N.; Paz-Kagan, T.; Panov, N.; Peeters, A.; Tsoar, A.; Karnieli, A. Time series analysis of vegetation-cover response to environmental factors and residential development in a dryland region. *Gisci. Remote Sens.* **2019**, *56*, 362–387. [CrossRef]
54. Song, X.P.; Hansen, M.C.; Stehman, S.V.; Potapov, P.V.; Tyukavina, A.; Vermote, E.F.; Townshend, J.R. Author Correction: Global land change from 1982 to 2016. *Nature* **2018**, *563*, E26. [CrossRef]
55. Wang, Z.; Deng, X.; Song, W.; Li, Z.; Chen, J. What is the main cause of grassland degradation? A case study of grassland ecosystem service in the middle-south Inner Mongolia. *Catena* **2017**, *150*, 100–107. [CrossRef]
56. Xu, G.; Zhang, J.; Li, P.; Li, Z.; Lu, K.; Wang, X.; Wang, F.; Cheng, Y.; Wang, B. Vegetation restoration projects and their influence on runoff and sediment in China. *Ecol. Indic.* **2018**, *95*, 233–241. [CrossRef]
57. Jiang, G.; Han, X.; Wu, J. Restoration and management of the Inner Mongolia grassland require a sustainable strategy. *AMBIO A J. Hum. Environ.* **2006**, *35*, 269–270. [CrossRef]

Technical Note

Extraction of Water Body Information from Remote Sensing Imagery While Considering Greenness and Wetness Based on Tasseled Cap Transformation

Chao Chen [1], Huixin Chen [1], Jintao Liang [1], Wenlang Huang [2,*], Wenxue Xu [3], Bin Li [4] and Jianqiang Wang [5]

[1] Marine Science and Technology College, Zhejiang Ocean University, Zhoushan 316022, China; chenchao@zjou.edu.cn (C.C.); s20070700014@zjou.edu.cn (H.C.); liangjintao@zjou.edu.cn (J.L.)
[2] Physical and Military Training Education Department, Zhejiang Ocean University, Zhoushan 316022, China
[3] Center for Marine Surveying and Mapping Research, First Institute of Oceanography, Ministry of Natural Resources, Qingdao 266061, China; xuwx@fio.org.cn
[4] Beijing Vminfull Limited, Beijing 100086, China; li.bin@pku.edu.cn
[5] Zhejiang Institute of Hydrogeology and Engineering Geology, Ningbo 315012, China; joson@bolts-nut.com
* Correspondence: 010015@zjou.edu.cn

Citation: Chen, C.; Chen, H.; Liang, J.; Huang, W.; Xu, W.; Li, B.; Wang, J. Extraction of Water Body Information from Remote Sensing Imagery While Considering Greenness and Wetness Based on Tasseled Cap Transformation. *Remote Sens.* **2022**, *14*, 3001. https://doi.org/10.3390/rs14133001

Academic Editors: Qiusheng Wu, Xinyi Shen, Jun Li and Chengye Zhang

Received: 7 March 2022
Accepted: 19 June 2022
Published: 23 June 2022

Publisher's Note: MDPI stays neutral with regard to jurisdictional claims in published maps and institutional affiliations.

Abstract: Water, as an important part of ecosystems, is also an important topic in the field of remote sensing. Shadows and dense vegetation negatively affect most traditional methods used to extract water body information from remotely sensed images. As a result, extracting water body information with high precision from a wide range of remote sensing images which contain complex ground-based objects has proved difficult. In the present study, a method used for extracting water body information from remote sensing imagery considers the greenness and wetness of ground-based objects. Ground objects with varied water content and vegetation coverage have different characteristics in their greenness and wetness components obtained by the Tasseled Cap transformation (TCT). Multispectral information can be output as brightness, greenness, and wetness by Tasseled Cap transformation, which is widely used in satellite remote sensing images. Hence, a model used to extract water body information was constructed to weaken the influence of shadows and dense vegetation. Jiangsu and Anhui provinces are located along the Yangtze River, China, and were selected as the research area. The experiment used the wide-field-of-view (WFV) sensor onboard the Gaofen-1 satellite to acquire remotely sensed photos. The results showed that the contours and spatial extent of the water bodies extracted by the proposed method are highly consistent, and the influence of shadow and buildings is minimized; the method has a high Kappa coefficient (0.89), overall accuracy (92.72%), and user accuracy (88.04%). Thus, the method is useful in updating a geographical database of water bodies and in water resource management.

Keywords: remote sensing imagery; water body extraction; Tasseled Cap transformation; WFV onboard Gaofen-1; accuracy evaluation

1. Introduction

Water is an important resource for human survival and economic development [1–4]. Whether water is treated as an environmental factor, as a resource, or as a cause of flood disasters, the monitoring and investigation of the nature of water bodies have great significance in the use of natural resources; it is also important in land use planning, development and protection of the environment, and in flood protection and mitigation [5–7]. Remote sensing satellite observations can effectively overcome all kinds of the limitations that may be encountered in ground mapping, with the graphics of the landscape recorded numerically and processed by the computers [8–10]. Remote sensing satellite images cover the entire Earth with high resolution, repeatable observation, and multispectral images, which can accurately record rivers, lakes, coastlines, tidal conditions, as well as related ground information and allow researchers to determine the range of a water body quickly and accurately; the cost of land-based surveys are saved while remotely sensed images provide many economic and social benefits [11–14].

Since the 1970s, researchers have conducted a considerable amount of research on the extraction of boundaries between water bodies and other ground-based objects [15–18]. From the earliest efforts in edge detection and threshold segmentation to the application of deep learning, the methods for extracting water body information have been developing over time with continuous progress [19–21]. The methods used to extract water body information from remote sensing imagery can be divided into three categories: (i) the single band threshold-band method, (ii) the spectrum photometric-based method, and (iii) the water body index-based method [22–25]. (i) The single band threshold-based method is an early commonly used method that uses a single band of a remote sensing image, the reflectivity of water is significantly lower or higher than other features, and a water body can be automatically extracted by setting the threshold [26–28]. The process of extracting water bodies using the single band threshold-based method is relatively simple, and the effects of extracting local water body information are clearly visible; nevertheless, dense vegetation, mountain shadows, and the water body spectrum cannot always be correctly distinguished while small water bodies cannot be extracted using this method [29–31]. (ii) The spectrum photometric-based method extracts water bodies by searching for the difference between the characteristics of the spectral curve of a water body and other features; this method can extract water bodies as well as distinguish water bodies from shadows, making it suitable for the extraction of water bodies in mountain plateaus [32–35]. For plains, this method can extract the wider part of lakes, larger rivers, and smaller rivers, but a phenomenon known as staggered buildings can create problems; however, a threshold can be used to judge the conditions for the extraction of water bodies from small rivers and those of larger urban areas [36–38]. In addition to buildings, this method is also affected by clouds. (iii) The water body index-based method is performed by using normalized difference processing of specific wavelengths of ground-based objects to highlight water body information in remote sensing images [39–42]. The method is very precise, has wide applicability, is simple to operate, and is currently the most widely used and developed method [43–45]. This method can effectively eliminate shadow pixels and improve the accuracy of shadow extraction or of other dark surface areas; however, the reflective surfaces of urban areas, such as ice, snow, and reflective roofs, may be accidentally classified as water.

Tasseled Cap transformation (TCT) is a special type of principal component transformation, which combines complex factors into several components while introducing multi-faceted variables, and it simplifies the problem and, at the same time, obtains more scientifically sound and effective data information [46]. In this paper, we aim to provide a detailed guide on how to develop an approach capable of providing estimates of water body information from the wide-field-of-view (WFV) image of the Gaofen-1 satellite with a wide range. This approach both considers the greenness and wetness of ground-based objects obtained by TCT to weaken the influence of shadows and dense vegetation. This paper is of great significance for updating basic geographic information and water resources investigation.

2. Materials and Methods

2.1. Study Area

The study area is located at the intersection of Jiangsu and Anhui provinces, China, which is a part of the Yangtze River Delta. The Yangtze River forms the Yangtze River Delta, an alluvial plain, before entering the sea [47–49]. The delta covers parts of two provinces, Zhejiang and Jiangsu, and the city of Shanghai [50,51]. This densely populated low-lying plain also supports a well-developed area of agriculture; the plain is covered by a dense river network, agricultural areas, and more than 200 lakes. The research area includes a variety of terrain such as plains, lakes, and mountains with numerous urban areas (Figure 1).

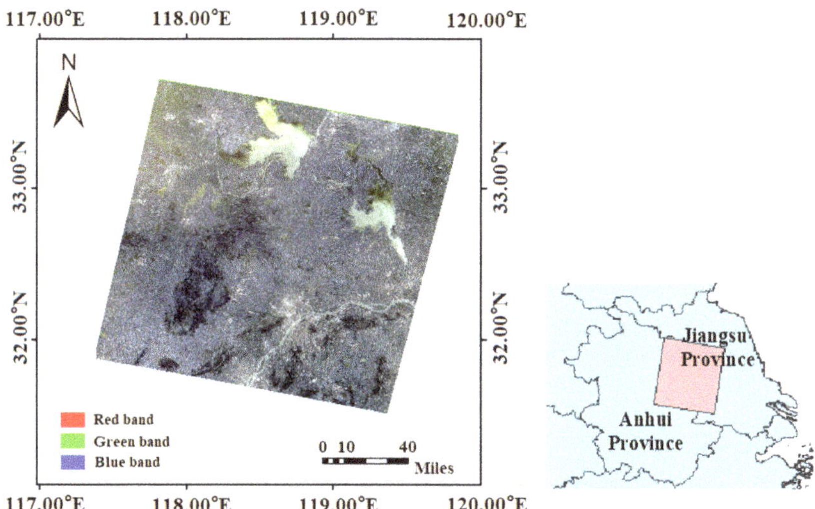

Figure 1. Remotely sensed photo of the study area. An inset map shows the location of the study area within the Anhui and Jiangsu provinces.

2.2. Data

In the study, the image of the covered study area was acquired on 31 December 2016 by the WFV camera onboard the Gaofen-1 satellite. Gaofen-1 (in the Chinese language, gao fen means high resolution) is the first of a series of high-resolution optical Earth observation satellites of the China National Space Administration, Beijing, China, which is configured with two 2 m panchromatic/8 m resolution multispectral cameras and a set of four 16 m resolution multispectral medium-resolution wide-field cameras [52]. The WFV sensor with 16 m spatial resolution consists of three visible light bands (0.45–0.52 μm, 0.52–0.59 μm, 0.63–0.69 μm) and a near-infrared band (0.77–0.89 μm) [53–55]. With the satellite's four-day repetition cycle, the combination of high temporal and spatial resolution is achieved; this combination has played an important role in land and resources investigation and in the dynamic monitoring of the environment and climate change as well as in supporting precision agriculture and urban planning [56–58].

For the Gaofen-1 satellite data, no definite coefficient for TCT has yet to be developed; however, by comparing the parameters of the WFV cameras sensor of Gaofen-1 with known satellite sensors with Tasseled Cap transform coefficients, it was found that the WFV sensor and the IKONOS satellite sensor have a similar number of bands and spectral range of bands [59]. As the TCT conversion coefficient is fixed and determined by the band number and band range of a sensor, the conversion coefficient of the IKONOS satellite was selected in this paper (Table 1) [60,61].

It can be seen from Table 1 that (1) All band coefficients used to calculate the brightness component are positive, each band makes a positive contribution to the brightness component, and the contribution sequence is near-infrared band > red band > green band > blue band. (2) The near-infrared band has a positive contribution to the greenness component, while the other bands have a negative contribution, and the contribution sequence is near-infrared band > green band > red band > blue band. (3) The red band has a positive contribution to the wetness component, while the other bands have a negative contribution, and the contribution sequence is red band > blue band > green band > near-infrared band. (4) The green band has a positive contribution to the yellowness component, while the other bands have a negative contribution, and the contribution sequence is green band > blue band > red band > near-infrared band.

Table 1. The weights of the Tasseled Cap transformation for the WFV sensor onboard the Gaofen-1 satellite.

	Blue Band	Green Band	Red Band	Near-Infrared Band
Brightness component	0.326	0.509	0.560	0.567
Greenness component	−0.311	−0.356	−0.325	0.819
Wetness component	−0.612	−0.312	0.722	−0.081
Yellowness component	−0.650	0.719	−0.243	−0.031

2.3. Method

Tasseled Cap transformation (TCT), also known as Kauth–Thomas transformation, was first reported in 1976 by using maximum segment size data on the growth of crops and vegetation; it consisted of four bands in the maximum segment size analysis [62–64]. In the four-dimensional space, the spectral data points of vegetation are regularly distributed, forming a hat-like shape; therefore, this transformation is named a Tasseled Cap transformation [65,66]. TCT has been widely used in many fields, such as crop type recognition [67], crop growth monitoring [68], remote sensing ecological assessment [69,70], and monitoring of land surface changes [71,72]. It also has important application value in forest type classification [73], distinguishing forest density [74], and biomass inversion [75,76].

For satellite remote sensing imagery, a TCT can compress a multispectral image into a sum of a set of components, each of which corresponds to a weighted index [66]. The weighted index can reflect each pixel in the original multispectral image.

The TCT is as Equation (1) [62–64]:

$$y = cx + a \tag{1}$$

where y is the component in multispectral space after the transformation, x corresponds to the multispectral bands before the transformation, c is the transformation coefficient, which is related to the sensor onboard a satellite, and a is a constant used to avoid negative values.

In this new spectral space after the TCT occurs, the first four components (brightness, greenness, wetness, and yellowness components) are important. The brightness component reflects the comprehensive effect of total reflectivity; the greenness component is related to vegetation cover, leaf base cover index, and biomass; the wetness component reflects the moisture condition of the ground-based objects, and the yellowness component indicates the maturity of vegetation [49,77,78].

The size of the component index corresponding to each pixel can reflect the correlation between its corresponding features and these components (Figure 2). Since different features have different correlations with the four components, these pixels are based on the component values in a 4-dimensional space. The distribution forms different settlements. The pixels in each settlement correspond to the same kind of features, and the boundaries are obvious. It is easy to find the difference between different objects and extract the water information.

Tasseled Cap transformation uses principal component analysis, and its conversion coefficient is fixed, which is determined by the number of frequency bands and spectral range of frequency bands. Therefore, it is independent of individual images so that the soil brightness and greenness can be compared between different images. Compared with other ground-based objects, water bodies have unique characteristics in their greenness and wetness components. Therefore, based on the characteristics of the TCT, a method for extracting water body information from remote sensing imagery has been proposed for use in processing remote sensing imagery; water body information can be extracted accurately and the influence of shadows and dense vegetation can be weakened.

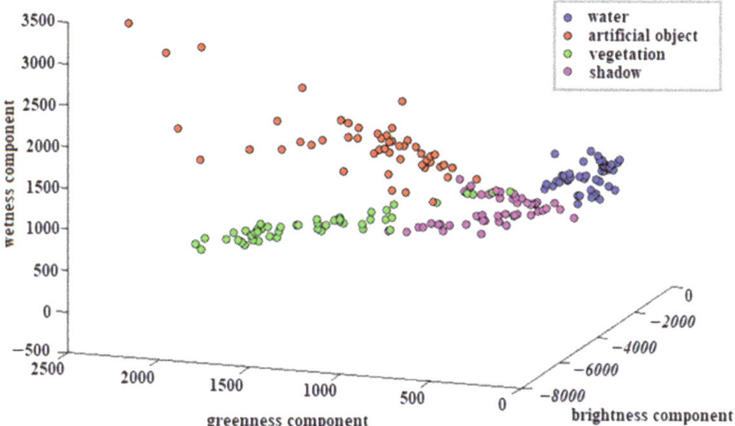

Figure 2. Scatter plot of ground-based objects in the space of TCT.

The variation in water content and vegetation coverage of the ground-based objects will cause water bodies to show unique characteristics in the feature space after TCT. The overall workflow of the method for extracting water body information from remote sensing imagery considering greenness and wetness can be divided into four steps: (a) pre-processing, (b) TCT, (c) extraction of water body information, and (d) accuracy assessment (Figure 3).

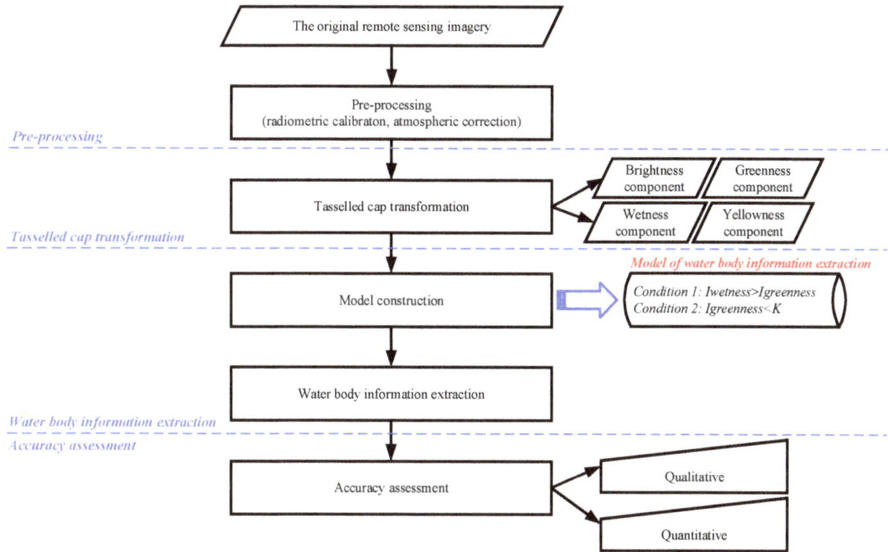

Figure 3. Flowchart of the proposed method.

First, the original remote sensing image data were pre-processed with radiometric calibration and atmospheric correction to eliminate the deviation caused by atmosphere absorption and scattering of light. Second, the correlated remote sensing image information was converted into uncorrelated linear information using TCT; four components were obtained: brightness, greenness, wetness, and yellowness. Third, based on the greenness and wetness components, the difference between the water body and other ground-based objects (vegetation, man-made object, shadow) was analyzed and the model was constructed to extract water body information. Finally, an accuracy assessment was carried out by calculating the Kappa Coefficient, overall accuracy, and user accuracy.

3. Results

3.1. Water Body Information Extraction from Remote Sensing Imagery While Considering Greenness and Wetness

The original remote sensing image was pre-processed with radiometric calibration and atmospheric correction to eliminate most of the deviation caused by the atmosphere, and Environment for Visualizing Images (ENVI) version 4.8 software has been used to perform statistical and visual analyzes on satellite imagery. Then, a TCT was performed to obtain the four components, brightness, greenness, wetness, and yellowness, using the weights in Table 1. The results are shown in Figure 4.

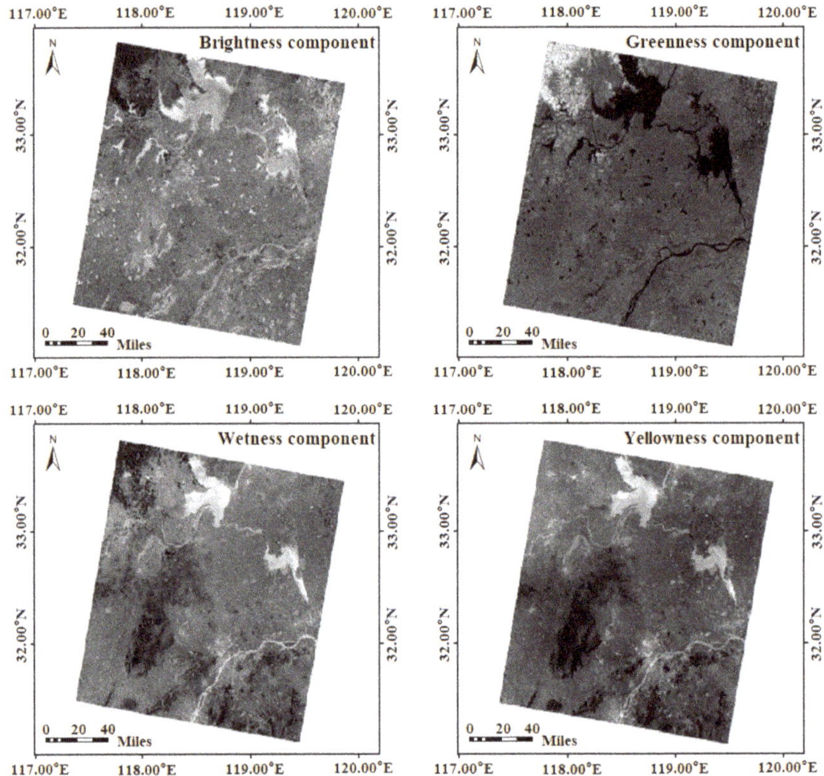

Figure 4. The TCT components of the WFV cameras onboard the Gaofen-1 satellite.

In the feature map (Figure 5), the water bodies show different distribution characteristics from those of vegetation (including dry field), man-made features (including roads, residential, and industrial land), and shadow (mainly mountain and building shadows).

As can be seen from Figures 4 and 5, the water bodies correspond to relatively high and low values in the wetness and greenness components, respectively; vegetation has a relatively high value in the greenness component; meanwhile, shadows and man-made objects have more diverse reflectivity due to the variety of their species. By adding an auxiliary line $I_{wetness} = I_{greenness}$ to Figure 5, it is easy to see that the corresponding pixels of water bodies are all located below the auxiliary line, and the vegetation and shadows are all located above the line; therefore, the condition of "$I_{wetness} > I_{greenness}$" can easily be used to divide the water bodies from the shadows and vegetation. However, some man-made objects still fall below the auxiliary line; one can see a clear gap exists between the greenness component of these man-made objects and the greenness component of the water body. Therefore, the thresholding condition "$I_{greenness} < K$" was set to distinguish the water bodies from the man-made object.

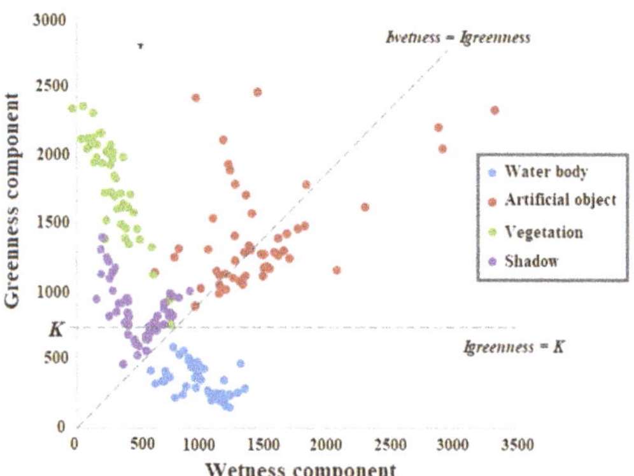

Figure 5. The distribution characteristics of ground-based objects in the feature map generated by Tassel Cap transformation components.

Equation (2) provides the formula for extracting water body information as follows:

$$\begin{cases} I_{wetness} > I_{greenness} \\ I_{greenness} < K \end{cases} \tag{2}$$

where $I_{wetness}$ and $I_{greenness}$ are the wetness and greenness components obtained by TCT, respectively, and K is the empirical constant of the greenness component, usually 750.

The greenness and wetness components obtained from the TCT were used to extract water body information with the support of Equation (1). The extraction results are shown in Figure 6.

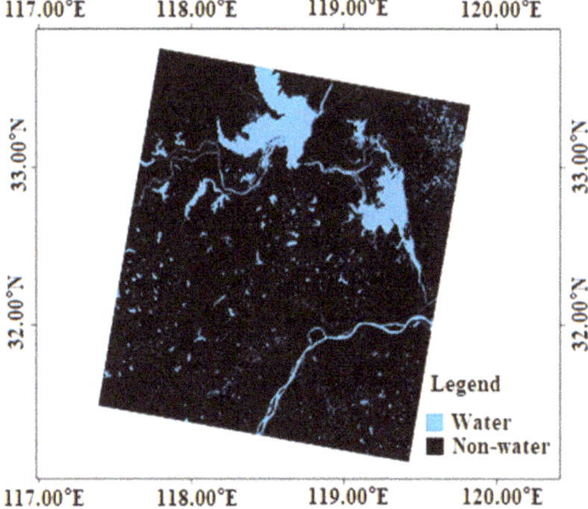

Figure 6. The water body information extracted using the proposed method and ENVI 4.8 software has been used to perform statistical and visual analyzes on satellite imagery.

By visually observing the difference between the extraction results and the original remote sensing image, Figures 1 and 6 show that the extracted water body information is consistent with the visual results regardless of the size and contour of water bodies and can effectively remove the shadows of mountains, buildings, and clouds. However, man-made buildings cause some errors.

3.2. Accuracy Assessment

Accuracy assessment is an important part of information extraction from remote sensing imagery, and using the confusion matrix to calculate the Kappa coefficient, overall accuracy (OA), and user accuracy (UA) is a common method for accuracy assessment [79–81]. The real boundary of the water bodies was obtained by visual interpretation from the Gaofen-1 satellite image, and the Kappa coefficient, OA, and UA were calculated based on visual interpretation and the extracted results. The water bodies were obtained by the proposed method and the results were obtained by single band threshold-based method, spectrum photometric-based method, normalized difference water index-based method (NDWI-based method), water ratio index-based method (WRI-based method), and automated water extraction index-based method (AWEIsh-based method) which were compared to perform a quantitative evaluation of the results [82–84], and the experiment was carried out by ENVI 4.8 software.

Table 2 shows that the Kappa coefficient, OA, and UA of the proposed method are all higher than those of the traditional methods. This indicates that the water body information extracted using the proposed method is consistent with the visual interpretation results, and the method considering the greenness and wetness of ground-based objects is reliable and accurate.

Table 2. Quantitative evaluation of extracting water body information.

Method	Formula	Kappa Coefficient	OA (%)	UA (%)
Single-band threshold-based method	$\rho_{NIR} < 0$	0.77	86.63	80.28
Spectrum photometric-based method	$\rho_{green} + \rho_{red} > 2 \times \rho_{NIR}$	0.86	90.05	83.51
NDWI-based method	$(\rho_{green} - \rho_{NIR})/(\rho_{green} + \rho_{NIR}) > 0$	0.88	93.35	87.69
WRI-based method	$(\rho_{green} + \rho_{red})/(2 \times \rho_{NIR}) > 1$	0.83	87.08	84.92
AWEI$_{sh}$-based method	$(\rho_{blue} + 2.5 \times \rho_{green} - 3.25 \times \rho_{NIR}) > 0$	0.78	80.34	78.59
The proposed method	$I_{wetness} > I_{greenness}$ and $I_{greenness} < 750$	0.91	97.02	90.81

Note: ρ is the reflectance of remote sensing imagery; *red, green, NIR* are red, green, and near-infrared bands, respectively; $I_{wetness}$ and $I_{greenness}$ are wetness and greenness obtained by TCT, respectively.

4. Discussion

The components obtained by the TCT are associated with water content and vegetation coverage of ground-based objects, which is helpful in extracting water body information quickly and simply by using remote sensing images in a wide range. Following remote sensing image pre-processing, TCT, and model construction considering greenness and wetness, the water body information can be extracted accurately. Unfortunately, there are some factors affecting the extraction accuracy of the water body information. First, a pixel in the remote sensing image measures the combined reflectance of all ground objects contained on the ground of the same size as the spatial resolution [4,28,49]. Therefore, mixed pixels cause boundaries between land and water body to become vague and affect the accurate statistics of the extracted water body information. Second, the threshold values in the constructed model are obtained by the statistics of the sampling points and are representative of the study area [3,10,49]. Therefore, the use of the model in other regions may produce uncertainty errors. Third, the water surface can be shaded by vegetation covering the bank zone, and then the surface morphology of extracted water body information is affected [85–87]. Therefore, considering the characteristics of smooth edges of banks, edge restoration of extracted water body information is also worth discussing.

5. Conclusions

The present study developed a method for extracting water body information from remote sensing imagery by considering the greenness and wetness of ground-based objects using TCT. This was applied to reduce the influence caused by shadows and dense vegetation to achieve simple, fast, and accurate extraction of water body information. The experiment using the WFV sensor onboard the Gaofen-1 satellite was carried out along the border between Jiangsu and Anhui of the Yangtze River Delta, China, to demonstrate and validate this method. Based on the qualitative and quantitative evaluations, the proposed method is accurate, and the validity and practicality of the proposed method have been verified.

The proposed method has high precision and is simple to operate. However, some weakness remained that has not been eliminated in this experiment. The influence of mixed pixels caused the boundaries of water body information extracted from remote sensing imagery to be inconsistent with the actual boundaries of water bodies in the real world. In addition, using the coefficient of IKONOS to perform the TCT of data collected by the WFV sensor onboard the Gaofen-1 satellite also leads to an uncertain effect in the extraction of water body information. Therefore, developing a mixed pixel decomposition method that can be used to extract the boundaries of water body information more accurately will need to be a follow-up research study. Obtaining the coefficient of the WFV imagery acquired using the Gaofen-1 satellite to reduce the uncertain influence of water body information extraction is another direction that will need continuing efforts.

Author Contributions: Conceptualization, C.C. and W.H.; methodology, C.C. and W.X.; software, C.C., H.C., J.L. and W.H.; validation, C.C., H.C., J.L. and B.L.; formal analysis, C.C., H.C. and J.L.; investigation, C.C. and J.W.; resources, H.C. and J.L.; data curation, C.C.; writing—original draft preparation, C.C., H.C. and J.L.; writing—review and editing, C.C., H.C., W.H., W.X., B.L. and J.W.; project administration, C.C.; funding acquisition, C.C., W.H., B.L. and J.W. All authors have read and agreed to the published version of the manuscript.

Funding: This study was funded by the National Key R&D Program of China (Grant Number: 2021YFB3901305), in part by the National Natural Science Foundation of China (Grant Number: 42171311), in part by the Fundamental Research Funds for Zhejiang Provincial Universities and Research Institutes (Grant Numbers: 2021R005, 2021JZ001), in part by the Project of Beijing VMinFull Limited (Grant Number: VMF2021RS), in part by the Training Program of Excellent Master Thesis of Zhejiang Ocean University.

Data Availability Statement: The satellite remote sensing images of Gaofen-1 were provided by the Chinese Academy of Science (www.ids.ceode.ac.cn). All Gaofen-1 data used in the study is not applicable.

Acknowledgments: The authors would like to thank the editors and the anonymous reviewers for their outstanding comments and suggestions, which greatly helped them to improve the technical quality and presentation of this manuscript.

Conflicts of Interest: The authors declare no conflict of interest.

References

1. Jain, S.K.; Saraf, A.K.; Goswami, A.; Ahmad, T. Flood inundation mapping using NOAA AVHRR data. *Water Resour. Manag.* **2006**, *20*, 949–959. [CrossRef]
2. Duan, Z.; Bastiaanssen, W. Estimating water volume variations in lakes and reservoirs from four operational satellite altimetry databases and satellite imagery data. *Remote Sens. Environ.* **2013**, *134*, 403–416. [CrossRef]
3. Sun, W.; Peng, J.; Yang, G.; Du, Q. Correntropy-Based Sparse Spectral Clustering for Hyperspectral Band Selection. *IEEE Geosci. Remote Sens. Lett.* **2020**, *17*, 484–488. [CrossRef]
4. Wang, L.; Chen, C.; Xie, F.; Hu, Z.; Zhang, Z.; Chen, H.; He, X.; Chu, Y. Estimation of the value of regional ecosystem services of an archipelago using satellite remote sensing technology: A case study of Zhoushan Archipelago, China. *Int. J. Appl. Earth Obs. Geoinf.* **2021**, *105*, 102616. [CrossRef]
5. Sethre, P.R.; Rundquist, B.C.; Todhunter, P.E. Remote Detection of Prairie Pothole Ponds in the Devils Lake Basin, North Dakota. *GIScience Remote Sens.* **2005**, *42*, 277–296. [CrossRef]

6. Masocha, M.; Dube, T.; Makore, M.; Shekede, M.D.; Funani, J. Surface water bodies mapping in Zimbabwe using Landsat 8 OLI multispectral imagery: A comparison of multiple water indices. *Phys. Chem. Earth Parts A/B/C* **2018**, *106*, 63–67. [CrossRef]

7. Pôças, I.; Calera, A.; Campos, I.; Cunha, M. Remote sensing for estimating and mapping single and basal crop coefficients: A review on spectral vegetation indices approaches. *Agric. Water Manag.* **2020**, *233*, 106081. [CrossRef]

8. Chen, H.; Chen, C.; Zhang, Z.; Lu, C.; Wang, L.; He, X.; Chu, Y.; Chen, J. Changes of the spatial and temporal characteristics of land-use landscape patterns using multi-temporal Landsat satellite data: A case study of Zhoushan Island, China. *Ocean Coast. Manag.* **2021**, *213*, 105842. [CrossRef]

9. Fu, J.; Chen, C.; Chu, Y. Spatial–temporal variations of oceanographic parameters in the Zhoushan sea area of the East China Sea based on remote sensing datasets. *Reg. Stud. Mar. Sci.* **2019**, *28*, 100626. [CrossRef]

10. Chen, C.; Wang, L.; Zhang, Z.; Lu, C.; Chen, H.; Chen, J. Construction and application of quality evaluation index system for remote-sensing image fusion. *J. Appl. Remote Sens.* **2021**, *16*, 012006. [CrossRef]

11. Han, C.; Zhang, B.; Chen, H.; Wei, Z.; Liu, Y. Spatially distributed crop model based on remote sensing. *Agric. Water Manag.* **2019**, *218*, 165–173. [CrossRef]

12. Ranjan, S.; Sarvaiya, J.N.; Patel, J.N. Integrating Spectral and Spatial features for Hyperspectral Image Classification with a Modified Composite Kernel Framework. *PFG—J. Photogramm. Remote Sens. Geoinf. Sci.* **2019**, *87*, 275–296. [CrossRef]

13. Chen, Y.; Ming, D.; Lv, X. Superpixel based land cover classification of VHR satellite image combining multi-scale CNN and scale parameter estimation. *Earth Sci. Inform.* **2019**, *12*, 341–363. [CrossRef]

14. Sun, W.; Yang, G.; Peng, J.; Du, Q. Lateral-Slice Sparse Tensor Robust Principal Component Analysis for Hyperspectral Image Classification. *IEEE Geosci. Remote Sens. Lett.* **2020**, *17*, 107–111. [CrossRef]

15. Gautam, V.K.; Gaurav, P.K.; Murugan, P.; Annadurai, M. Assessment of Surface Water Dynamicsin Bangalore Using WRI, NDWI, MNDWI, Supervised Classification and K-T Transformation. *Aquat. Procedia* **2015**, *4*, 739–746. [CrossRef]

16. Huang, X.; Xie, C.; Fang, X.; Zhang, L. Combining Pixel- and Object-Based Machine Learning for Identification of Water-Body Types from Urban High-Resolution Remote-Sensing Imagery. *IEEE J. Sel. Top. Appl. Earth Obs. Remote Sens.* **2015**, *8*, 2097–2110. [CrossRef]

17. He, X.; Chen, C.; Liu, Y.; Chu, Y. Inundation Analysis Method for Urban Mountainous Areas Based on Soil Conservation Service Curve Number (SCS-CN) Model Using Remote Sensing Data. *Sensors Mater.* **2020**, *32*, 3813. [CrossRef]

18. Valderrama-Landeros, L.; Flores-de-Santiago, F. Assessing coastal erosion and accretion trends along two contrasting subtropical rivers based on remote sensing data. *Ocean. Coast. Manag.* **2019**, *169*, 58–67. [CrossRef]

19. Teodoro, A.C.; Goncalves, H.; Veloso-Gomes, F.; Goncalves, J.A. Modeling of the Douro River Plume Size, Obtained Through Image Segmentation of MERIS Data. *IEEE Geosci. Remote Sens. Lett.* **2009**, *6*, 87–91. [CrossRef]

20. Malahlela, O.E. Inland waterbody mapping: Towards improving discrimination and extraction of inland surface water features. *Int. J. Remote Sens.* **2016**, *37*, 4574–4589. [CrossRef]

21. Su, H.; Peng, Y.; Xu, C.; Feng, A.; Liu, T. Using improved DeepLabv3+ network integrated with normalized difference water index to extract water bodies in Sentinel-2A urban remote sensing images. *J. Appl. Remote Sens.* **2021**, *15*, 018504. [CrossRef]

22. Jawak, S.D.; Kulkarni, K.; Luis, A.J. A Review on Extraction of Lakes from Remotely Sensed Optical Satellite Data with a Special Focus on Cryospheric Lakes. *Adv. Remote Sens.* **2015**, *4*, 196–213. [CrossRef]

23. Fisher, A.; Flood, N.; Danaher, T. Comparing Landsat water index methods for automated water classification in eastern Australia. *Remote Sens. Environ.* **2016**, *175*, 167–182. [CrossRef]

24. Sarp, G.; Ozcelik, M. Water body extraction and change detection using time series: A case study of Lake Burdur, Turkey. *J. Taibah Univ. Sci.* **2017**, *11*, 381–391. [CrossRef]

25. Li, D.; Wu, B.; Chen, B.; Xue, Y.; Zhang, Y. Review of water body information extraction based on satellite remote sensing. *J. Tsinghua Univ. (Sci. Technol.)* **2020**, *60*, 147–161.

26. Lira, J. Segmentation and morphology of open water bodies from multispectral images. *Int. J. Remote Sens.* **2006**, *27*, 4015–4038. [CrossRef]

27. Grodsky, S.; Reverdin, G.; Carton, J.; Coles, V.J. Year-to-year salinity changes in the Amazon plume: Contrasting 2011 and 2012 Aquarius/SACD and SMOS satellite data. *Remote Sens. Environ.* **2014**, *140*, 14–22. [CrossRef]

28. Chen, C.; Fu, J.Q.; Sui, X.X.; Lu, X.; Tan, A.H. Construction and application of knowledge decision tree after a disaster for water body information extraction from remote sensing images. *J. Remote Sens.* **2018**, *22*, 792–801.

29. Wilson, E.H.; Sader, S.A. Detection of forest harvest type using multiple dates of Landsat TM imagery. *Remote Sens. Environ.* **2002**, *80*, 385–396. [CrossRef]

30. Sharma, D.; Singhai, J. An Object-Based Shadow Detection Method for Building Delineation in High-Resolution Satellite Images. *PFG—J. Photogramm. Remote Sens. Geoinf. Sci.* **2019**, *87*, 103–118. [CrossRef]

31. Vanama, V.S.K.; Mandal, D.; Rao, Y.S. GEE4FLOOD: Rapid mapping of flood areas using temporal Sentinel-1 SAR images with Google Earth Engine cloud platform. *J. Appl. Remote Sens.* **2020**, *14*, 034505. [CrossRef]

32. Rogers, A.S.; Kearney, M.S. Reducing signature variability in unmixing coastal marsh Thematic Mapper scenes using spectral indices. *Int. J. Remote Sens.* **2004**, *25*, 2317–2335. [CrossRef]

33. Jain, S.K.; Singh, R.D.; Jain, M.; Lohani, A.K. Delineation of Flood-Prone Areas Using Remote Sensing Techniques. *Water Resour. Manag.* **2005**, *19*, 333–347. [CrossRef]

34. Yang, Z.; Wang, L.; Sun, W.; Xu, W.; Tian, B.; Zhou, Y.; Yang, G.; Chen, C. A New Adaptive Remote Sensing Extraction Algorithm for Complex Muddy Coast Waterline. *Remote Sens.* **2022**, *14*, 861. [CrossRef]
35. Ahmed, A.; Drake, F.; Nawaz, R.; Woulds, C. Where is the coast? Monitoring coastal land dynamics in Bangladesh: An integrated management approach using GIS and remote sensing techniques. *Ocean. Coast. Manag.* **2018**, *151*, 10–24. [CrossRef]
36. Guttler, F.; Niculescu, S.; Gohin, F. Turbidity retrieval and monitoring of Danube Delta waters using multi-sensor optical remote sensing data: An integrated view from the delta plain lakes to the western–northwestern Black Sea coastal zone. *Remote Sens. Environ.* **2013**, *132*, 86–101. [CrossRef]
37. Wan, W.; Long, D.; Hong, Y.; Ma, Y.; Yuan, Y.; Xiao, P.; Duan, H.; Han, Z.; Gu, X. A lake data set for the Tibetan Plateau from the 1960s, 2005, and 2014. *Sci. Data* **2016**, *3*, 160039. [CrossRef]
38. Wei, B.; Li, Y.; Suo, A.; Zhang, Z.; Xu, Y.; Chen, Y. Spatial suitability evaluation of coastal zone, and zoning optimisation in Ningbo, China. *Ocean Coast. Manag.* **2021**, *204*, 105507. [CrossRef]
39. McFeeters, S.K. The use of the Normalized Difference Water Index (NDWI) in the delineation of open water features. *Int. J. Remote Sens.* **1996**, *17*, 1425–1432. [CrossRef]
40. Xu, H. Modification of normalised difference water index (NDWI) to enhance open water features in remotely sensed imagery. *Int. J. Remote Sens.* **2006**, *27*, 3025–3033. [CrossRef]
41. Feyisa, G.L.; Meilby, H.; Fensholt, R.; Proud, S.R. Automated Water Extraction Index: A new technique for surface water mapping using Landsat imagery. *Remote Sens. Environ.* **2014**, *140*, 23–35. [CrossRef]
42. Ahmed, N.; Howlader, N.; Hoque, M.A.-A.; Pradhan, B. Coastal erosion vulnerability assessment along the eastern coast of Bangladesh using geospatial techniques. *Ocean Coast. Manag.* **2021**, *199*, 105408. [CrossRef]
43. Zhang, G.; Yao, T.; Xie, H.; Zhang, K.; Zhu, F. Lakes' state and abundance across the Tibetan Plateau. *Chin. Sci. Bull.* **2014**, *59*, 3010–3021. [CrossRef]
44. Liao, H.-Y.; Wen, T.-H. Extracting urban water bodies from high-resolution radar images: Measuring the urban surface morphology to control for radar's double-bounce effect. *Int. J. Appl. Earth Obs. Geoinf.* **2020**, *85*, 102003. [CrossRef]
45. Xia, Z.; Guo, X.; Chen, R. Automatic extraction of aquaculture ponds based on Google Earth Engine. *Ocean Coast. Manag.* **2020**, *198*, 105348. [CrossRef]
46. El Din, E.S. A novel approach for surface water quality modelling based on Landsat-8 tasselled cap transformation. *Int. J. Remote Sens.* **2020**, *41*, 7186–7201. [CrossRef]
47. Wu, Q.; Miao, S.; Huang, H.; Guo, M.; Zhang, L.; Yang, L.; Zhou, C. Quantitative Analysis on Coastline Changes of Yangtze River Delta Based on High Spatial Resolution Remote Sensing Images. *Remote Sens.* **2022**, *14*, 310. [CrossRef]
48. Yang, C.; Xia, R.; Li, Q.; Liu, H.; Shi, T.; Wu, G. Comparing hillside urbanizations of Beijing-Tianjin-Hebei, Yangtze River Delta and Guangdong–Hong Kong–Macau greater Bay area urban agglomerations in China. *Int. J. Appl. Earth Obs. Geoinf.* **2021**, *102*, 102460. [CrossRef]
49. Chen, C.; Liang, J.; Xie, F.; Hu, Z.; Sun, W.; Yang, G.; Yu, J.; Chen, L.; Wang, L.; Wang, L.; et al. Temporal and spatial variation of coastline using remote sensing images for Zhoushan archipelago, China. *Int. J. Appl. Earth Obs. Geoinf.* **2022**, *107*, 102711. [CrossRef]
50. Jia, M.; Wang, Z.; Mao, D.; Ren, C.; Wang, C.; Wang, Y. Rapid, robust, and automated mapping of tidal flats in China using time series Sentinel-2 images and Google Earth Engine. *Remote Sens. Environ.* **2021**, *255*, 112285. [CrossRef]
51. Yang, X.; Zhu, Z.; Qiu, S.; Kroeger, K.D.; Zhu, Z.; Covington, S. Detection and characterization of coastal tidal wetland change in the northeastern US using Landsat time series. *Remote Sens. Environ.* **2022**, *276*, 113047. [CrossRef]
52. Liu, K.; Su, H.; Li, X.; Wang, W.; Yang, L.; Liang, H. Quantifying spatial–temporal pattern of urban heat island in Beijing: An improved assessment using land surface temperature (LST) time series observations from LANDSAT, MODIS, and Chinese new satellite GaoFen-1. *IEEE J. Sel. Top. Appl. Earth Obs. Remote Sens.* **2016**, *9*, 2028–2042. [CrossRef]
53. Xu, L.; Wu, Z.; Zhang, Z.; Wang, X. Forest classification using synthetic GF-1/WFV time series and phenological parameters. *J. Appl. Remote Sens.* **2021**, *15*, 042413. [CrossRef]
54. Sun, Q.; Zhang, P.; Sun, D.; Liu, A.; Dai, J. Desert vegetation-habitat complexes mapping using Gaofen-1 WFV (wide field of view) time series images in Minqin County, China. *Int. J. Appl. Earth Obs. Geoinf.* **2018**, *73*, 522–534. [CrossRef]
55. Tian, L.; Wai, O.W.H.; Chen, X.; Li, W.; Li, J.; Li, W.; Zhang, H. Retrieval of total suspended matter concentration from Gaofen-1 Wide Field Imager (WFI) multispectral imagery with the assistance of Terra MODIS in turbid water—case in Deep Bay. *Int. J. Remote Sens.* **2016**, *37*, 3400–3413. [CrossRef]
56. Orimoloye, I.R.; Mazinyo, S.P.; Kalumba, A.M.; Nel, W.; Adigun, A.I.; Ololade, O.O. Wetland shift monitoring using remote sensing and GIS techniques: Landscape dynamics and its implications on Isimangaliso Wetland Park, South Africa. *Earth Sci. Inform.* **2019**, *12*, 553–563. [CrossRef]
57. Zhu, S.; Wan, W.; Xie, H.; Liu, B.; Li, H.; Hong, Y. An Efficient and Effective Approach for Georeferencing AVHRR and GaoFen-1 Imageries Using Inland Water Bodies. *IEEE J. Sel. Top. Appl. Earth Obs. Remote Sens.* **2018**, *11*, 2491–2500. [CrossRef]
58. Liu, Q.; Liu, G.; Huang, C.; Xie, C. Comparison of tasselled cap transformations based on the selective bands of Landsat 8 OLI TOA reflectance images. *Int. J. Remote Sens.* **2015**, *36*, 417–441. [CrossRef]
59. Yang, F.; Fan, M.; Tao, J. An Improved Method for Retrieving Aerosol Optical Depth Using Gaofen-1 WFV Camera Data. *Remote Sens.* **2021**, *13*, 280. [CrossRef]

60. Cheng, W.-C.; Chang, J.-C.; Chang, C.-P.; Su, Y.; Tu, T.-M. A Fixed-Threshold Approach to Generate High-Resolution Vegetation Maps for IKONOS Imagery. *Sensors* **2008**, *8*, 4308–4317. [CrossRef]
61. Lee, S.; Jin, C.; Choi, C.; Lim, H.; Kim, Y.; Kim, J. Absolute radiometric calibration of the KOMPSAT-2 multispectral camera using a reflectance-based method and empirical comparison with IKONOS and QuickBird images. *J. Appl. Remote Sens.* **2012**, *6*, 063594. [CrossRef]
62. Kauth, R.J.; Thomas, G.S. The tasselled cap–A graphic description of the spectral-temporal development of agricultural crops as seen by Landsat. In *LARS Symposia, Proceedings of the Symposium on Machine Processing of Remotely Sensed Data, West Lafayette, IN, USA, 29 June–1 July 1976*; The Institute of Electrical and Electronics Engineers, Inc.: New York, NY, USA, 1976; p. 159.
63. Crist, E.P.; Kauth, R.J. The tasseled cap de-mystified. *Photogramm. Eng. Remote Sens.* **1986**, *52*, 81–86.
64. Crist, E.P. A TM Tasseled Cap equivalent transformation for reflectance factor data. *Remote Sens. Environ.* **1985**, *17*, 301–306. [CrossRef]
65. Sheng, L.; Huang, J.-F.; Tang, X.-L. A tasseled cap transformation for CBERS-02B CCD data. *J. Zhejiang Univ. Sci. B* **2011**, *12*, 780–786. [CrossRef]
66. Chen, C.; Fu, J.; Zhang, S.; Zhao, X. Coastline information extraction based on the tasseled cap transformation of Landsat-8 OLI images. *Estuarine, Coast. Shelf Sci.* **2019**, *217*, 281–291. [CrossRef]
67. Tatsumi, K.; Yamashiki, Y.; Morante, A.K.M.; Fernández, L.R.; Nalvarte, R.A. Pixel-based crop classification in Peru from Landsat 7 ETM+ images using a Random Forest model. *J. Agric. Meteorol.* **2016**, *72*, 1–11. [CrossRef]
68. Gilbertson, J.K.; van Niekerk, A. Value of dimensionality reduction for crop differentiation with multi-temporal imagery and machine learning. *Comput. Electron. Agric.* **2017**, *142*, 50–58. [CrossRef]
69. Song, C.; Ren, H.; Huang, C. Estimating soil salinity in the Yellow River Delta, Eastern China—An integrated approach using spectral and terrain indices with the generalized additive model. *Pedosphere* **2016**, *26*, 626–635. [CrossRef]
70. Hu, X.; Xu, H. A new remote sensing index for assessing the spatial heterogeneity in urban ecological quality: A case from Fuzhou City, China. *Ecol. Indic.* **2018**, *89*, 11–21. [CrossRef]
71. Rahman, S.; Mesev, V. Change Vector Analysis, Tasseled Cap, and NDVI-NDMI for Measuring Land Use/Cover Changes Caused by a Sudden Short-Term Severe Drought: 2011 Texas Event. *Remote Sens.* **2019**, *11*, 2217. [CrossRef]
72. Zanchetta, A.; Bitelli, G.; Karnieli, A. Monitoring desertification by remote sensing using the Tasselled Cap transform for long-term change detection. *Nat. Hazards* **2016**, *83*, 223–237. [CrossRef]
73. Santra, A.; Mitra, S.S. A Comparative Study of Tasselled Cap Transformation of DMC and ETM+ Images and their Application in Forest Classification. *J. Indian Soc. Remote Sens.* **2014**, *42*, 373–381. [CrossRef]
74. Ahmed, O.S.; Franklin, S.E.; Wulder, M.A. Interpretation of forest disturbance using a time series of Landsat imagery and canopy structure from airborne lidar. *Can. J. Remote Sens.* **2014**, *39*, 521–542. [CrossRef]
75. Frazier, R.J.; Coops, N.C.; Wulder, M.; Kennedy, R. Characterization of aboveground biomass in an unmanaged boreal forest using Landsat temporal segmentation metrics. *ISPRS J. Photogramm. Remote Sens.* **2014**, *92*, 137–146. [CrossRef]
76. Kazar, S.A.; Warner, T.A. Assessment of carbon storage and biomass on minelands reclaimed to grassland environments using Landsat spectral indices. *J. Appl. Remote Sens.* **2013**, *7*, 073583. [CrossRef]
77. Kakooei, M.; Baleghi, Y. A two-level fusion for building irregularity detection in post-disaster VHR oblique images. *Earth Sci. Inform.* **2020**, *13*, 459–477. [CrossRef]
78. Singh, H.; Garg, R.D.; Karnatak, H.C. Online image classification and analysis using OGC web processing service. *Earth Sci. Inform.* **2019**, *12*, 307–317. [CrossRef]
79. Kakooei, M.; Baleghi, Y. Shadow detection in very high resolution RGB images using a special thresholding on a new spectral–spatial index. *J. Appl. Remote Sens.* **2020**, *14*, 016503. [CrossRef]
80. Yue, H.; Li, Y.; Qian, J.; Liu, Y. A new accuracy evaluation method for water body extraction. *Int. J. Remote Sens.* **2020**, *41*, 7311–7342. [CrossRef]
81. Prošek, J.; Gdulová, K.; Barták, V.; Vojar, J.; Solský, M.; Rocchini, D.; Moudrý, V. Integration of hyperspectral and LiDAR data for mapping small water bodies. *Int. J. Appl. Earth Obs. Geoinf.* **2020**, *92*, 102181. [CrossRef]
82. Wang, L.; Chen, C.; Zhang, Z.; Gan, W.; Yu, J.; Chen, H. Approach for estimation of ecosystem services value using multitemporal remote sensing images. *J. Appl. Remote Sens.* **2021**, *16*, 012010. [CrossRef]
83. Li, Y.; Dang, B.; Zhang, Y.; Du, Z. Water body classification from high-resolution optical remote sensing imagery: Achievements and perspectives. *ISPRS J. Photogramm. Remote Sens.* **2022**, *187*, 306–327. [CrossRef]
84. Chen, C.; Fu, J.; Gai, Y.; Li, J.; Chen, L.; Mantravadi, V.S.; Tan, A. Damaged Bridges Over Water: Using High-Spatial-Resolution Remote-Sensing Images for Recognition, Detection, and Assessment. *IEEE Geosci. Remote Sens. Mag.* **2018**, *6*, 69–85. [CrossRef]
85. Bode, C.A.; Limm, M.P.; Power, M.E.; Finlay, J.C. Subcanopy Solar Radiation model: Predicting solar radiation across a heavily vegetated landscape using LiDAR and GIS solar radiation models. *Remote Sens. Environ.* **2014**, *154*, 387–397. [CrossRef]
86. Burrell, T.K.; O'Brien, J.M.; Graham, S.E.; Simon, K.S.; Harding, J.S.; McIntosh, A.R. Riparian shading mitigates stream eutrophication in agricultural catchments. *Freshw. Sci.* **2014**, *33*, 73–84. [CrossRef]
87. Kałuża, T.; Sojka, M.; Wróżyński, R.; Jaskuła, J.; Zaborowski, S.; Hämmerling, M. Modeling of river channel shading as a factor for changes in hydromorphological conditions of small lowland rivers. *Water* **2020**, *12*, 527. [CrossRef]

Article

Accessing the Time-Series Two-Dimensional Displacements around a Reservoir Using Multi-Orbit SAR Datasets: A Case Study of Xiluodu Hydropower Station

Qi Chen [1,2], Heng Zhang [1,2], Bing Xu [3,*], Zhe Liu [4] and Wenxiang Mao [3]

[1] China Centre for Resources Satellite Data and Application, Beijing 100094, China
[2] China Siwei Surveying and Mapping Technology Co., Ltd., Beijing 100048, China
[3] School of Geosciences and Info-Physics, Central South University, Changsha 410083, China
[4] Hanzhong Vocational and Technical College, Hanzhong 723002, China
* Correspondence: xubing@csu.edu.cn

Citation: Chen, Q.; Zhang, H.; Xu, B.; Liu, Z.; Mao, W. Accessing the Time-Series Two-Dimensional Displacements around a Reservoir Using Multi-Orbit SAR Datasets: A Case Study of Xiluodu Hydropower Station. *Remote Sens.* **2023**, *15*, 168. https://doi.org/10.3390/rs15010168

Academic Editors: Qiusheng Wu, Xinyi Shen, Jun Li and Chengye Zhang

Received: 15 November 2022
Revised: 12 December 2022
Accepted: 19 December 2022
Published: 28 December 2022

Abstract: The construction of large-scale hydropower stations could solve the problem of China's power and energy shortages. However, the construction of hydropower stations requires reservoir water storage. Artificially raising the water level by several tens of meters or even hundreds of meters will undoubtedly change the hydrogeological conditions of an area, which will lead to surface deformation near the reservoir. In this paper, we first used SBAS-InSAR technology to monitor the surface deformation near the Xiluodu reservoir area for various data and analyzed the surface deformation of the Xiluodu reservoir area from 2014 to 2019. By using the 12 ALOS2 ascending data, the 100 Sentinel-1 ascending data, and the 97 Sentinel-1 descending data, the horizontal and vertical deformations of the Xiluodu reservoir area were obtained. We found that the Xiluodu reservoir area is mainly deformed along the vertical shore, with a maximum deformation rate of 250 mm/a, accompanied by vertical deformation, and the maximum deformation rate is 60 mm/a. Furthermore, by analyzing the relationship between the horizontal deformation sequence, the vertical deformation sequence, and the impoundment, we found the following: (1) Since the commencement of Xiluodu water storage, the vertical shore direction displacement has continued to increase, indicating that the deformation caused by the water storage is not due to the elastic displacement caused by the load, but by irreversible shaping displacement. According to its development trend, we speculate that the vertical shore direction displacement will continue to increase until it eventually stabilizes; (2) Vertical displacement increases rapidly in the initial stage of water storage; after two water-storage cycles, absolute settlement begins to slow down in the vertical direction, but its deformation still changes with the change in the storage period.

Keywords: SBAS; two-dimensional deformation sequence; water-level adjustment

1. Introduction

As a clean and recyclable resource, hydroelectricity is receiving increasing attention from all countries [1]. Inevitably, the rapid development of China's economy has led to a huge demand for energy. Southwest China has abundant water resources for the construction of hydropower stations, so it has obtained a reputation as the "Asian battery" [2]. However, the impounding of a reservoir changes the conditions of local hydrogeology and the regional gravity field, resulting in ground deformation [3]. Furthermore, fluctuations in water level accelerate the deformation, causing local geohazards. In 1961, a landslide with a volume of about 165×10^4 m^3 occurred in the Zixi Reservoir in Zishui, Hunan Province, causing 40 deaths [4]. In 1963, a landslide occurred in the Vajont Reservoir in Italy; a huge volume of landslide soil (about 270×10^6 m^3) rushed into the reservoir, and the surge caused 1925 deaths [5].

The Three Gorges Dam is the largest hydropower station in the world. After the water level of the Three Gorges Dam reached 175 m in 2008, the number of resurrection landslides was tens of times the number from 1992 to 1995 [6]. In order to decrease such geological disasters, scholars are paying more attention to the ground deformation around hydropower stations.

Currently, the tools to monitor ground deformation include the spirit leveling method [7], the Global Positioning System (GPS) [8], and optical remote sensing [9]. However, the spirit leveling and GPS methods are costly and a have low spatial resolution for capturing the detail of a whole deformation pattern [10]. Optical remote sensing is susceptible to the air conditions of clouds and rain.

Synthetic aperture radar interferometry (InSAR), a satellite-based geodetic technique, can overcome the influence of cloud and rain on optical remote sensing and can work both day and night, demonstrating advantages in ground-deformation monitoring [11]. InSAR has been used in the geohazard monitoring of landslides [12], earthquakes [13], volcanos [14], and ground subsidence [15], among others. Additionally, in order to bypass the effects of decorrelation [16] and atmospheric delay [17], scholars have proposed time-series InSAR techniques, such as Persistent Scatterer InSAR (PS-InSAR) [18] and Short Baseline Subset InSAR (SBAS-InSAR) [19]. The PS-InSAR or SBAS-InSAR techniques can recover long-temporal time-series ground deformations with accuracy levels of mm to cm. Recently, scholars have applied time-series InSAR to monitor the landslide at hydropower stations, such as the Three Gorges Dam [10] and the Wudongde Hydropower Station [20].

The Xiluodu Hydropower Station is located upstream of the Jinsha River in China. As the world's third-largest hydropower station, the Xiluodu Hydropower Station has experienced frequent landslides and other geohazards since its impoundment in 2013. In 2013, the Huangping landslide, with an inflow of earthworks amounting to 20×104 m^3, caused 12 deaths. The Ganhaizi landslide, with an influx of earthworks of 7800×104 m^3, was the largest landslide that occurred at the Xiluodu Hydropower Station [21]. LIANG Guohe [22] and ZHOU Zhifang [23] used the leveling points and valley-shrink surveying lines, respectively, to monitor subsidence and valley-shrink deformation around the shoreside of the Xiluodu reservoir and the Xiluodu arch dam, and their results showed that the Xiluodu arch dam and the shoreside of the reservoir were under valley-shrinkage deformation of 50 mm, accompanied by about 25 mm subsidence. Additionally, ZHOU Zhifang [23] pointed out that valley-shrinkage deformation is highly correlated with the water level. In 2017, LI Lingjing [24] analyzed the Yizicun landslide with ALOS, ASAR, and TerraSAR datasets and found that there was a push-type landslide before the water impoundment and that the landslide changed to a pull-type after the water impoundment.

Currently, the research on the Xiluodu Hydropower Station has focused on valley-shrinkage deformation of the shoreside and the arch dam and their subsidence. In 2019, Li et al. [25] used D-InSAR technology to determine the vertical and horizontal displacement of the landslide near the Xiluodu Hydropower Station and further analyzed its deformation characteristics. In 2022, Zhu et al. [26] obtained the horizontal and vertical displacement of the landslide of the Xiluodu Water Station by SBAS technology, using ALOS2 ascending and descending data, and explained the cause of slope instability by combining geomorphic data and lithologic characteristics.

However, these studies were mainly based on traditional measurements and could not completely reflect the ground deformations occurring near Xiluodu. Although the InSAR technique was used to monitor the Xiluodu reservoir area, it was limited to a specific landslide. Furthermore, the hydrogeological conditions have changed after water impoundment. Whether the change in the hydrogeological conditions will lead to wide-range deformation in Yongshan and Leibo counties and change the relationship between the water level and deformation, needs to be analyzed.

2. Geological Setting

2.1. Geological Setting of Xiluodu

As the backbone project of "Power Transmission from West to East" in China, Xiluodu is the largest hydropower station on the Jinsha River. The Xiluodu Hydropower Station is located in the U-shaped river valley upstream of the Jinsha River. On its left side is Leibo County of Sichuan Province, and on its right side is Yongshan County of Yunnan Province. The construction of Xiluodu Hydropower Station started in April 2007, and Xiluodu Hydropower Station carried out the first stage of water storage trial operation in May 2013. In June 2014, Xiluodu Hydropower Station officially began operating, and at the end of September, the water level of the reservoir reached 600 m for the first time. The Xiluodu Hydropower Station is mainly designed to generate power, control floods, fix sand, and navigation [27].

As shown in Figure 1, the reservoir area is located on the southwestern margin of the Yangtze platform, which is the transitional zone between the Yunnan–Guizhou Plateau and the Sichuan Basin. The terrain is generally high in the west and low in the east, belonging to the strong erosion of the high-mountain landform [28]. The Xiluodu Hydropower Station is located at the junction of the first and second steps of China. Rapid changes in the terrain may easily cause geological disasters. Moreover, with the completion of the hydropower station, significant fluctuations in water levels will significantly affect the fragile geological environment.

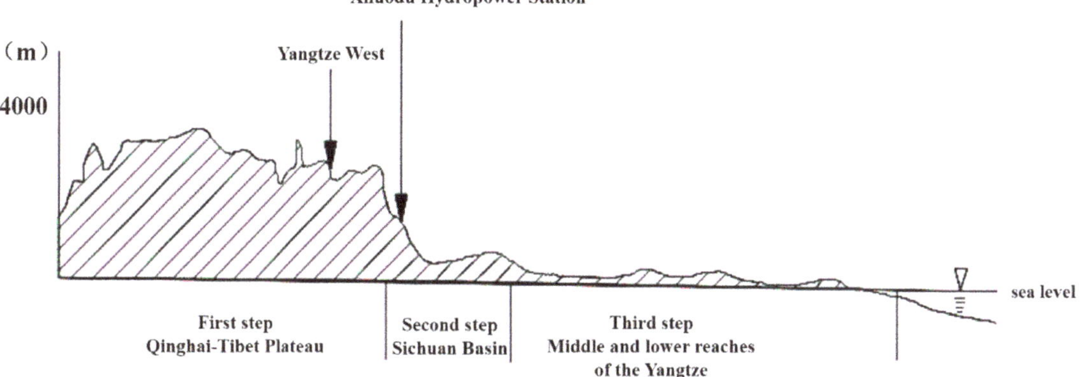

Figure 1. Topography profile of Western—Eastern China.

As shown in Figure 2, the Xiluodu area is surrounded by the yellow mud slope anticline in the north, the wall rock syncline in the south, the Majinzi fault (F1) in the west, and the Jiziba fault (F2) in the east, forming a relatively stable zone. However, a thrust—nappe structure fault (named Majiahe, F3) is located 3 km from the dam area, and the number of thrust outliers is influenced by the Majiahe fault. When the water level of the Xiluodu reservoir increases, most of the thrust outliers are submerged, and subsequently, the fault becomes unstable, affecting the safety of the dam area [29].

2.2. Impoundment of Xiluodu

As shown in Figure 3a, the water level reached its highest level of 600 m in October 2014 for the first time. The water level dropped to 590 m in March 2015. In June 2015, the water level rapidly dropped to 540 m, and then a new round of water storage was started. In July 2015, the water level rose to 555 m. In August 2015, the water level rose to 560 m. In October 2015, it reached 600 m again. Since the water impoundment in October 2014, four complete water impoundment cycles have been completed. Because the annual rainfall is different, the water storage mode is adjusted accordingly.

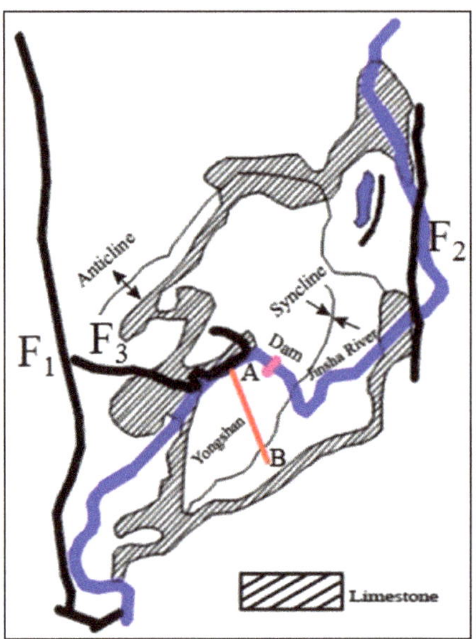

Figure 2. Basic geological structural map of Xiluodu area. F1 is the Majingzi fault, F2 is the Jiziba fault, and F3 is the Majiahe fault.

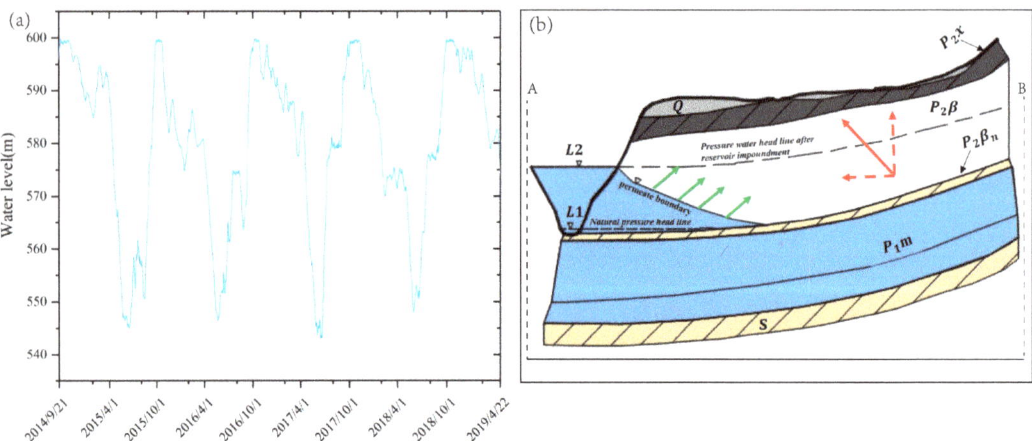

Figure 3. (**a**) Water level of Xiluodu reservoir; (**b**) distribution of aquifers (along the red line in Figure 2) of Xiluodu after water impoundment. A is the river reservoir, B is the area of Yongshan County. S is the Rammell aquifuge layer; P1 m is the limestone permeable bed layer; P2βn is the Rammell aquifuge layer; P2β is the basalt permeable bed layer; P2x is the sandy shale layer; Q is the Quaternary loose accumulation layer; the green arrow indicates the direction of water flow; the red arrow indicates the direction of deformation along the Jinsha River valley. L1 is the water level before constructing Xiluodu Hydropower Station, and L2 is the water level after its construction.

The distribution of the Xiluodu aquifers changed significantly after water impoundment. The geological setting mainly comprises three water-repellent layers (i.e., S, P2βn, and P2x) and three water-permeable layers (i.e., P1 m, P2β, and Q) [23]; see Figure 3b. The highest water level is 600 m, which is lower than the elevation of the sandy shale layer (P2x). Therefore, only the P1 m and P2β permeable bed layers are directly connected with the reservoir water during the process of impoundment. The potential energy of Xiluodu water increases with the water level. For the P1 m limestone permeable bed layers in a pressurized state, the water permeates the P1 m layer through the exposed limestone area, increasing the P1 m pressure potential energy. For the P2β basalt permeable bed layers in a state of low pressure, the water will directly flow into the P2β layer through the submerged basalt area (see the green arrow in Figure 3b), causing the P2β layer to be saturated gradually.

When the water level rises to 600 m, the rising value of the water level (the difference in L2 and L1) is about 220 m. It leads to an increase in the water's potential energy. Due to the existence of confining beds (Rammell layer P2βn), the groundwater pressure potential energy under P2βn will generate an oblique upward pressure (see the solid red arrow in Figure 3b), which will cause the uplift and horizontal displacement perpendicular to the shoreline (see the red dash arrow in Figure 3b). Moreover, as the P2β layer becomes increasingly saturated, the saturated soil will consolidate under the action of gravity, causing vertical subsidence.

3. Methods

3.1. Theoretical Basis of Two-Dimensional Deformation Decomposition

The InSAR measures a one-dimensional deformation along the line of sight (LOS) direction, and such measurements can lead to misinterpretation, as the true deformations may consist of horizontal and vertical displacements [30]. As is known, the LOS deformation is the sum of projection results of horizontal (including north–south and east–west components) and vertical displacement in the LOS direction. Given that there are three or more LOS measurements from different viewing geometries, we can decompose the LOS measurements into north–south, east–west, and vertical displacements by least-squares adjustment [31]. However, for this specific study, we know that displacement along a certain direction is zero or can be ignored based on certain assumptions. For the Xiluodu Reservoir, which is a valley reservoir, the displacement along the direction parallel to the shoreline can be ignored. Thus, the horizontal displacement in a direction perpendicular to the shoreline and the displacement in a vertical direction are sufficient for interpretation. Moreover, due to the small number of viewing angles in the study area, it lacks any redundant observation for us to check for errors if we decomposed the three-dimensional deformations. Thus, we proposed an approach to decompose the InSAR measurements in the LOS direction into two-dimensional displacements, i.e., the horizontal displacement in a direction perpendicular to the shoreline and the vertical displacement.

The specific geometry for decomposing LOS deformation into two-dimensional displacements is shown in Figure 4.

The LOS deformation measurement can be expressed as

$$d_{los} = d_u cos\theta + d_p cos(\delta - \varphi)sin\theta \tag{1}$$

where δ is the azimuth angle of a predefined direction, i.e., the clockwise angle with respect to the north direction; θ is the incident angle of the radar; φ is the heading angle of the satellite; h and u are the horizontal deformation along predefined direction and vertical deformation, respectively; and d_{los} is the InSAR measured deformation in the LOS direction.

Given that there are K ($K >= 2$) LOS measurements for some specific time periods, there is a set of observation equations written in matrix form:

$$\begin{bmatrix} d_{los1} \\ d_{los2} \\ \vdots \\ d_{losK} \end{bmatrix} = \begin{bmatrix} cos(\delta - \varphi_1)sin\theta_1 & cos\theta_1 \\ cos(\delta - \varphi_2)sin\theta_2 & cos\theta_2 \\ \vdots & \vdots \\ cos(\delta - \varphi_K)sin\theta_K & cos\theta_K \end{bmatrix} \times \begin{bmatrix} h \\ u \end{bmatrix} \tag{2}$$

$$Ax = b \tag{3}$$

where A is the design matrix with a size of $K \times 2$, x is the unknown parameter vector, and b is the observation vector.

When solving Formula (3), because there are random errors in the interferogram, such as noise and some unremoved clean atmospheric delay phase, a random model needs to be added to constrain some of the errors. In this paper, interferogram coherence is used for weighting; then, we can obtain the two-dimensional deformations by using the least-squares adjustment method:

$$x = \left(A^{T}PA\right)^{-1}A^{T}Pb \tag{4}$$

where P is the weight matrix composed of interferogram coherence.

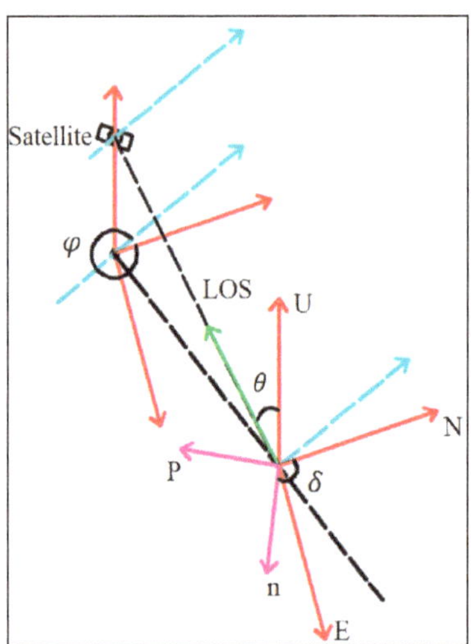

Figure 4. Schematic diagram of LOS decomposition. The red arrow represents the local east, north, up (ENU) coordinates; the blue arrow represents the flight direction of the satellite; the green arrow indicates the LOS direction; φ and θ are the azimuth and incident angle of the satellite, respectively; δ is the azimuth angle of a predefined direction; n and p directions are defined local coordinate systems.

3.2. InSAR Two-Dimensional Deformation Method

Recently, the decomposition of three-dimensional deformation sequences has been developed. Samsonov and D'Oreye proposed the MSBAS method to decompose the time-series East–West and vertical deformations [32] by ignoring the North–South deformations; Pepe et al. proposed the minimum-acceleration (MinA) method, which is used to recover three-dimensional (3D) time-series deformations [33]; Xuguo Shi et al. recovered three-dimensional deformation sequences by using cubic spline interpolation [34]. However, rather than obtaining the 3D deformations or the East–West and vertical deformations by ignoring the North–South deformations, we obtained the vertical and horizontal deformation, called two-dimensional (2D) deformations hereinafter, along a specified direction in reality, which benefits the interpretation.

As shown in Figure 5, given that we have three tracks of SAR data, that is, the Sentinel-1 ascending and descending and ALOS2, covering the study area. The SAR datasets are processed to obtain the time-series LOS deformations. The deformations are geocoded into a geographic coordinate system and further resampled to a common geographic grid. Assuming that the number of images of the three tracks SAR data is N_1, N_2, and N_3, and the acquisition times are $T_1 = \left[t_1^{(1)}, t_2^{(1)}, \ldots, t_{N_1}^{(1)}\right]$, $T_2 = \left[t_1^{(2)}, t_2^{(2)}, \ldots, t_{N_2}^{(2)}\right]$, and $T_3 = \left[t_1^{(3)}, t_2^{(3)}, \ldots, t_{N_3}^{(3)}\right]$, respectively.

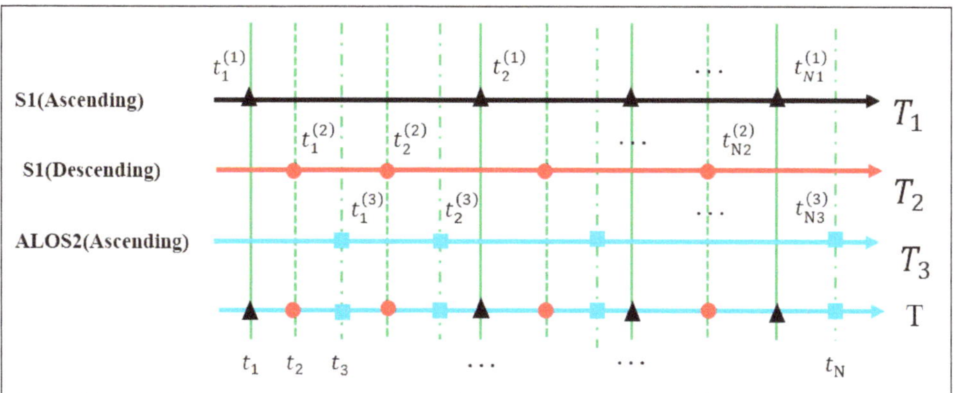

Figure 5. Time series of Sentinel-1 ascending and descending data, and ALOS2 ascending data.

Assuming that the acquisition times are unique for different pieces of SAR data, the total number of images is $N = N_1 + N_2 + N_3$. The times of all of the data acquisitions are $T = T_1 \cup T_2 \cup T_3$. The corresponding LOS time-series deformations sequence are $d_1 = \left[d_1^{(1)}, d_2^{(1)}, \ldots, d_{N_1}^{(1)}\right]$, $d_2 = \left[d_1^{(2)}, d_2^{(2)}, \ldots, d_{N_2}^{(2)}\right]$, and $d_3 = \left[d_1^{(3)}, d_2^{(3)}, \ldots, d_{N_3}^{(3)}\right]$; then, the merged LOS time-series deformations are $d = d_1 \cup d_2 \cup d_3$.

According to Equation (2), the LOS time-series deformations d can be written as a set of equations with respect to the time-series 2-D deformations, and expressed in matrix form:

$$Gx = d \tag{5}$$

where $d = \begin{bmatrix} d_1 & d_2 & \cdots & d_N \end{bmatrix}^T$ is the observation vector, and the unknown parameters vector $x = \begin{bmatrix} h_1 & u_1 & h_2 & u_2 & \cdots & h_N & u_N \end{bmatrix}^T$ with a size of $2N \times 1$. The design matrix has a size of $N \times 2N$, and its form is as follows:

$$G = \begin{bmatrix} A_1 & & & & \\ & A_2 & & & \\ & & A_3 & & \\ & & & \ddots & \\ & & & & A_N \end{bmatrix}$$

where $A_1 = \left[cos(\delta - \varphi^{(1)})sin\theta^{(1)} \quad cos\theta^{(1)}\right]$, $A_2 = \left[cos(\delta - \varphi^{(2)})sin\theta^{(2)} \quad cos\theta^{(2)}\right]$, and $A_3 = \left[cos(\delta - \varphi^{(3)})sin\theta^{(3)} \quad cos\theta^{(3)}\right]$. The superscript in the angles of φ and θ is dependent on which subset of d_i it is from.

Obviously, Equation (5) is rank deficient as the number of equations is smaller than the number of parameters to be estimated, i.e., $N < 2N$. To solve the problem of rank deficiency, we add smoothing constraints to the parameters by assuming that the deformation rates of adjacent displacements are equal, i.e.,

$$\frac{h_{i+2} - h_{i+1}}{t_{i+2} - t_{i+1}} = \frac{h_{i+1} - h_i}{t_{i+1} - t_i} \tag{6}$$

$$\frac{u_{i+2} - u_{i+1}}{t_{i+2} - t_{i+1}} = \frac{u_{i+1} - u_i}{t_{i+1} - t_i} \tag{7}$$

where $i \in [1, N-2]$. Then, we obtain a set of equations with respect to unknown parameters vector, and it is expressed by

$$Rx = 0 \tag{8}$$

where R is the roughness matrix with a size of $(N - 2) \times N$, and

$$R = \begin{pmatrix} t_3 - t_2 & 0 & t_1 - t_3 & 0 & t_2 - t_1 & 0 & \cdots & \cdots & \cdots & \cdots & 0 & 0 \\ 0 & t_3 - t_2 & 0 & t_1 - t_3 & 0 & t_2 - t_1 & \cdots & \cdots & \cdots & \cdots & 0 & 0 \\ \cdots & \cdots & \cdots & \cdots & \cdots & \cdots & \cdots & \cdots & \cdots & \cdots & \cdots & \cdots \\ 0 & 0 & 0 & 0 & 0 & \cdots & t_N - t_{N-1} & 0 & t_{N-2} - t_N & 0 & t_{N-1} - t_{N-2} & 0 \\ 0 & 0 & 0 & 0 & 0 & \cdots & 0 & t_N - t_{N-1} & 0 & t_{N-2} - t_N & 0 & t_{N-1} - t_{N-2} \end{pmatrix}.$$

Combining Equations (5) and (8), we solve the regularized least-squares problem [35]:

$$min \left\| \begin{bmatrix} G \\ \alpha R \end{bmatrix} x - \begin{bmatrix} d \\ 0 \end{bmatrix} \right\|_2^2 \tag{9}$$

The Tikhonov regularization solution is determined by

$$x = \left(G^T G + \alpha^2 R^T R \right)^{-1} G^T d \tag{10}$$

where α is the smoothing parameter and is unknown beforehand. We determine it by using the L-curve method [35].

Figure 6 presents a flowchart of the proposed method.

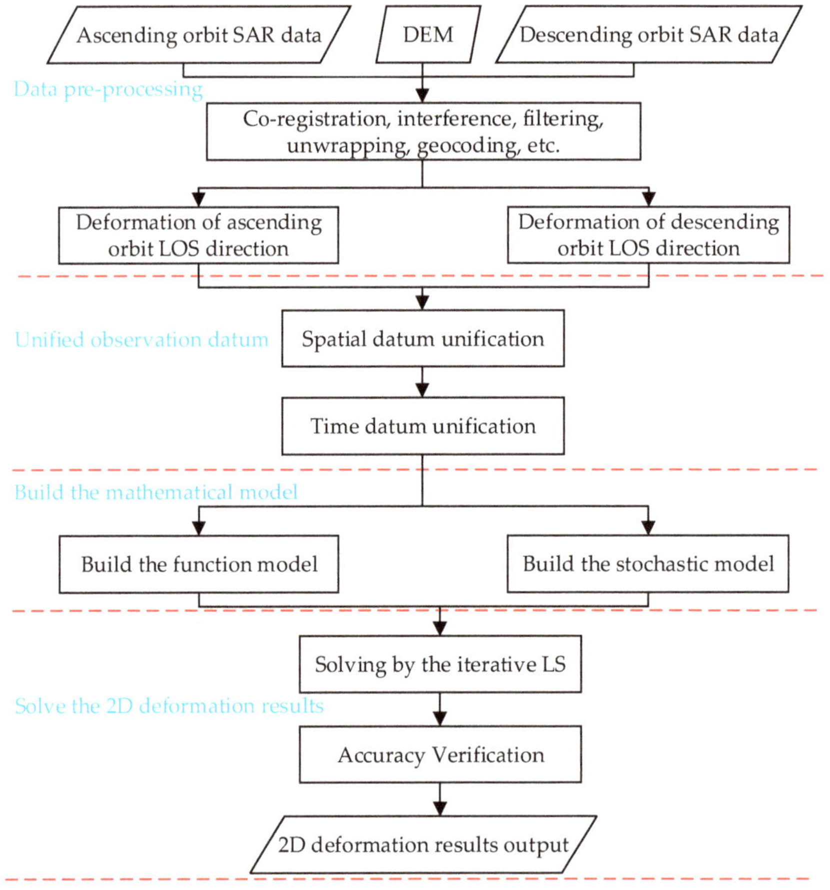

Figure 6. Flowchart of the proposed method.

4. Data Processing and Results

4.1. SAR Data Used

In order to study surface deformation near the Xiluodu reservoir area, we used the ascending and descending data of Sentinel-1 and the ascending data of ALOS2: from October 2014 to March 2019, there was a total of 100 ascending Sentinel-1 data, and the incident angle of the image center was 39.29°. From October 2014 to April 2019, there were 97 descending Sentinel-1 data, and the image center incident angle was 33.9°. From September 2014 to December 2018, there was a total of 12 pieces of ascending ALOS2 data, and the image center incident angle was 31.4°; the coverage area is shown in Figure 7. See Table 1 for specific information. At the same time, we chose the Shuttle Radar Topography Mission (SRTM) with a 30 m x 30 m resolution to remove the terrain phase.

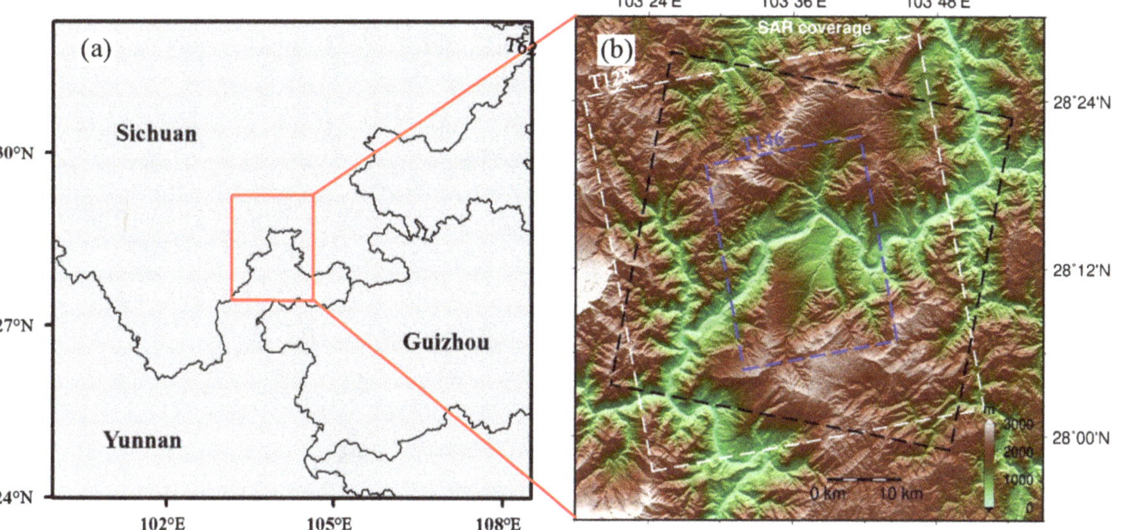

Figure 7. (**a**) Study area; (**b**) SAR data coverage area. The white dashed box represents the Sentinel-1 ascending data, and the black dashed box represents the Sentinel-1 descending data. The blue dashed box represents the ALOS2 ascending data.

Table 1. Three kinds of data for SBAS interference pair information.

Data	Orbit Type	Track	No. of Images	No. of Int. Pairs	Time Span
S1 A	Ascending	T128	100	197	October 2014–March 2019
S1 A	Descending	T62	97	191	October 2014–April 2019
ALOS2	Ascending	T146	12	21	September 2014–December 2018

4.2. Data Processing

We first performed SBAS processing on each piece of data to obtain the deformation sequence of Sentinel-1 data and ALOS2 data. In the data-processing phase, first, in order to ensure the consistency of the three sets of data results, we performed 8 × 2 multi-look on the Sentinel-1 data and 3 × 7 multi-look on the ALOS2 data, thus obtaining image results of about 15 m resolution. Second, because the Sentinel-1 data and ALOS2 data orbital control were accurate, the spatial baseline was short, so the influence of spatial decoherence was not considered. At the same time, in order to overcome the time decoherence, we adopted the two-connected interference pairing method (as shown in Figure 8: 1 and 2 constitute the interference pair, 1 and 3 constitute the interference pair, 2 and 3 constitute the interference pair, and so on), so that any scene image was in a triangular closed loop, which overcame the time decoherence and could test for errors. Therefore, we obtained 21 interference pairs from the ALOS2 ascending data, 197 interference pairs from the Sentinel-1 ascending data obtained, and 191 interference pairs from the Sentinel-1 descending data. Third, we used the SRTM data of 30 × 30 m resolution to obtain the DEM in the SAR coordinate system by geocoding, thus obtaining the simulated terrain phase, and then using the generated interferogram subtracts in the simulated interference phase to remove the terrain phase. In order to eliminate the phase residual and improve the signal-to-noise ratio of the interferogram, we used a 'Goldstein' filter on the differential interference phase with a filter factor of 0.5 and a filter window size of 64. As the low-coherence region could be affected by phase noise, we averaged the coherence coefficient map when unwrapping and then ensured that the region with coherence below 0.52 did not participate in the unwrapping, thus ensuring the reliability of the wrap. In the phase unwrapping stage, we selected the unwrapping reference point in the stable region away from the city. After unwrapping with the minimum cost flow, we refined the baseline of the distracted differential interferogram, thereby effectively removing the track phase. Through the above process, we obtained the exact differential interference phase after unwrapping and then converted it into a settlement sequence result. Finally, we inversely geocoded the resulting sedimentation results and resampled them to the same geographic coordinate range to obtain the deformation sequence results for the three images at 15 m resolution.

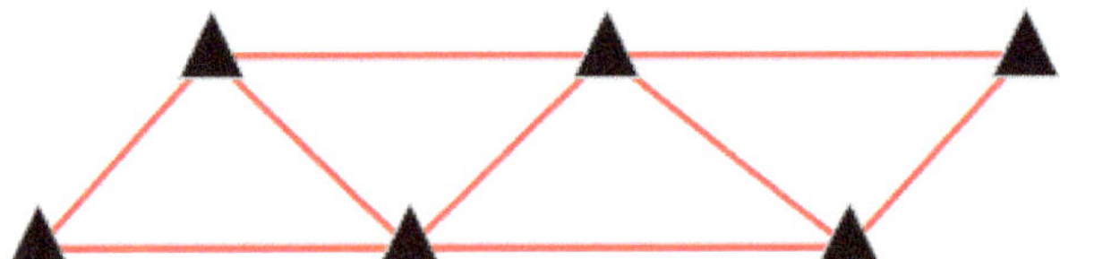

Figure 8. Two-connection network mode. The black triangles represent the SAR images in chronological order from left to right, and the red lines represent different interference pairs.

On this basis, using the algorithm described in Section 3.2, the deformation sequence was decomposed to obtain the deformation sequence results in the vertical shore direction and the vertical direction, and the deformation rate in the horizontal direction and the vertical direction was calculated by using the two-dimensional deformation sequence. The results are provided in Section 4.3.

4.3. Results

4.3.1. LOS Deformation

Figure 9 shows the average rate map of the Sentinel-1 data and ALOS2 data obtained using the SBAS technique. The reference point was selected in the same stable area farther from the deformation zone. For the Yongshan County area, the ALOS2 ascending data LOS has a maximum average speed of −90 mm/a; the ascending Sentinel-1 data LOS has a maximum average speed of −110 mm/a; and the descending Sentinel-1 data LOS has a maximum average rate of 60 mm/a (positive values represent proximity to the satellite and negative values represent the distance from the satellite). For the ALOS2 ascending data and the Sentinel-1 ascending data, the incident angles are 31.4° and 39.29°, respectively. The reasons for the different deformation rates may be as follows: (1) the data incident angles are different, resulting in different observational geometry; (2) Xiluodu horizontal shifts may occur in the reservoir area, resulting in different projections onto the LOS backward direction. For the Sentinel-1 ascending data and the descending Sentinel-1 data, the deformation rate results are opposite, indicating that the horizontal displacement must occur in the Yongshan County area. Therefore, we needed to achieve this in order to decompose and obtain horizontal displacement and vertical displacement.

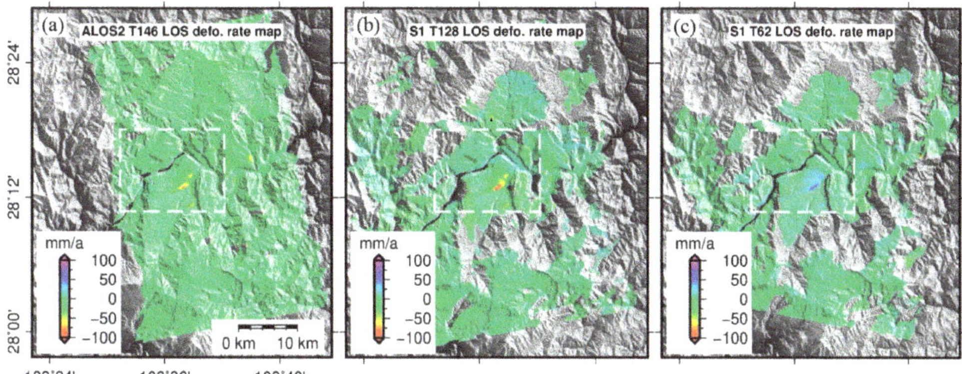

Figure 9. (**a**) Mean velocity map of ALOS2 ascending; (**b**) mean velocity map of Sentinel-1 ascending; (**c**) mean velocity map of Sentinel-1 descending.

4.3.2. Two-Dimensional Deformation Rate

We used the Section 3.2 formula to decompose the deformation sequence to obtain the deformation sequence results in the horizontal direction and the vertical direction and then obtained the mean velocity map perpendicular to the shore direction and the vertical direction, as shown in Figure 10. As can be seen from Figure 10, the maximum deformation rate is 250 mm/a perpendicular to the outward direction of the slope, and the maximum deformation rate is 60 mm/a in the vertical downward direction. It can be seen that the deformation of the Xiluodu reservoir area is mainly in the horizontal direction, accompanied by vertical sedimentation, which is consistent with the previous results of Zhou et al. [23]. Furthermore, we sought to determine the relationship between water storage and horizontal and vertical directions by analyzing the deformation sequence.

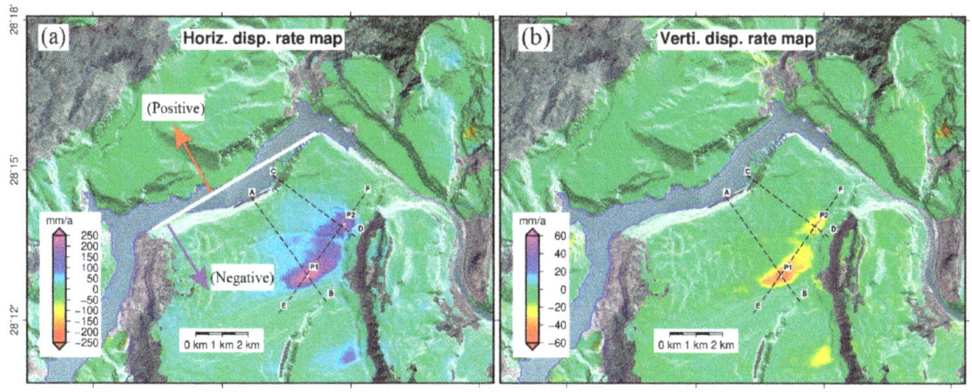

Figure 10. (**a**) Mean rate perpendicular to the shore direction; (**b**) vertical mean rate.

In order to more intuitively display the surface deformation of the Xiluodu reservoir area, because the number of images was too large, we drew some horizontal and vertical deformation sequence images, as shown in Figure 10. It can be seen from Figure 11 that, since October 2014, the horizontal deformation of the Yongshan County area has been accelerating, and the maximum deformation in April 2019 was about 1000 mm; at the same time, the vertical direction deformation has also slowly increased, and by April 2019, the maximum deformation was about 200 mm. Since the reservoir was filled with water in 2014, such a huge deformation occurred in the vicinity of the Xiluodu reservoir area. According to previous studies, the reservoir impoundment may lead to instability of the reservoir slope [36]. Therefore, we suspect that due to the huge water level after the impoundment, the water level difference caused by the change threatens the Xiluodu surface, which could lead to the deformation of the Earth's surface.

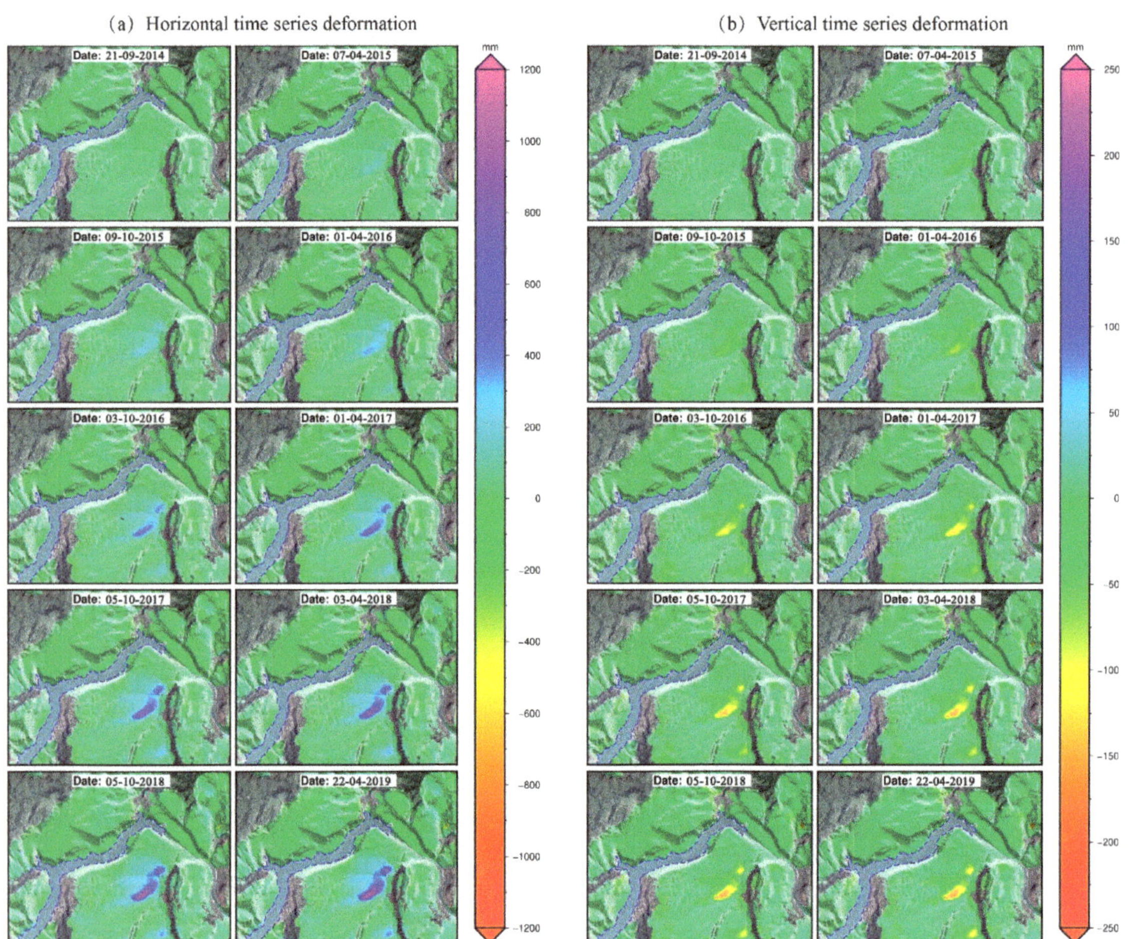

Figure 11. (**a**) Horizontal time-series deformation; (**b**) vertical time-series deformation.

4.3.3. Two-Dimensional Time Series

It can be seen from Figure 10 that the closer the bank is, the smaller the deformation rate, and the farther away it is from the shore, the greater the deformation rate. At the same time, the study area is obviously two distinct regions. In order to prevent the formation of the trailing edge of the study area from creating a landslide and to prevent the formation of a ground crack in the middle of the study area, we selected the section line AB, CD perpendicular to the shore direction and the section line EF parallel to the shore for analysis. Figure 12a–d are the deformation rate profile lines of the vertical shore, and Figure 12e,f are the deformation rate profile lines of the parallel shore. (1) The AB section line reaches a peak at 3.7 km from the shore, while the horizontal deformation rate is greater than 100 mm and the range is about 0.75 km, and the vertical deformation rate is greater than −40 mm.

The range is about 0.5 km. (2) The CD profile line reaches a peak at 3.4 km from the shore, while the horizontal deformation rate is greater than 100 mm, which is about 0.65 km; the vertical deformation rate is greater than −20 mm, which is about 0.6 km. (3) The EF section achieved peaks at 1.8 km and 4.2 km from point E. It can be seen from the results that these two areas are more dangerous, and we will continue to analyze them in combination with optical images.

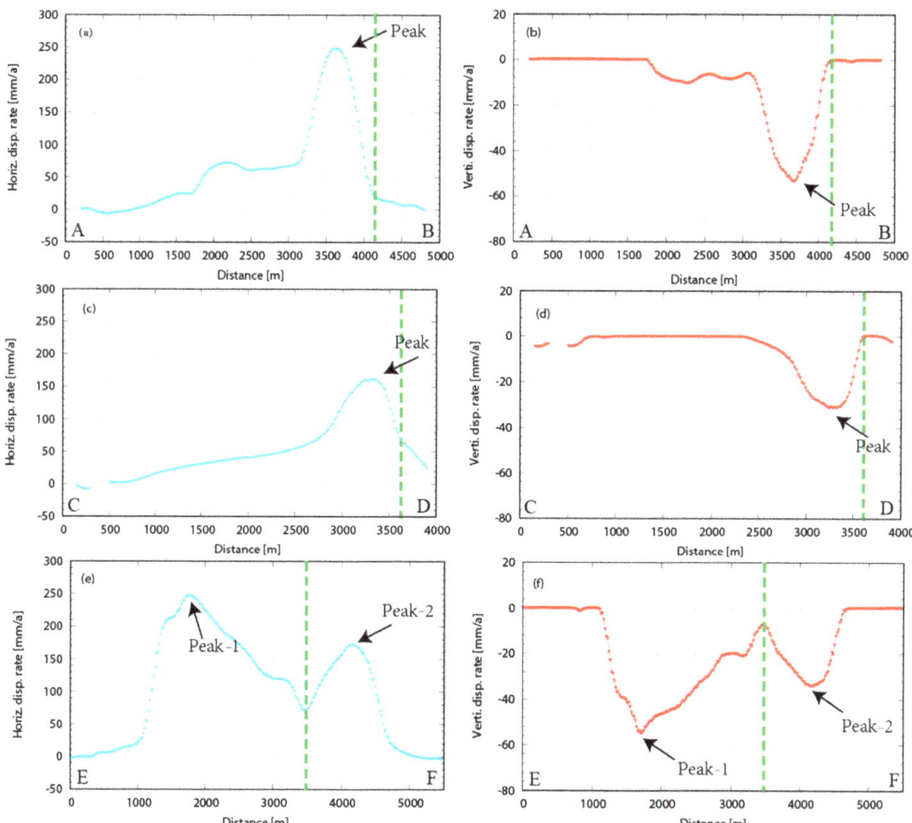

Figure 12. (**a**) is the horizontal displacement rate of profile line A–B; (**b**) is the vertical displacement rate of profile line A–B; (**c**) is the horizontal displacement rate of profile line C–D; (**d**) is the vertical displacement rate of profile line C–D; (**e**) is the horizontal displacement rate of profile line E–F; (**f**) is the vertical displacement rate of profile line E–F. The green line represents the suspected fault.

5. Analysis and Discussion

5.1. Precision Analysis

The accuracy of this experiment was evaluated by calculating the RMS of the residual of the deformation sequence. The results are shown in Figure 13. Furthermore, we statistically analyzed the RMS of the deformation sequence (see Table 2 for details). We found that the mean root mean square error of the region is 4.7 mm, RMS < 10 mm is about 95.36% of the total, and RMS < 15 mm is about 98.85% of the total. The root error average is 4.1 mm, RMS < 10 mm accounts for 95.40% of the total, and RMS < 15 mm accounts for 99.31% of the total. The mean square root error of the non-deformation significant region is 4.8 mm, RMS < 10 mm accounts for 95.35% of the total, and RMS < 15 mm is approximately 98.79% of the total. It can be seen from Figure 14 that the deformation region mainly occurs in the county area, and the terrain is relatively flat, so the accuracy of the InSAR monitoring result is high, while the other regions have steep terrain and relatively low precision. Therefore, we think that the deformation accuracy could reach 10 mm within the 95% confidence interval, so the accuracy of this experiment is reliable.

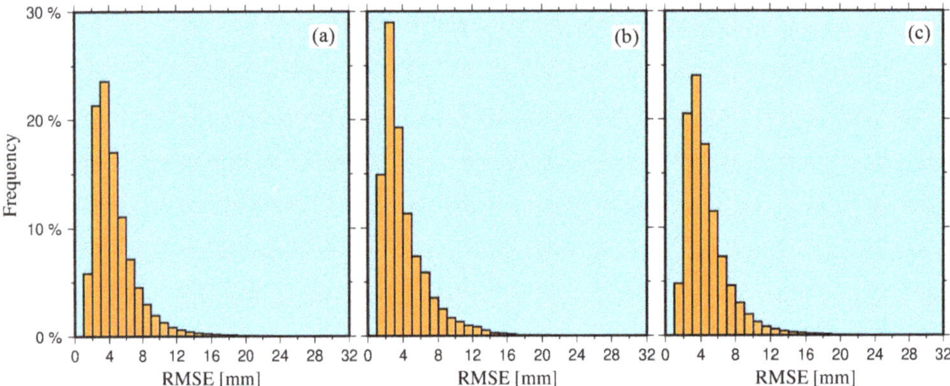

Figure 13. Root mean square error chart. (**a**) Total mean square error; (**b**) deformation significant regional mean square error; (**c**) deformation non-significant area mean square error.

Table 2. RMS data statistics.

	Average of RMSE (mm)	Percentage of RME < 10 mm	Percentage of RMS < 15 mm
Total	4.7	95.36%	98.85%
Deformation significant area	4.1	95.40%	99.31%
Non-deformation significant area	4.8	95.35%	98.79%

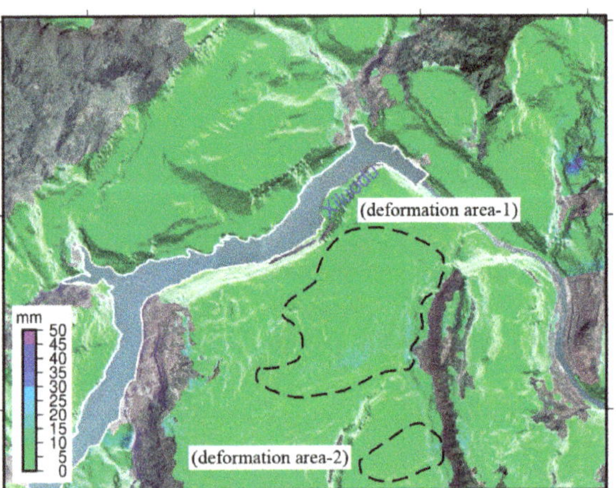

Figure 14. Overall root mean square error map.

5.2. Relationships between Displacement and Water Storage

According to the literature [23], the elevation of the Jinsha River riverbed near the Xiluodu reservoir area was about 400 m before water storage. When the Xiluodu Hydropower Station began to store water, the lowest water level in the reservoir area rose to 540 m, and the highest water level rose to 600 m. It leads to a continuous increase in surface deformation near the reservoir area. At the same time, the fluctuation of the Xiluodu water level of about 60 m per year will also cause the surface deformation near the reservoir area to fluctuate accordingly. To this end, we selected two feature points, P1 and P2, in the two significant areas of deformation, to analyze the relationship between water storage and deformation and provide a basis for future reservoir water storage.

5.2.1. Relationship between Horizontal Displacement and Water Storage

As shown in Figure 15, the water of Xiluodu has been stored for the first time since October 2014, and the water level dropped to 540 m in June of the following year. At present, it has experienced four complete water storage–drainage cycles, of which the cumulative shape variable of the direction of the P1 level reached 110 mm, and the cumulative shape variable of the horizontal direction of the P2 level reached 800 mm. At the same time, the horizontal deformation rate obviously accelerated in each round of water storage, and when the water level of Xiluodu rises to 600 m, the horizontal deformation rate begins to decrease. Over time, in the third round and the fourth round of water storage, although the absolute deformation variable increased in the horizontal direction of the Xiluodu reservoir area, the relative deformation rate began to slow down, indicating that the horizontal deformation of the Xiluodu reservoir area tends to be stable. According to Zhou et al.'s literature analysis [23], as the water level continues to rise, the pore water pressure inside the permeable layer of Yangxin limestone increases continuously, which leads to a reduction in effective stress, which leads to the elastic expansion deformation of Yangxin limestone. At the same time, since the upper layer of the Yangxin limestone is covered with mud shale as a water-blocking layer, the shale also undergoes elastic deformation as the pressure increases. Since the Yongsheng syncline basin generally tends toward the Jinsha River, the horizontal direction of the Xiluodu reservoir area will be deformed outwardly along the slope under the superposition of two elastic deformations. However, as the water level of Xiluodu continues to rise, the water level of the reservoir enters from the permeable layer of basalt, resulting in a stable horizontal deformation.

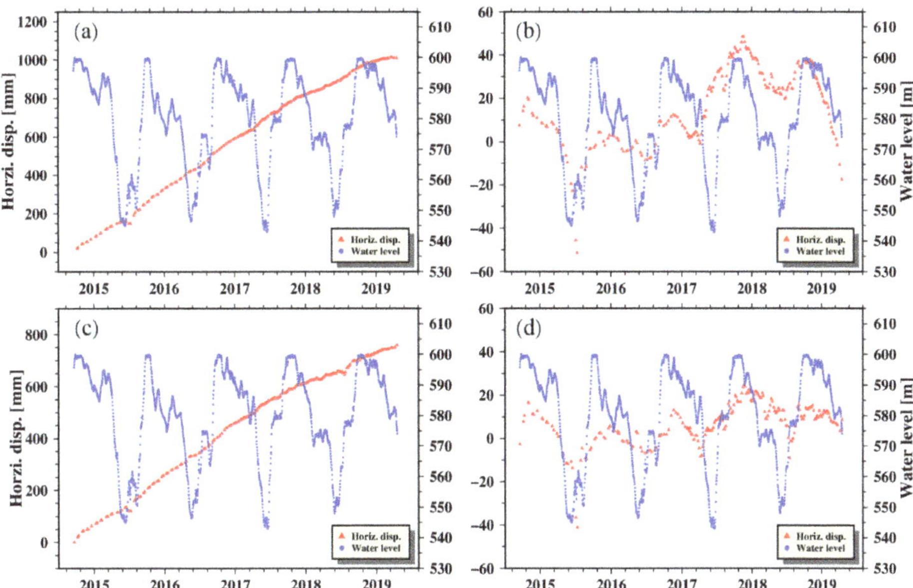

Figure 15. Relationship between horizontal displacement and impoundment. (**a**) is the relationship between the deformation sequence and water level in the horizontal direction of point P1 in Figure 10; (**b**) is the deformation sequence in (**a**) that removes the linear trend; (**c**) is the relationship between the deformation sequence and water level in the horizontal direction of point P2 in Figure 10; (**d**) is the deformation sequence in (**c**) that removes the linear trend.

5.2.2. Relationship between Vertical Displacement and Water Storage

As shown in Figure 16, from October 2014 to April 2019, the cumulative shape variables at the P1 point and the vertical point of the P2 point were −175 mm and −110 mm, respectively. At the same time, when the water level of the reservoir rapidly decreases, the deformation in the vertical direction accelerates. When the water level of the reservoir rises, the deformation in the vertical direction begins to rebound and finally, the deformation in the vertical direction changes with the fluctuation of the water level. Therefore, we speculate that when the water in the Jinsha River continues to infiltrate, the floating force is generated, causing the Xiluodu reservoir area to move upward in a vertical direction. However, the elevation of the water level leads to an increase in the gravitational potential energy, and the soil is consolidated under the action of gravity. The gravity is much greater than the buoyancy. Therefore, the Xiluodu reservoir area has a vertical downward deformation. Finally, as the water-storage cycle continues to increase, the natural consolidation state of the soil becomes saturated so that the cyclical changes caused by the water storage are not offset; thus, the vertical deformation changes with the water level fluctuation.

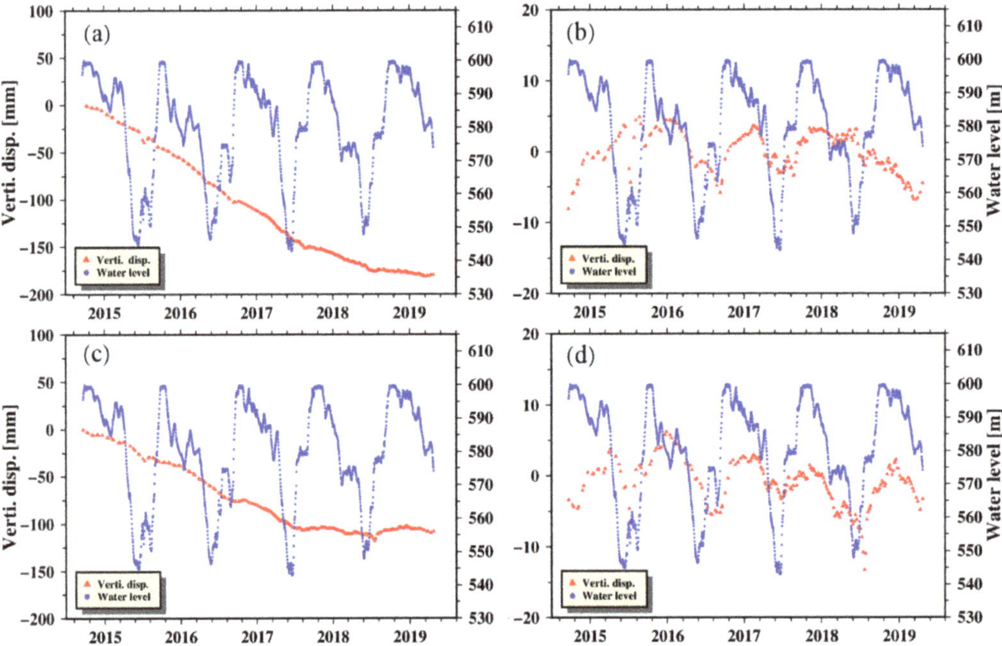

Figure 16. Relationship between vertical displacement and water storage. (**a**) is the relationship between the deformation sequence and water level in the vertical direction of point P1 in Figure 10; (**b**) is the deformation sequence in (**a**) that removes the linear trend; (**c**) is the relationship between the deformation sequence and water level in the vertical direction of point P2 in Figure 10; (**d**) is the deformation sequence in (**c**) that removes the linear trend.

5.3. Potential Geological Hazards

According to the National Geological Hazard Bulletin, there were 2966 geological disasters in 2018, resulting in 105 deaths and seven missing people. The direct economic losses amounted to RMB 1.47 billion. Among them, the southwestern region suffered the most serious disasters, accounting for 35.2% of national geological disasters. Therefore, monitoring geological disasters is vital. As a wide-ranging monitoring method, InSAR can undoubtedly provide necessary observation of geological disasters. However, due to the limitation of satellite side-view imaging, the SAR satellite of a single platform may have overlapping, shadow, and top–bottom inversion in complex mountainous areas, which may easily cause blind spots of identification, resulting in misjudgment [37]. Therefore, through a variety of data (ascending and descending data), the true strength of InSAR technology can be realized in complex mountainous areas. In this paper, we used the data of multiple orbits to decompose the study area horizontally and vertically, thus restoring the deformation of the entire study area. According to the analysis in Section 5.1, combined with the optical image (as shown in Figure 17a), we found that the deformation of the trailing edge (fault-1) of the significant region of deformation was large, and then through the analysis in Section 5.2, we speculate that the region will continue to move along the coast, which may lead to cracks at the trailing edge, and the formation of geological disasters such as ground fissures and collapses, which require following up. At the same time, the left side of the area has a distinct block in the deformed area on the right. As the deformation in the horizontal direction continues to accumulate, the unequal horizontal forces in the two areas may cause relative shear, causing the middle area to form a fault (Fault-2), which may lead to geological disasters such as landslides and ground fissures. As can be seen from the foregoing, the geological conditions in the area are relatively fragile. Therefore, in order to avoid large casualties, the area should be continuously monitored.

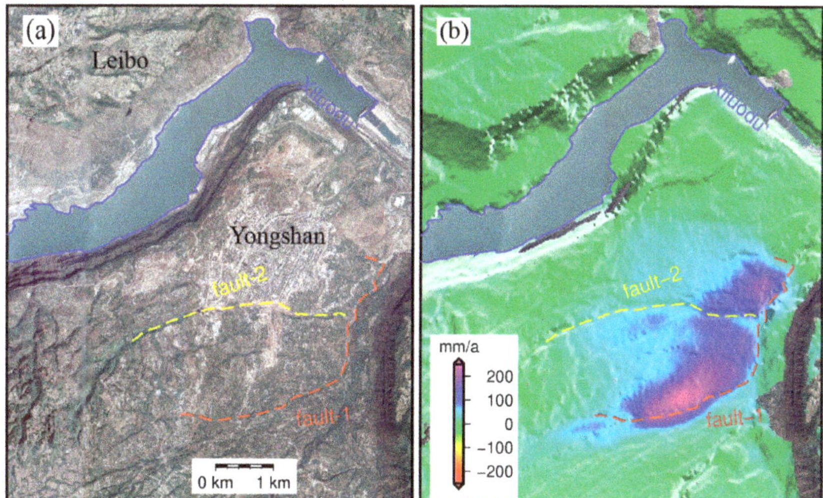

Figure 17. (**a**) Optical image; (**b**) potential fault.

6. Conclusions

As China's demand for power resources continues to increase, hydropower has become increasingly important as a clean and recyclable energy source. However, the artificial construction of a reservoir undoubtedly changes the hydrogeological conditions of the hydropower station's attached structures, resulting in deformation of the surface near the hydropower station, which can lead to landslides, mudslides, earthquakes, etc. In this paper, using time-series InSAR technology, through a variety of data (ALOS2 ascending, Sentinel-1 ascending and descending), large-scale surface deformation monitoring was carried out near the Xiluodu reservoir area after water storage and power generation. The conclusions are as follows:

(1) For a variety of data, the space is not synchronized, and the deformation sequence decomposition cannot be directly performed. A new solution was proposed, and the horizontal and vertical deformation values of the entire time series were obtained. The results show that the maximum average velocity is perpendicular to the bank edge, the maximum average velocity is 250 mm/a, and the maximum vertical velocity is 60 mm/a, which indicates that the surface deformation near the Xiluodu reservoir area is mainly horizontal and the vertical direction is supplemented. At the same time,

we used the residuals to analyze the accuracy of the deformation solution. The results show that the deformation accuracy can reach 10 mm within the 95% confidence interval, indicating that the accuracy of this experiment is reliable.

(2) By analyzing the relationship between horizontal displacement and vertical displacement and water level, it was found that after the Xiluodu water storage, the vertical bank direction displacement continued to increase. This indicates that the deformation caused by the water storage was not due to the elastic displacement caused by the load but caused instead by the irreversible shaping displacement. According to its development trend, we speculate that the vertical shore direction displacement will continue to increase and finally stabilize. According to its development trend, we speculate that the displacement in the vertical shore direction will continue to increase and finally stabilize; the displacement in the vertical direction increases rapidly at the initial stage of water storage. After two water-storage cycles, the vertical deformation begins to stabilize, and the vertical deformation will change with the change in storage period.

(3) Through the analysis of the results and potential faults, we speculate that there may be two faults at this location. As the horizontal deformation continues to accumulate, it may cause cracking, which will aggravate the geological vulnerability of the site.

This experiment failed to obtain GPS data, so it is impossible to verify the decomposition results of InSAR data with external data. At the same time, the acquisition date of this experimental data was after the Xiluodu Hydropower Station started operating. In subsequent analysis, we can try to obtain the InSAR data before the Xiluodu Reservoir was stored and thus gain a clearer understanding of the landmark deformation near the Xiluodu Reservoir.

Author Contributions: Conceptualization, Q.C., B.X. and Z.L.; data curation, H.Z.; formal analysis, W.M. and Z.L.; funding acquisition, B.X.; investigation, W.M., Z.L. and B.X.; methodology, Z.L. and Q.C.; supervision, B.X.; writing—original draft, Q.C. and Z.L.; writing—review and editing, all authors. All authors have read and agreed to the published version of the manuscript.

Funding: This work was partly supported by the National Science Fund for Distinguished Young Scholars (No. 41925016), and the National Natural Science Foundation of China (No. 41804008).

Data Availability Statement: The Sentinel-1 data have been made freely available by the European Space Agency and distributed and archived by the Alaska Satellite Facility (https://www.asf.alaska.edu/sentinel/, 25 October 2019). The ALOS2 SAR data were provided by JAXA under the second Research Announcement on the Earth Observations (EO-RA2) for its Earth observation satellite projects (Project No. ER2 A2 N167).

Acknowledgments: Several figures were plotted by General Mapping Tools (GMT v6.0).

Conflicts of Interest: The authors declare no conflict of interest.

References

1. Bartle, A. Hydropower potential and development activities. *Energy Policy* **2002**, *30*, 1231–1239. [CrossRef]
2. Hennig, T.; Wang, W.; Feng, Y.; Xiaokun, O.U.; Daming, H.E. Review of Yunnan's hydropower development. Comparing small and large hydropower projects regarding their environmental implications and socio-economic consequences. *Renew. Sustain. Energy Rev.* **2013**, *27*, 585–595. [CrossRef]
3. Wang, L.; Chao, C.; Rong, Z.; Du, J. Surface gravity and deformation effects of water storage changes in China's Three Gorges Reservoir constrained by modeled results and in situ measurements. *J. Appl. Geophys.* **2014**, *108*, 25–34. [CrossRef]
4. Xiao Shirong, L.D.; Hu, Z. Engineering geologic study of three actual dip bedding rockslides associated with reservoirs in world. *J. Eng. Geol.* **2010**, *18*, 52–59.
5. Bosa, S.; Petti, M. Shallow water numerical model of the wave generated by the Vajont landslide. *Environ. Model. Softw.* **2011**, *26*, 406–418. [CrossRef]
6. Zhang, L.; Liao, M.; Balz, T.; Shi, X.; Jiang, Y. Monitoring Landslide Activities in the Three Gorges Area with Multi-frequency Satellite SAR Data Sets. In *Modern Technologies for Landslide Monitoring and Prediction*; Springer: Berlin/Heidelberg, Germany, 2015; pp. 181–208.

7. Pingue, F.; Obrizzo, F.; Serio, C. Vertical ground movements in the Colli Albani area (central Italy) from recent precise levelling. *Appl. Geomat.* **2013**, *5*, 203–214. [CrossRef]
8. Fastellini, G.; Radicioni, F.; Stoppini, A. The Assisi landslide monitoring: A multi-year activity based on geomatic techniques. *Appl. Geomat.* **2011**, *3*, 91–100. [CrossRef]
9. Barazzetti, L.; Gianinetto, M.; Scaioni, M. A New Approach to Satellite Time-series Co-registration for Landslide Monitoring. In *Modern Technologies for Landslide Monitoring and Prediction*; Springer: Berlin/Heidelberg, Germany, 2015.
10. Liu, P.; Li, Z.; Hoey, T.; Kincal, C.; Zhang, J.; Zeng, Q.; Muller, J.P. Using advanced InSAR time series techniques to monitor landslide;movements in Badong of the Three Gorges region, China. *Int. J. Appl. Earth Obs. Geoinf.* **2013**, *21*, 253–264. [CrossRef]
11. Massonnet, D.; Feigl, K.L.; Massonnet, D.; Feigl, K.L. Radar interferometry and its application to changes in the Earth's surface. *Rev. Geophys.* **1998**, *36*, 441–500. [CrossRef]
12. Fruneau, B.; Achache, J.; Delacourt, C. Observation and Modelling of the Saint-Etienne-de-Tinée Landslide Using SAR Interferometry. *Tectonophysics* **1996**, *265*, 181–190. [CrossRef]
13. Simons, M.; Fialko, Y.; Rivera, L. Coseismic Deformation from the 1999 Mw 7.1 Hector Mine, California, Earthquake as Inferred from InSAR and GPS Observations. *Bull. Seismol. Soc. Amer.* **2002**, *92*, 1390–1402. [CrossRef]
14. Hooper, A.; Zebker, H.; Segall, P.; Kampes, B. A new method for measuring deformation on volcanoes and other natural terrains using InSAR persistent scatterers. *Geophys. Res. Lett.* **2004**, *31*, L23611. [CrossRef]
15. Bing, X.; Feng, G.; Li, Z.; Wang, Q.; Wang, C.; Xie, R. Coastal Subsidence Monitoring Associated with Land Reclamation Using the Point Target Based SBAS-InSAR Method: A Case Study of Shenzhen, China. *Remote Sens.* **2016**, *8*, 652.
16. Zebker, H.A.; Villasenor, J. Decorrelation in interferometric radar echoes. *IEEE Trans. Geosci. Remote Sens.* **1992**, *30*, 950–959. [CrossRef]
17. Li, Z.W.; Xu, W.B.; Feng, G.C.; Hu, J.; Wang, C.C.; Ding, X.L.; Zhu, J.J. Correcting atmospheric effects on InSAR with MERIS water vapour data and elevation-dependent interpolation model. *Geophys. J. Int.* **2012**, *189*, 898–910. [CrossRef]
18. Ferretti, A.; Prati, C.; Rocca, F. Nonlinear subsidence rate estimation using permanent scatterers in differential SAR interferometry. *IEEE Trans. Geosci. Remote Sens.* **2000**, *38*, 2202–2212. [CrossRef]
19. Mora, O.; Lanari, R.; Mallorquí, J.J.; Berardino, P. A new algorithm for monitoring localized deformation phenomena based on small baseline differential SAR interferograms. In Proceedings of the IEEE International Geoscience & Remote Sensing Symposium, Toronto, ON, Canada, 24–28 June 2002.
20. Zhao, C.Y.; Kang, Y.; Zhang, Q.; Zhu, W.; Li, B. Landslide detection and monitoring with InSAR technique over upper reaches of Jinsha River, China. In Proceedings of the IGARSS 2016, Beijing, China, 10–15 July 2016.
21. Yin, Y.; Sun, P.; Zhu, J.; Yang, S. Research on catastrophic rock avalanche at Guanling, Guizhou, China. *Landslides* **2011**, *8*, 517–525. [CrossRef]
22. Liang, G.H.Y.; Fan, Q.; Li, Q. Analysis on valley deformation of Xiluodu high arch dam during impoundment and its influence factors. *J. Hydroelectr. Eng.* **2016**, *35*, 101–110.
23. Zhou, Z.L.M.; Zhuang, C.; Guo, Q. Impact factors and forming conditions of valley deformation of Xiluodu Hydropower Station. *J. Hohai Univ. (Nat. Sci.)* **2018**, *46*, 497–505.
24. Li, L.Y.X.; Zhou, Z.; Feng, X.; Liu, X. The deformation characteristics of a large landslide before and after impoundment in the xiluodu area based on insar technology. In Proceedings of the 2017 National Engineering Geology Academic Annual Meeting, Guilin, China, 28–29 October 2017; p. 5.
25. Li, L.; Yao, X.; Yao, J.; Zhou, Z.; Feng, X.; Liu, X. Analysis of deformation characteristics for a reservoir landslide before and after impoundment by multiple D-InSAR observations at Jinshajiang River, China. *Nat. Hazards* **2019**, *98*, 719–733. [CrossRef]
26. Zhu, Y.; Yao, X.; Yao, L.; Zhou, Z.; Ren, K.; Li, L.; Yao, C.; Gu, Z. Identifying the Mechanism of Toppling Deformation by InSAR: A Case Study in Xiluodu Reservoir, Jinsha River. *Landslides* **2022**, *19*, 2311–2327. [CrossRef]
27. Zhongli, R. Main Environmental Problems at Xiluodu Waterpower Project. *Sichuan Water Power* **1994**, 40–47 + 94.
28. Hongyan, D.W.C. Analysis on Engineering Geological Characteristics and Genesis Mechanism of Ku'an Lao Landslide in Xiluodu Reservoir Area. *Soil Water Conserv. China* **2011**, 59–62.
29. Wu, D.C.; Li, Y.S.; Liu, W.L.; Deng, J.H.; Wang, D.Y.; Xiao, Y.F. Activity of the Majiahe Dam fault and stability of engineering works in the Xiluodu Hydropower Station in the lower reaches of the Jinsha River, southwestern China. *Geol. Bull. China* **2006**, *25*, 506–511.
30. Hu, J.; Li, Z.W.; Ding, X.L.; Zhu, J.J.; Zhang, L.; Sun, Q. Resolving three-dimensional surface displacements from InSAR measurements: A review. *Earth-Sci. Rev.* **2014**, *133*, 1–17. [CrossRef]
31. Hu, J.; Li, Z.W.; Zhu, J.J.; Zhang, L.; Sun, Q. 3D coseismic Displacement of 2010 Darfield, New Zealand earthquake estimated from multi-aperture InSAR and D-InSAR measurements. *J. Geod.* **2012**, *86*, 1029–1041. [CrossRef]
32. Samsonov, S.V.; D'Oreye, N. Multidimensional Small Baseline Subset (MSBAS) for Two-Dimensional Deformation Analysis: Case Study Mexico City. *Can. J. Remote Sens.* **2017**, *43*, 318–329. [CrossRef]
33. Pepe, A.; Solaro, G.; Calo, F.; Dema, C. A Minimum Acceleration Approach for the Retrieval of Multiplatform InSAR Deformation Time Series. *IEEE J. Sel. Top. Appl. Earth Obs. Remote Sens.* **2016**, *9*, 3883–3898. [CrossRef]
34. Xuguo Shi, L.Z.; Zhou, C.; Li, M.; Liao, M. Retrieval of time series three-dimensional landslide surface displacements from multi-angular SAR observations. *Landslides* **2018**, *15*, 1015–1027.
35. Aster, R.C.; Borchers, B.; Thurber, C.H. *Parameter Estimation and Inverse Problems*; Elsevier: Amsterdam, The Netherlands, 2018.

36.   Li, M.; Zhang, L.; Shi, X.; Liao, M.; Yang, M. Monitoring active motion of the Guobu landslide near the Laxiwa Hydropower Station in China by time-series point-like targets offset tracking. *Remote Sens. Environ.* **2019**, *221*, 80–93. [CrossRef]
37.   Zhao, C.; Liu, X.; Zhang, Q.; Peng, J.; Xu, Q. Research on Loess Landslide Identification, Monitoring and Failure Mode with InSAR Technique in Herfangtai, Gansu. *Geomat. Inf. Sci. Wuhan Univ.* **2019**, *44*, 996–1007.

Article

Evolution Simulation and Risk Analysis of Land Use Functions and Structures in Ecologically Fragile Watersheds

Yafei Wang [1,2], Yao He [1,2], Jiuyi Li [1,3] and Yazhen Jiang [1,4,*]

[1] Key Laboratory of Regional Sustainable Development Modeling, Institute of Geographic Sciences and Natural Resources Research, Chinese Academy of Sciences, Beijing 100101, China
[2] College of Resources and Environment, University of Chinese Academy of Sciences, Beijing 100049, China
[3] Key Laboratory of Water Cycle and Related Land Surface Processes, Institute of Geographic Sciences and Natural Resources Research, Chinese Academy of Sciences, Beijing 100101, China
[4] State Key Laboratory of Resources and Environment Information System, Institute of Geographic Sciences and Natural Resources Research, Chinese Academy of Sciences, Beijing 100101, China
* Correspondence: jiangyz@lreis.ac.cn

Abstract: The evolution of land use functions and structures in ecologically fragile watersheds have a direct impact on regional food security and sustainable ecological service supply. Previous studies that quantify and simulate land degradation in ecologically fragile areas from the perspective of long-term time series and the spatial structure of watersheds are rare. This paper takes the Huangshui Basin of the Qinghai-Tibet Plateau in China as a case study and proposes a long-time series evolution and scenario simulation method for land use function using the Google Earth Engine platform, which realizes the simulation of land use function and structure in ecologically fragile areas by space–time cube segmentation and integrated forest-based prediction. This allows the analysis of land degradation in terms of food security and ecological service degradation. The results show that: (1) the land use function and structure evolution of the Huangshui watershed from 1990 to 2020 have a significant temporospatial variation. In the midstream region, the construction land expanded 151.84% from 1990 to 2004, driven by urbanization and western development policy; in the middle and downstream region, the loss of farmland was nearly 12.68% from 1995 to 2005 due to the combined influence of the policy of returning farmland to forest and urban expansion. (2) By 2035, the construction land in the watershed will be further expanded by 28.47%, and the expansion intensity will be close to the threshold in the upstream and midstream areas and will continue to increase by 33.53% over 2020 in downstream areas. (3) The evolution of land use function and structure will further induce land degradation, causing a 15.30% loss of farmland and 114.20 km^2 of occupation of ecologically vulnerable areas, seriously threatening food security and ecological protection. Accordingly, this paper proposes policy suggestions to strengthen the spatial regulation for land degradation areas and the coordination of upstream, midstream, and downstream development.

Keywords: GEE; long-term time series images; land use classification; spatial modeling; land degradation; Huangshui watershed

Citation: Wang, Y.; He, Y.; Li, J.; Jiang, Y. Evolution Simulation and Risk Analysis of Land Use Functions and Structures in Ecologically Fragile Watersheds. *Remote Sens.* **2022**, *14*, 5521. https://doi.org/10.3390/rs14215521

Academic Editors: Qiusheng Wu, Xinyi Shen, Jun Li and Chengye Zhang

Received: 29 September 2022
Accepted: 31 October 2022
Published: 2 November 2022

Publisher's Note: MDPI stays neutral with regard to jurisdictional claims in published maps and institutional affiliations.

1. Introduction

How human societies utilize, manage, and interact with land are key to addressing current sustainable development issues, including sustainable livelihoods, food security, biodiversity, climate change, and sustainable energy, which have been mentioned in high-level political agreements such as the 2030 agenda for sustainable development, the Paris climate agreement, and the convention on biological diversity [1,2]. Meanwhile, the global climate keeps showing a warming trend, which causes a long-term effect on the human–land system, especially in areas with vulnerable natural conditions [3,4]. Ecologically fragile areas are vulnerable to the combined effects of climate change and human activities due to the poor stability of their own functional and structural systems [5]. These areas are susceptible to exceeding critical thresholds and path dependence, leading to negative social or environmental impacts, and the impacts are difficult to reverse, are receiving more and more attention and becoming a hot topic. Retracing, monitoring, and analyzing the long-term time series changes of land use function and structure in ecologically fragile areas, revealing their process and mechanism, and modeling their potential degradation risks are of importance to global climate change monitoring and the sustainable development of ecologically fragile areas [6].

The existing dominant models for land use and land cover change include the logistic regression [7], Markov [8], system dynamics (SD) [9], cellular automata (CA) [10], CLUE-S [11], agent-based model (ABM) [12], and future land use simulation (FLUS) [13]. With the advantages of wide applicability, easy operation, and high openness to integration with other models, CA is widely used and has proven effective in various areas [14]. The key to these models is to set different development paths according to the land use evolution and development policies, predict the total amount and structure of land use in different paths, and use spatial simulation models to predict the land use patterns in the corresponding scenarios [13,15]. However, these methods are usually constrained by the temporal frequency of the historical land use data and lower classification accuracy, which lead to higher errors and uncertainties in prediction parameters, while seriously affecting the accuracy and reliability of simulation results.

With the open application of remote sensing big data platforms such as the Google Earth Engine, it is possible to use long-term free data and high-performance platforms for long-term land use classification [16]. At present, the realization of large-scale, long-term land use classification and change detection with the help of the Google Earth Engine (GEE) platform has become a hot issue, and a series of breakthroughs have been made. From a methodological view, these studies mainly include three categories: one is to identify and monitor the land use expansion through the construction of long-term remotely sensed ecological environment indexes such as NDVI [17], NDWI [18], etc. [19,20]; second is to realize the change detection of land use functions and structures [21,22] through full-element classification and mapping with long-term Landsat or Sentinel series images; third is to directly classify and extract change information by using multi-source images with the fusion method and dense stacking of time series data as input into the classifier, which prove that the accuracy of change detection can be significantly improved [23,24]. Based on long-term remote sensing classification results, some researchers explored the impacts caused by global issues, such as climate change [25,26], international protocols [27,28], and so on. However, these studies basically only focused on classification and change monitoring, and it is rarely shown how to reveal the evolution laws and patterns contained in long-term classification data and further apply them to the spatial simulation of future scenarios, which is our exploration in this study.

Located in the transition zone between the Qinghai-Tibet Plateau and Loess Plateau, the Huangshui Basin has fragile natural conditions and features of an alpine zone and soil erosion area, which causes great ecological vulnerability and significant ecological importance in the basin. Therefore, the Huangshui Basin is a typically ecologically fragile area. Meanwhile, it is the area with the highest population density and the most concentrated economy in the Qinghai-Tibet Plateau, and the human–land relationship is relatively adversarial and fragile. In 2020, the population of the Huangshui Basin was ~3.43 million, accounting for 56% of that of Qinghai Province; the population density was 95.62 persons/km^2, which was about 13 times that of Qinghai Province. In terms of agricultural production, the Huangshui Basin concentrates about 50% of it's farmland in the Qinghai Province and produces about 65% of its grain; the industrial production there is more than 70% of that of the Qinghai Province. Under the background of global climate change, the changes in natural environment and the expansion of human activities in the past 30 years have had a relatively significant impact on the Huangshui Basin. In particular, the average annual precipitation shows a slight increase, the irrigation water diversion projects are continuously networked, the population and heavy chemical enterprises are further concentrated in the Huangshui Valley, the urban construction land has expanded by 3.6 times, and the tourist and economy in farming and pastoral areas of the upstream have continued to grow [29]. All these have led to great changes in the spatial structure of the Huangshui Basin, resulting in increasingly acute contradictions between urban construction, agricultural and animal husbandry production and ecologically functional land, and an imbalance in the spatial evolution of the upstream and downstream territorial structures. For example, the water pollution in the Xining-Minhe section was once in the lower five categories, and the downstream areas suffered from serious soil erosion and poverty problems [30]. However, China plans to achieve modernization by 2035, which means the intensity of human activities in Huangshui basin tends to maintain a high increase. In the future, with the development of the western region in the new period, the integration of Lanzhou-Xining urban agglomeration and the construction of the Qinghai-Tibet Plateau national park group, the population and economy of the Huangshui Basin will be further agglomerated, and the contradiction between humans and the land will become more prominent.

Therefore, this study took the Huangshui watershed of the Qinghai-Tibet plateau in China as an example and used the GEE platform to realize long-term land use classification. Specifically, a space-time cube that meets the prediction accuracy was constructed based on fully exploiting the advantage of long-term classification, the functional land parameters of future scenarios were measured and calculated by using space-time cube segmentation and prediction methods based on integrated forests, and the land-use function and structure in 2035 were simulated based on adaptive cellular automata. Based on the spatial simulation, relevant indicators were selected, and the risk of land degradation was analyzed from the two aspects of food security and ecological security. Finally, this study verified the reliability of the simulation and risk analysis results by combining field investigation visits to sit in typical areas of the watershed.

2. Study area and Data

2.1. Study Area

The Huangshui Basin ($36°02'\sim37°28'$N, $100°42'\sim103°04'$E) is located in northeastern Qinghai, China, which is between the Qinghai-Tibet plateau and the Loess plateau. The basin comprises the mainstream area of Huangshui and the tributary Datong river basin, a plume-shaped water system with high terrain in the northwest and low terrain in the southeast. Its area is about 16,200 km^2 and its average elevation is 2300 m. The basin is divided into Chuanshui area, Qianshan area, Naoshan area and Shishan forest area from low to high, and the elevation of the dividing line is 2200 m, 2800 m, and 3200 m, respectively. From a sub-basin division perspective, the basin is divided into upstream, midstream and downstream and the points of demarcation are located in the confluences of Xina river and Yinsheng river (Figure 1), respectively. The climate here is a plateau arid and semi-arid continental climate, with high cold and little rainfall, and obvious vertical differences in climate and meteorology with an average temperature of 2.5~7.5 °C and large temperature differences between day and night. The average annual precipitation is 394.6 mm, about 70% of which is concentrated between June and September, resulting in frequent droughts and floods in the basin and a staggering trend. Water resources are scarce in the basin, the ecosystem is fragile, and droughts occur frequently, with typical characteristics of plateau basins and arid and semi-arid basins. The Huangshui basin involves Xining City, Haidong City, and Haibei Prefecture, including 13 districts.

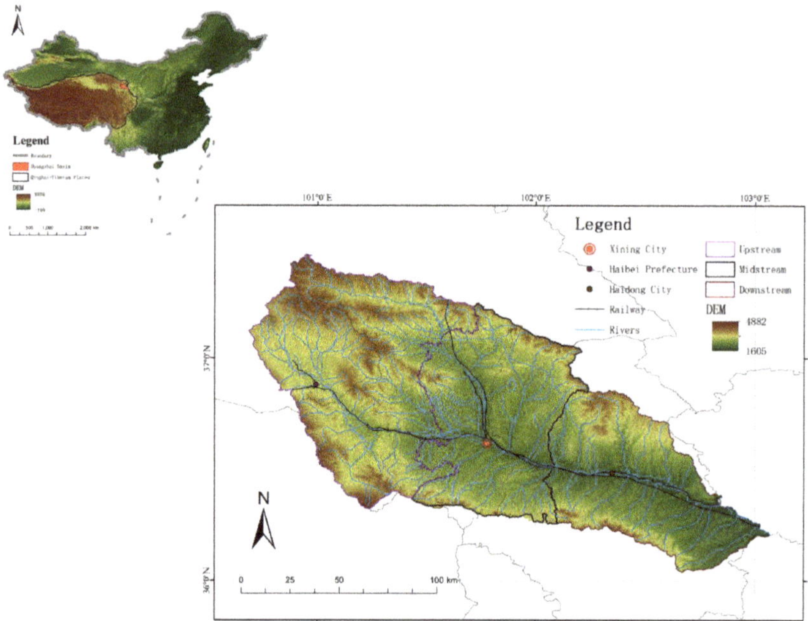

Figure 1. Overview map of the study area (WGS84, 1:1,000,000).

2.2. Data

2.2.1. Remote Sensing Data

The remote sensing data applied in this study were mainly Landsat series image data, obtained from GEE collections "LANDSAT/LT05/C02/T1_L2" and "LANDSAT/LC08/C01/T1_SR", including TM/ETM+/OLI, and periods are from June to October (vegetation flourishing period) and from November to March of the next year (vegetation withering period). In order to obtain as much observational data as possible with cloud cover <10% from 1987 to 2019, we utilized the pixel mosaic-based method [31] and >9000 images were obtained, obtaining 6 years images more than that of the traditional method. After image mosaic, image synthesis, and other operations, we obtained a long-term Landsat image dataset of the Huangshui watershed with a time step of 1 year. Spectral indices such as NDVI, NDWI, and NDBI were then calculated. Auxiliary data include VIIRS night light data (Nighttime Day/Night Band Composites Version 1) and digital elevation data (SRTM). The VIIRS nighttime light data was applied to distinguish urban and non-urban areas and is the monthly average radiation of nighttime data. The digital elevation data (SRTM) is obtained from the US Land Distributed Activity Archive and was gap-filled using open data (ASTER GDEM2, GMTED2010 and NED) [32].

2.2.2. Climate Data

The climate data in this study were FLDAS (Famine Early Warning Systems Network (FEWS NET) Land Data Assimilation System) data, including water content, humidity, evapotranspiration, soil temperature, and precipitation [33]. In addition, the Chinese meteorological dataset was obtained from 1915 stations in the Resource and Environmental Science and Data Center of the Chinese Academy of Sciences (http://www.resdc.cn/ (accessed on 7 July 2021)). After altitude correction, the annual average temperature (Ta), annual average precipitation (Pa), and accumulated temperature $\geq 10\ °C$ (TaDEM), are obtained and then interpolated to obtain the nationwide dataset using the inverse distance-weighted average method.

2.2.3. Land Classification Sample Data

We selected a sample dataset of land-use function classification in the Huangshui basin year by year, including the data for cultivated land, shrubs, forests, grasslands, water bodies, urban construction, and unused land, by combining the visual interpretation of historical images and the selection of field samples. The selection of the dataset was mainly based on the 2019 high-resolution images of the same period as the selected Landsat image, combined with the field sample points to obtain the 2019 sample point data; after that, 2019 was used as the base year, and the reference was gradually selected forward. The image was interpreted sample point by sample point, and the sample points that changed the category were eliminated, and the corresponding sample points were filled to ensure the sample size. By repeating this process, a total of 1065 sample points covering 24-year were finally selected.

3. Methodology

The Methodology (Figure 2) in our study mainly included: (1) using a random forest and spatiotemporal consistency algorithm based on the GEE platform to realize long-term land use classification; (2) constructing a spatiotemporal cube and using spatiotemporal cube segmentation and integrated forest-based prediction methods to calculate the future scenario of the functional land parameters, and using the adaptive cellular automata to simulate land use function and structure; (3) selecting related indicators, such as the area and quality of food loss, the importance, and vulnerability of ecological service functions, analyzing the risks of land degradation from the perspective of food security and ecological security, and verifying the results with field investigations in typical areas of the watershed.

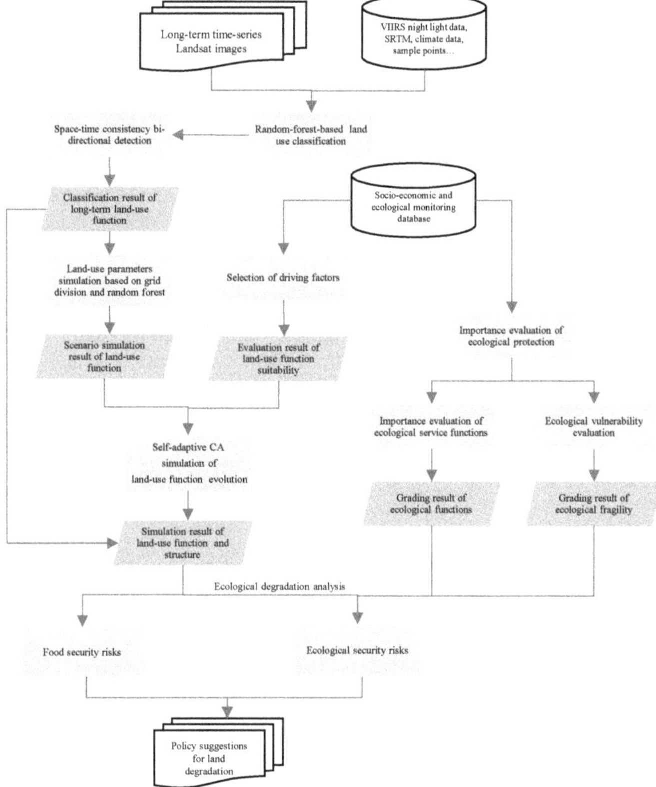

Figure 2. The main technical flows adopted in the methodology.

3.1. Long-Term Functional Classification of Land Use

In this study, we used the GEE platform and the random forest algorithm to initially classify long-term images year by year. We first obtained sample points of 2019 manually. Then to improve the accuracy and strengthen the consistency, we checked whether the features changed and provided category replacement based on historical images of 2018. The process was conducted annually until sample points of 1990 were obtained [29,34]. Next, the classifier was trained by continuously adjusting parameters such as sample point distribution, optimal feature vector combination, texture, optimal window size, and climate factors. After the training, we used the random forest classifier to classify the images of the corresponding years to obtain the initial classification results. The training was as follows: Firstly, taking 80% of the sample points as the training sample points and 20% as the verification sample points, and using the GEE's built-in random function to randomly distribute the training sample points 30 times, we took the result with the highest accuracy as the final training sample [35]. Secondly, based on long-term remote sensing image data, the characteristic variables of each pixel, including the normalized vegetation index NDVI, normalized building index NDBI, normalized difference water index NDWI, and the night light data VIIRS, were calculated and combined year by year. Considering the significant influence of altitude and terrain factors on land-use function in the plateau area [26], the slope and aspect data generated by DEM were introduced as the classification basis, in which DEM adopted the SRTM dataset. At the same time, the parameters in the gray level co-occurrence matrix (GLCM) were used as texture feature variables, the window size and parameter combination with the highest verification accuracy were selected as the final texture feature, through the experiments of various window size and multi-type parameter combination. These texture feature variables included a total of 17 index parameters proposed by Haralick [36] and Conners [37], which are angular second moment (ASM), correlation (CORR), entropy (SENT), variance (VAR), inverse difference moment (IDM), average (SAVG), sum variance (SVAR), sum entropy (SENT), information entropy (ENT), difference variance (DVAR), difference entropy (DENT), relevant information measure 1 (IMCORR1), relevant information measure 2 (IMCORR2), maximum correlation coefficient (MAXCORR), dissimilarity (DISS), inertia (INERTIA), cluster shadow (SHADE), and cluster protrusion (PROM). In addition, considering the natural background characteristics of the Huangshui basin located on the Qinghai-Tibet plateau is sensitive to hydrothermal conditions, meteorological elements were introduced for classification. Among them, the temperature was characterized by annual average temperature, annual accumulated temperature and soil average temperature, and the precipitation was characterized by annual precipitation, soil moisture and soil moisture content. Finally, the final training samples, feature variables, and terrain, texture and climate features were input into the random forest classifier.

Due to the interannual differences in the quality of the Landsat images used in the long-term classification results, we implemented a two-way consistency detection algorithm by constructing a sliding window and a seed window to correct the consistency of the classification results [29]. The main processing included: (1) setting up the bidirectional sliding window w_m and the seed window w_s by selecting the five years with the highest average precision as the starting position of the sliding window, and the third year as the reference year of the seed window to construct a 3×3 seed window; (2) placing the sliding window w_m immediately outside the seed window w_s; (3) determining the dominant classification f_m in the sliding window w_m by investing the frequency ϱ of a function classification: if it was >0.7, the function classification was regarded as the dominant classification f_m of the sliding window w_m, and if the classification frequency of all functions was <0.7, it was considered that there was no dominant classification in the sliding window; (4) determining whether the object type *fs* in the seed window w_s was consistent with the dominant classification f_m in the adjacent sliding detection window w_m: if they were consistent, we set all the classifications in w_m to f_s, and moved the position of the seed window to the grid position where the last dominant classification f_s appeared in the sliding window, and went to step (2) until the seed window reached the beginning and end of the time series; if it was inconsistent, we moved the seed window one bit inward, and went to step (2), and repeated until the seed window reached the beginning and end of the time series to complete the consistency correction.

3.2. Scenario Simulation of Land Use Function

Simulation analysis of land use function mainly included: (1) land use scale simulation based on grid division and random forest, (2) spatial structure simulation based on adaptive cellular automata, which is the simulation of the spatial structure of land use functions in 2035 based on land use function classification data, quantitative impact factor parameters and the result of land use scale simulation.

The Huangshui basin had the characteristics of significant spatial differentiation of natural background conditions, the factors that dominated the evolution of land-use functions in different regions were significantly different, and it was difficult to obtain simulation results with high confidence. Therefore, we constructed a 2 km × 2 km grid division unit, downscaled the long-term land use function classification data, and calculated the area proportion of each functional type $P_t^{m,q}$ as a function parameter year by year. The $P_t^{m,q}$ was expressed as follows:

$$P_t^{m,q} = \frac{s_t^{m,q}}{s}$$

where $s_t^{m,q}$ represents the area occupied by the land-use function q in the grid unit m at the t-th time, and s is the area of the grid unit.

Then, based on the existing functional parameter data, the Lagrangian interpolation method was used to interpolate the functional parameters of 8 time nodes that lacked land use functional classification data grid by grid, and the long time series data of functional parameters with a step of 1 year were constructed. After that, based on the long-term time series data of grid cell functional parameters, the random forest algorithm was used to predict the functional parameters for 2035. We implemented the algorithm through the tool Forest-based Forecast, developed and provided in ArcGIS Pro 10.8 by ESRI. The size of the decision tree forest constructed in this study was 300, and the length of the decision tree training sample was 7 time nodes. To improve prediction accuracy, the data of three time nodes were selected as the verification samples, and the decision tree with the root mean square error (RMSE), calculated by simulated value and observed value, less than 10^{-4} was retained by simulating the verification samples based on the training samples. Finally, the simulation results of functional parameters were normalized by functional classification, and the total area of each functional classification in the Huangshui basin was counted to obtain the simulation results of land use functional classification in the Huangshui basin in 2035.

Based on the principle of functional genetics, we selected corresponding indicators by considering natural carrying capacity, development status, and future potential. Critical indicators such as land construction suitability, water resource convenience, and geological disaster risk were selected for natural carrying capacity. Critical indicators such as population density and economic agglomeration were selected for development status. The future potential focused on the quantification of location conditions, with traffic dominance as the main indicator including traffic accessibility and distance from the central city. Then, based on the attributes of functional classification, the quantitative parameters of different influencing factors were linear combination by category to calculate the functional suitability probability S_k of land-use. The functional suitability probability S_k of land-use was expressed as follows:

$$S_k = \sum_x w_{x,k} * I_x$$

where x is the impact factor, $w_{x,k}$ is the impact weight of the impact factor x on the functional classification k, and I_x is the quantitative parameter of the impact factor x.

Using S_k as the driving factor, and the land scale simulation result of functional classification as the target parameter, the generation probability $TP_{p,k}^t$ of each functional classification was calculated by adaptive cellular automata, which is conducted in GeoSOS-FLUS V2.3, a software developed in 2017 [13], and the evolution simulation of functional classification was realized by generating floating-point random numbers [13].

$$TP_{p,k}^t = S_{p,k} * \Omega_{p,k}^t * Intertia_k^t$$

where k is the land type; p is the spatial location; t represents the t-th iteration; $TP_{p,k}^t$ represents the probability that the functional classification; $S_{p,k}$ is the functional suitability probability of land-use; $\Omega_{p,k}^t$ represents the neighborhood influence; and $Intertia_k^t$ represents the inertia coefficient which is used to realize the adaptive process of the cellular automaton.

3.3. Risk Analysis of Land Degradation

We analyzed the risk of land degradation from the aspects of food security and ecological security in this study. Food security risk mainly refers to the loss of farmland quantity and quality due to the superposition of human activities and climate change, which is mainly characterized by landscape indexes such as total farmland, farmland quality, average patch area, the proportion of large-scale patches, and fragmented patches in total patches. Ecological security risk refers to the decline of ecological service functions and the enhancement of ecological vulnerability due to human activities and climate change, which refers to the encroachment of other land-use on the ecological service function land, mainly characterized by the importance evaluation of ecological conservation. The importance evaluation (Z) was expressed by ecosystem service function importance (X) and ecological vulnerability (Y), both of which were measured by the quality index [38], InVEST model [39] and so on. Among them, the X was related to the importance of ecosystem services such as water conservation [40], soil and water conservation [41], biodiversity maintenance [42], and windbreak and sand fixation [43]; the Y took soil erosion [44] and desertification vulnerability [45] into consideration:

$$\begin{cases} Z = f(X, Y) \\ X = C_x(x_1, x_2, x_3, \dots, x_n) \\ Y = C_y(y_1, y_2, \dots y_n) \end{cases}$$

where Z represents the evaluation level of ecological protection importance, X and Y represent the importance of ecosystem service functions and ecological vulnerability, respectively, $x_1 \sim x_n$ represent different types of ecological service functions, and $y_1 \sim y_n$ represents different vulnerability types.

4. Results

4.1. Long-Term Land Use Functional Classification Result

The long-term classification results of land use in the Huangshui Basin are shown in Figure 3. The classification results show that the overall accuracy of the initial classification results was 76.02~88.92%, and the average accuracy was 83.06% with an improvement of 2.32%. Grassland and farmland were the main land-use types in this basin, which accounted for >77% of the total area of the basin. Grassland was mainly distributed in alpine mountain areas and river valley areas, among which the area of river valley grassland always remained higher than 4652.14 km². The farmland was mainly distributed in the river valley area, among which the area with a slope of <12° in the basin reached 2778.96 km² in 2019, accounting for 55.98% of the total area. The woodland was mainly (63.60%) distributed in mountainous areas with higher altitudes. Compared with grassland, woodland areas were generally located upstream with higher elevations and steeper slopes.

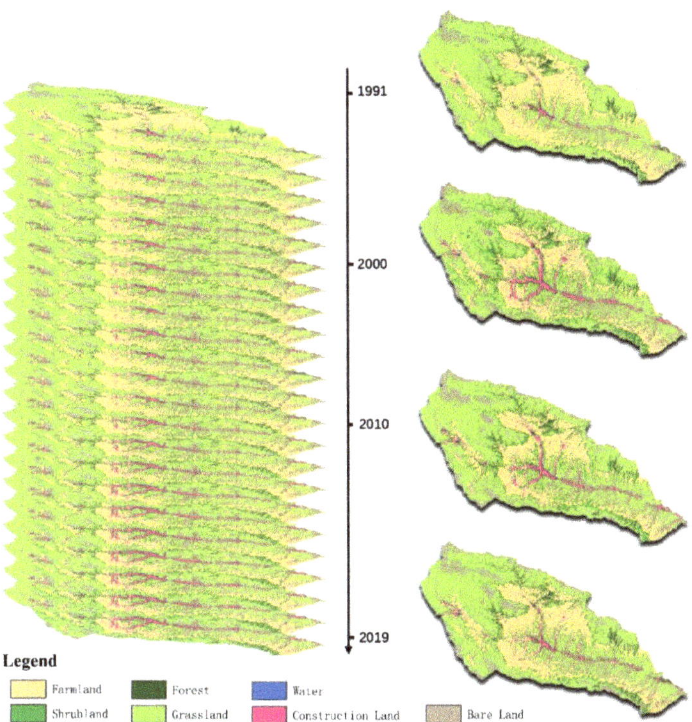

Figure 3. Results of year-by-year land use function classification from 1990–2019.

The spatial pattern of land use in the basin varied not only with terrain, but also with upper, middle, and lower streams. The land-use types dominated by human activities were mainly distributed in the middle and lower steams, while the original natural land types were mainly in upstream. In 2019, the farmland area in the midstream and downstream reached 3658.18 km^2, accounting for 73.70% of the total area of the watershed; for construction land dominated by human activities as well, it mainly (725.88 km^2, 62.72%) located in Xining and Haidong in midstream area. The bare land was mainly distributed upstream, with an area of 348.47 km^2, which was higher than the sum of that in the middle and lower stream, accounting for 57.18%.

We found that the largest changes were in construction land and farmland from 1990 to 2019. The construction land increased from 315.72 km^2 to 1091.83 km^2, with an expansion of 2.67 times; while the area of farmland decreased in a wide range, from 6084.17 km^2 to 4900.33 km^2, with a loss of 18.41%, which is the land type that lost the largest areas over the past 30 years. A major reason for the loss of farmland was the expansion of construction land, occupying a total area of 601.73 km^2, accounting for 77.53% of the total area of newly added urban construction land; another major reason for this was that the evolution to ecological land and the area of woodland, shrubs and other ecological land from evolution was 762.21 km^2, which was higher than the amount of urban construction encroaching. At the same time, the farmland showed a certain trend of migrating to the mountainous areas, which was revealed by the average slope of the farmland in 2019 as 12.08° and as 11.86° in 1990.

In addition, the land use change in this watershed had typical stage characteristics. First, the rate of urban expansion in different stages was significantly different and the main new urban construction land was concentrated in the period from 2000 to 2014, accounting for 56.54% of the total new area. Among them, from 2000 to 2010, the average annual growth rate was 3.84%; from 2010 to 2014, it was a high-speed expansion stage, with an average annual growth rate of 6.74%, which was more consistent with the region's response to the western development strategy. Second, the process of encroaching on farmland by construction land and forests and grasses had two stages: 2000–2006 and 2014–2018, with 362.30 km^2 and 160.21 km^2 of farmland being converted into woodland respectively, which is closely related to the project of "returning farmland to forest".

4.2. Scenario Simulation Result of Land Use Function

The predicted parameters of different land use functions and structures in the Huangshui basin for 2035 are shown in Table 1. During the period from 2020 to 2035, the construction land in this basin still maintains a trend of rapid expansion. The simulation results reached 1487.57 km^2, which increased by 330.31 km^2 compared to 2019, and the average annual expansion rate reached 1.69%. The ecologically degraded bare land is another type of area that increased significantly, which increased from 609.72 km^2 to 752.20 km^2, with an increase of 23.37%. Shrubs and water showed a slight expansion trend, expanding by 277.76 km^2 and 36.13 km^2, with an increase of 11.07% and 37.38%, respectively. However, a loss of farmland in the simulation is always serious. The simulated farmland was 4206.25 km^2, which was only 84.70% of that in 2019, and the average annual decline rate reached 1.10%. The woodland also had serious degeneration, which decreased from 568.51 km^2 in 2019 to 455.31 km^2, with a decline rate of 19.91%. The grassland did not change significantly, and the area change was only 87.07 km^2, accounting for 0.82% of the total area.

Table 1. Simulation scale result table (km^2) + specific gravity.

	Arable	Shrubs	Woodland	Grassland	Water	Construction	Bare Land
2019	4966.18	2231.49	568.51	10524.76	60.53	1157.88	609.72
2035	4206.25	2509.25	455.31	10611.83	96.66	1487.57	752.20

The land use spatial simulation of the Huangshui basin for 2035 with the adaptive cellular automata model is shown in Figure 4. From the perspective of terrain structure, there were obvious differences in land-use in this Basin. The construction land and farmland are concentrated in river valleys and shallow mountainous areas with an altitude of <2800 m, with an area of 1280.01 km^2 and 2938.26 km^2 respectively, accounting for 86.20% and 69.40% of their respective total area. The woodlands were mainly distributed in the origins of the main stream and main tributaries of the Huangshui river and the upper mountains, and were concentrated in the water conservation areas with an altitude of >2800 m. Shrubs were mainly distributed in the outer edge of the valley, concentrated on the edge of farmland; grassland was widely distributed in the whole basin; and bare land was densely distributed in the mountainous areas of the watershed or part of the valley slopes of the tributaries of Huangshui.

From the perspective of the main and tributary network of the river, the simulation results had obvious spatial differences between the upstream and downstream and the main and tributary. The overall function and structure of the upper streams of the watershed had a relatively small evolution, and were dominated by natural land such as forests and grasses. At the same time, there was also the expansion of urban construction land and the loss of farmland, but the evolution scale was small. The area of newly added urban construction land and the area of lost farmland were 61.38 km^2 and 121.33 km^2, with ranges of 23.24% and 9.29%, which were lower than the overall level of the basin. The middle stream had the areas with the highest intensity of human activities and growth rate, and construction land and farmland accounted for 51.75% and 56.64%. From 2019 to 2035, the new construction land area was 167.93 km^2, accounting for 50.85% of the total new expansion; in addition, the mountain bare land at the source of the tributaries in the middle stream had the most obvious expansion, with an increase of 81.19 km^2, accounting for

50.69% of the new bare land area, and the expansion rate reached 163.36%. The expansion of construction land in the lower stream is also extremely serious. The newly added urban construction land covered an area of 100.88 km^2, with an average annual growth rate of 1.95%, which was 0.25% higher than the average level. Similar to the midstream, the bare land in downstream had a substantial expansion as well, with an expansion area of 38.05 km^2 and an expansion rate of 54.54%.

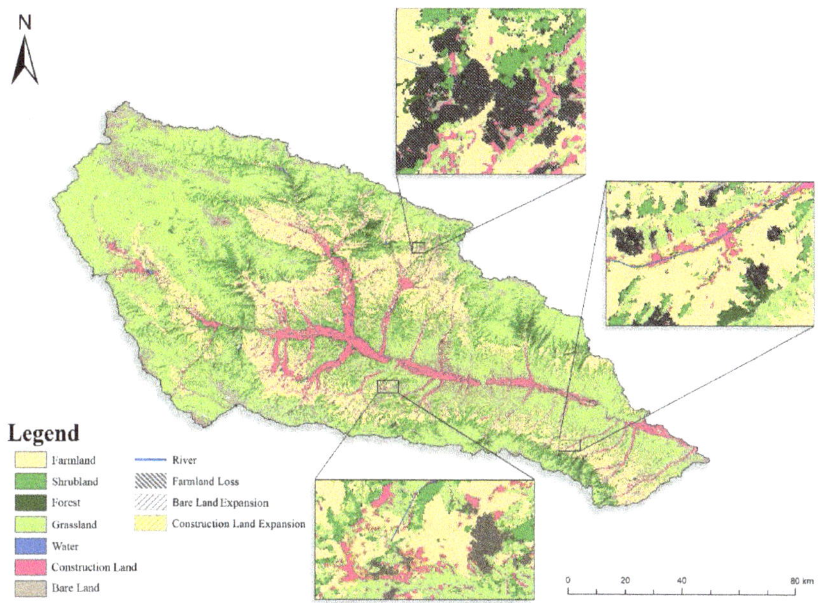

Figure 4. Simulation results of land use function and structure in Huangshui watershed in 2035.(WGS84, 1:1,000,000).

4.3. Risk Analysis Results of Land Degradation

The simulation results of land use in the Huangshui basin demonstrated that both construction land and bare land showed a significant expansion trend, which was accompanied by the loss of a large number of high-quality farmland and the loss of ecological service functions, threatening the food security and biodiversity of the watershed. The area of farmland decreased significantly, and the degree of fragmentation continued to increase. By 2035, 14.70% of the farmland was lost in this basin, and the average patch area was 0.326 km^2, with a decrease of 4.66%; and the broken farmland with an area of <1000 m^2 increased from 64.44% to 74.64%. By analyzing the evolution from farmland to other land use types (Figure 5), the expansion of construction land (typical of Duoba Town shown in Figure 5b), the conversion between woodland and grassland and farmland, and the degradation of farmland to bare land (typical of Hetaozhuang Town shown in Figure 5c) were the main reasons for the loss of farmland, accounting for 82% of the total farmland loss. The farmland replaced by woodlands and grasslands was mainly distributed in the sloping farmland with an altitude of >2800 m and a slope of >25% with a lower quality level. In addition, the degradation of farmland to bare land may also occur in the sloping areas of river valleys.

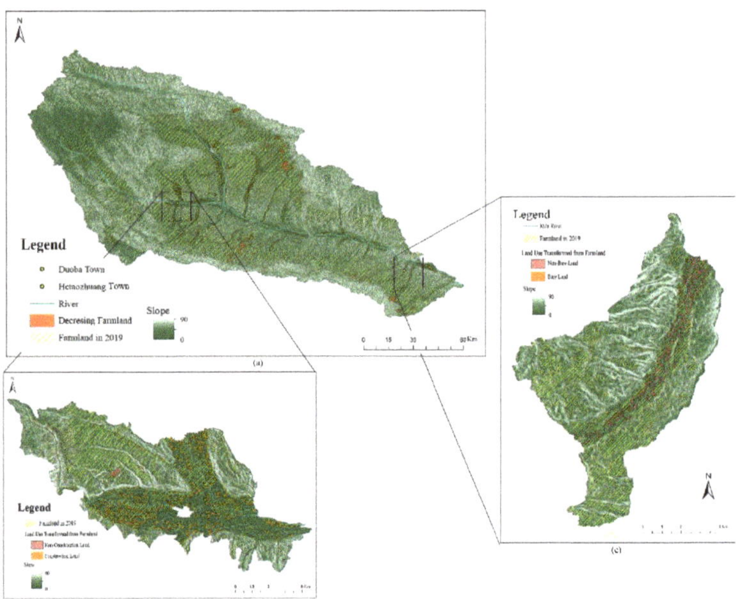

Figure 5. Results of risk analysis of farmland loss: (**a**) Huangshui Basin; (**b**) Duoba Town in Huangzhong District; (**c**) Hetaozhuang town in Minhe County. (WGS84, 1:975,000).

In addition to the loss of farmland, the risk of land degradation was also reflected in the partial weakening and loss of ecological services (Figure 6). The importance evaluation of the ecological service functions showed that the area with moderate or above ecological service function importance in the basin is 13,440.65 km^2, accounting for 66.78% of the total area of the basin. Among them, the proportion of soil and water conservation and biodiversity was relatively high, accounting for 88.09% and 97.37%, respectively, mainly distributed in the valley edge of this basin. This area is relatively flat, and the water and heat conditions are relatively abundant, which is suitable for the growth of various plants, and plays an important role in maintaining regional biodiversity and soil and water conservation. By 2035, the loss of areas with important ecological service functions increased to 111.64 km^2 (typical in Ledu District shown in Figure 6b), with an increase of 67.43% over the base year. The lost areas were mainly distributed in the transition zone from the middle and lower streams, accounting for 28.71% and 19.47% of the total area of important ecological service functions, respectively. The ecological vulnerability analysis showed that the total area above moderately vulnerable in this basin was 6792.41 km^2, accounting for 33.75%, mainly concentrated in the valley slopes and the watershed mountain areas of the basin. The area of bare land increased significantly, and the natural land degradation was relatively serious. Among them, 114.20 km^2 of newly added bare land was in areas with strong ecological vulnerability (typical of Lierbao Town shown in Figure 6c, where 52.99% bare land expansion was simulated in the area with strong ecological vulnerability), accounting for 71.29% of newly added bare land. The expansion of bare land in ecologically fragile areas mainly occurred on the slopes of tributary valleys in the middle stream. At the same time, by superimposing the results of the individual evaluation of ecological vulnerability and the simulated expansion of bare land, we found that 70.35% of the newly developed bare land was located in the middle and high-risk areas of soil erosion in this basin.

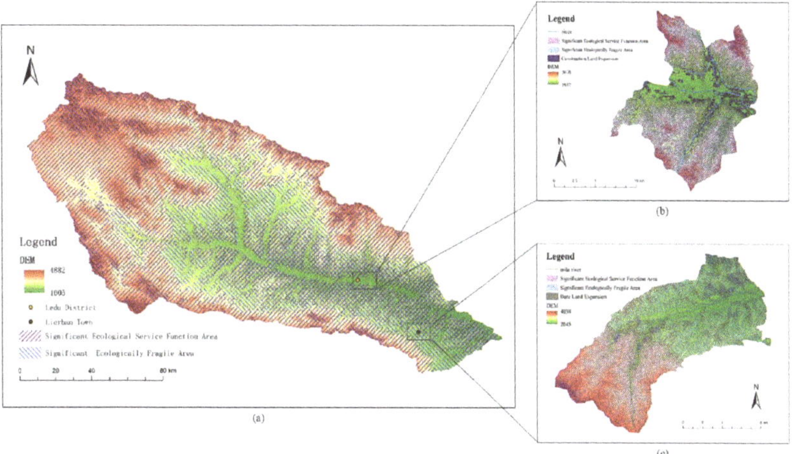

Figure 6. Results of ecological protection importance evaluation and risk analysis: (**a**) Huangshui Basin; (**b**) Ledu District; (**c**) Lierbao town in Minhe County. (WGS84, 1:975,000).

5. Discussion

The data used for the classification of land use functions in this study were long-series Landsat satellite images. Due to the aging of Landsat 5 sensors and the serious banding of Landsat 7 satellite images from 2007 to 2013, there are certain flaws in the reliability of the data [46]. The advantages of the method in this paper have been confirmed in previous studies, and a better classification has been achieved [47,48]. The overall accuracy (OA) of Random Forest classification results ranges from 76.38% to 89.07% and the Kappa range from 0.6881 to 0.9131. After bi-detection process, the OA averagely improved 5.60%, ranging from 84.20% to 91.74%.

Meanwhile, the method in this paper combines random forest (RF) and cellular automata (CA), both of which have proven stable and robust in land use research [49–52]. Moreover, a series of works have conducted a sensitivity analysis of the CA model, which found that the results are sensitive to time step, impact of driving factors, and neighborhood influence [13,53]. Because the CA model is not sensitive to the area that random forest simulates, the combined method is also stable and robust. To further assess accuracy, we used the indicator "figure of merit" (FoM) as a veridiction method. FoM is a ratio representing the proportion of samples that are observed and simulated correctly [54] and has been widely used in previous studies [55]. We simulated land use in 2018 using our method and the Markov-FLUS model, separately. The FoM using the method in this paper was 12.63%, which was 1.33% higher than FoM using Markov-FLUS model.

As a high-intensity human activity area with an important but vulnerable environment, the Huangshui basin is a hotspot of environmental protection policies and development plans, such as "returning farmland to forest", "protection of prime farmland", "Lanzhou-Xining Urban Area Development Plan", and so on. Effected by multiple policies, interconversion of land use has occurred in the Huangshui Basin, between farmland and shrubland, for instance. Taking policy factors and multiple special types of areas, such as agricultural and animal husbandry interlaced zones and ecologically fragile areas into consideration, the driving force is relatively complex, and the reliability of prediction and simulation needs to be verified. Therefore, this study conducted a field survey, and verified the conclusions of this study through a comprehensive practical investigation and visual interpretation of historical image data for the typical sites involved in the study.

Because of the loss of cultivated land and ecological degradation caused by the expansion of construction land, we selected Duoba Town in Huangzhong District Figure 7a and the urban area of Ledu District Figure 7b as the verification and inspection sites, both of which are urban expansion areas in the "Lanzhou-Xining Urban Area Development Plan". Combining historical images and land use function classification data, it was found that Duoba Town experienced a rapid expansion of urban construction land after 2012, and the loss of its cultivated land accelerated during the same period; in the simulation results, Duoba Town maintained the rapid expansion of urban construction land, its area reached 58.44 km^2 in 2035, with an increase of 15.73 km^2 compared to that in 2019; among them, 14.01 km^2 of new urban construction land was converted from cultivated land, accounting for 89.08% of the totally new area. The Ledu District had a similar situation to that of Duoba Town, and the overall urban area along the main banks of the Huangshui river showed a significant increase in 2013. The area of newly added urban construction land was 7.60 km^2, of which the area of the area with medium or above ecological service function importance accounted for 44.38%. The new urban area of Ledu District is flat, which is convenient for the construction implementation, and the transportation infrastructure is relatively superior, which reveals the extremely strong driving force for evolution into urban construction land. Considering that Ledu District is located at the entrance of the tributaries in the low stream, the weakening of the soil and water conservation service function can easily lead to the instability of the regional ecosystem, resulting in the risk of soil erosion in the upstream area.

Figure 7. Distribution of validation sites for risk analysis: (**a**) Duoba Town in Huangzhong District; (**b**) urban area of Ledu District; (**c**) Hetaozhuang town in Minhe County; (**d**) Donggou Town in Huzhu County. (WGS84, 1:975,000).

Given the degradation of farmland and ecological area into bare land, we selected Hetaozhuang town in Minhe County Figure 7c and Donggou Town in Huzhu County Figure 7d as verification sites. Several protection projects were conducted on these sites, such as "returning farmland to forest" and "protection of prime farmland". Based on the classification data of land use function, it was found that Hetaozhuang Town continued to maintain a certain decline of cultivated land after 1990, and the bare land also expanded slightly. In the simulation results, the cultivated land in Hetaozhuang Town continued to decline at significantly increased rates, which mainly evolved into grassland and bare land. The area of cultivated land in decline reached 4.91 km^2, accounting for 20.41% of the total cultivated land. From the field investigation, it was found that there is a significant vertical difference in some mountain land types in this area, which shows that the top mountain is land or bare land, and the bottom is cultivated land and vulnerability may induce bare land to encroach on cultivated areas. The bare land in Donggou Town also showed a large-scale expansion in the simulation results, with an expansion area of 12.24 km^2, and the simulated degradation area was 9.07 km^2, revealing potential risks of land degradation in areas with strong ecological vulnerability. The investigation shows that the vegetation coverage in this area is relatively good, the main natural vegetation is grassland, and the artificial development of the mountain is common with cultivated land at the bottom of the mountain. Further investigation demonstrated that this area consisted of bare land and low-coverage grassland before 2015, and there was an expansion trend, and the implementation of the wasteland treatment project and the conversion of farmland to forest has gradually restored the wasteland and part of the cultivated land into grassland and forest land after 2015. The stability of the ecosystem in Donggou Town has been enhanced, which has greatly curbed the expansion trend of bare land. With the auxiliary UAV images, we found that the grassland in Donggou town has regional heterogeneity and large-scale differences in coverage levels, which can prove that there are ecological restoration projects carried out in stages in this region.

After investigation and verification of typical sites in the Huangshui basin, it was found that: (1) there is a strong driving force for land-use conversion in each area based on the verification of the background conditions of urban construction land expansion and bare land degradation areas, revealing that the evolution of functional land is reasonable and the conclusions are reliable; (2) the degradation of bare land in the watershed mainly result from the regional topography, soil quality and vegetation coverage, the change of which has a great influence on the regional ecological effect. Destructive human activities lead to irreversible land degradation and bring serious ecological risks, while ecological management projects such as returning farmland to forests can curb the trend and improve regional ecological security.

6. Conclusions

This paper modeled the spatial evolution of land use function and structure in the Huangshui basin on the Qinghai-Tibet plateau using long-term time-series remote sensing land use/land cover classification data and analyzed land degradation analysis in terms of protection and degradation of ecological service functions. The study shows that Huangshui basin, as an ecologically fragile watershed, has undergone drastic changes in the past 30 years, manifested by the massive expansion of construction land and the massive loss of farmland. This change has been influenced not only by human activities such as urbanization and ecological immigration, but also by the combined effects of various policies such as the return of farmland to forests, farmland protection, and climate change. By continuing to follow past development paths, the Huangshui basin will suffer from further extensive land degradation, as evidenced by the massive expansion of built-up land and bare land, which will encroach further, leading to the loss of farmland, as well as the area of high ecological importance and vulnerability. Given the spatial management and control of specific functional areas and the coordination of upstream, midstream, and downstream development, we put forward specific policy recommendations in this study: (1) strengthen the protection of cultivated land, delineate the farmland protection redline, promote scientific planting pattern, and encourage the development of water-saving agriculture to improve planting efficiency; (2) strengthen ecological protection, ensure the ecological service function of the river basin, delineate ecological protection redline, ensure the ecological service function of the river basin, maintain the stability of the river basin ecosystem, and maximize the ecological benefits; (3) promote the unified planning and improve the collaborative management of the river basins.

Author Contributions: Conceptualization, Y.W. and Y.J.; formal analysis, Y.W. and Y.H.; funding acquisition, Y.W.; methodology, Y.W. and Y.H.; writing—original draft, Y.W., Y.H., J.L. and Y.J.; writing—review and editing, Y.W., Y.H. and Y.J. All authors have read and agreed to the published version of the manuscript.

Funding: This work was supported by the Second Tibetan Plateau Scientific Expedition and Research Pro-gram (2019QZKK0406), and the Strategic Priority Research Program of the Chinese Academy of Sciences (XDA20020301).

Data Availability Statement: Codes available for land cover/land use classification: https://code.earthengine.google.com/674cec07c398a7965fda4e0b1e8df58c (accessed on 7 July 2021). Other parts of the code or process is available from corresponding authors upon reasonable request.

Acknowledgments: We appreciate the critical and constructive comments and suggestions from the reviewers that helped improve the quality of this manuscript.

Conflicts of Interest: The authors declare no conflict of interest.

References

1. Desa, U. *Transforming Our World: The 2030 Agenda for Sustainable Development*; United Nations: New York, NY, USA, 2016.
2. Silvestre, B.S.; Ţîrcă, D.M. Innovations for Sustainable Development: Moving toward a Sustainable Future. *J. Clean. Prod.* **2019**, *208*, 325–332. [CrossRef]
3. Grabherr, G.; Gottfried, M.; Pauli, H. Climate Change Impacts in Alpine Environments. *Geogr. Compass* **2010**, *4*, 1133–1153. [CrossRef]
4. Orusa, T.; Borgogno Mondino, E. Exploring Short-Term Climate Change Effects on Rangelands and Broad-Leaved Forests by Free Satellite Data in Aosta Valley (Northwest Italy). *Climate* **2021**, *9*, 47. [CrossRef]
5. Huang, A.; Xu, Y.; Sun, P.; Zhou, G.; Liu, C.; Lu, L.; Xiang, Y.; Wang, H. Land Use/Land Cover Changes and Its Impact on Ecosystem Services in Ecologically Fragile Zone: A Case Study of Zhangjiakou City, Hebei Province, China. *Ecol. Indic.* **2019**, *104*, 604–614. [CrossRef]
6. De Lange, H.J.; Sala, S.; Vighi, M.; Faber, J.H. Ecological Vulnerability in Risk Assessment—A Review and Perspectives. *Sci. Total Environ.* **2010**, *408*, 3871–3879. [CrossRef] [PubMed]
7. Lin, Y.-P.; Chu, H.-J.; Wu, C.-F.; Verburg, P.H. Predictive Ability of Logistic Regression, Auto-Logistic Regression and Neural Network Models in Empirical Land-Use Change Modeling—A Case Study. *Int. J. Geogr. Inf. Sci.* **2011**, *25*, 65–87. [CrossRef]
8. Muller, M.R.; Middleton, J. A Markov Model of Land-Use Change Dynamics in the Niagara Region, Ontario, Canada. *Landsc. Ecol.* **1994**, *9*, 151–157.
9. Shen, Q.; Chen, Q.; Tang, B.; Yeung, S.; Hu, Y.; Cheung, G. A System Dynamics Model for the Sustainable Land Use Planning and Development. *Habitat Int.* **2009**, *33*, 15–25. [CrossRef]
10. Liu, X.; Ma, L.; Li, X.; Ai, B.; Li, S.; He, Z. Simulating Urban Growth by Integrating Landscape Expansion Index (LEI) and Cellular Automata. *Int. J. Geogr. Inf. Sci.* **2014**, *28*, 148–163. [CrossRef]
11. Verburg, P.H.; Soepboer, W.; Veldkamp, A.; Limpiada, R.; Espaldon, V.; Mastura, S.S.A. Modeling the Spatial Dynamics of Regional Land Use: The CLUE-S Model. *Environ. Manag.* **2002**, *30*, 391–405. [CrossRef]
12. Matthews, R.B.; Gilbert, N.G.; Roach, A.; Polhill, J.G.; Gotts, N.M. Agent-Based Land-Use Models: A Review of Applications. *Landsc. Ecol.* **2007**, *22*, 1447–1459. [CrossRef]
13. Liu, X.; Liang, X.; Li, X.; Xu, X.; Ou, J.; Chen, Y.; Li, S.; Wang, S.; Pei, F. A Future Land Use Simulation Model (FLUS) for Simulating Multiple Land Use Scenarios by Coupling Human and Natural Effects. *Landsc. Urban Plan.* **2017**, *168*, 94–116. [CrossRef]
14. Aburas, M.M.; Ho, Y.M.; Ramli, M.F.; Ash'aari, Z.H. The Simulation and Prediction of Spatio-Temporal Urban Growth Trends Using Cellular Automata Models: A Review. *Int. J. Appl. Earth Obs. Geoinf.* **2016**, *52*, 380–389. [CrossRef]
15. Zhou, L.; Dang, X.; Sun, Q.; Wang, S. Multi-Scenario Simulation of Urban Land Change in Shanghai by Random Forest and CA-Markov Model. *Sustain. Cities Soc.* **2020**, *55*, 102045. [CrossRef]
16. Gorelick, N.; Hancher, M.; Dixon, M.; Ilyushchenko, S.; Thau, D.; Moore, R. Google Earth Engine: Planetary-Scale Geospatial Analysis for Everyone. *Remote Sens. Environ.* **2017**, *202*, 18–27. [CrossRef]
17. Robinson, N.P.; Allred, B.W.; Jones, M.O.; Moreno, A.; Kimball, J.S.; Naugle, D.E.; Erickson, T.A.; Richardson, A.D. A Dynamic Landsat Derived Normalized Difference Vegetation Index (NDVI) Product for the Conterminous United States. *Remote Sens.* **2017**, *9*, 863. [CrossRef]
18. Tang, Z.; Li, Y.; Gu, Y.; Jiang, W.; Xue, Y.; Hu, Q.; LaGrange, T.; Bishop, A.; Drahota, J.; Li, R. Assessing Nebraska Playa Wetland Inundation Status during 1985–2015 Using Landsat Data and Google Earth Engine. *Environ. Monit. Assess.* **2016**, *188*, 654. [CrossRef]
19. Kong, D.; Zhang, Y.; Gu, X.; Wang, D. A Robust Method for Reconstructing Global MODIS EVI Time Series on the Google Earth Engine. *ISPRS J. Photogramm. Remote Sens.* **2019**, *155*, 13–24. [CrossRef]
20. Amani, M.; Ghorbanian, A.; Ahmadi, S.A.; Kakooei, M.; Moghimi, A.; Mirmazloumi, S.M.; Moghaddam, S.H.A.; Mahdavi, S.; Ghahremanloo, M.; Parsian, S.; et al. Google Earth Engine Cloud Computing Platform for Remote Sensing Big Data Applications: A Comprehensive Review. *IEEE J. Sel. Top. Appl. Earth Obs. Remote Sens.* **2020**, *13*, 5326–5350. [CrossRef]
21. Zurqani, H.A.; Post, C.J.; Mikhailova, E.A.; Schlautman, M.A.; Sharp, J.L. Geospatial Analysis of Land Use Change in the Savannah River Basin Using Google Earth Engine. *Int. J. Appl. Earth Obs. Geoinf.* **2018**, *69*, 175–185. [CrossRef]

22. Souza, C.M.; Shimbo, J.Z.; Rosa, M.R.; Parente, L.L.; Alencar, A.A.; Rudorff, B.F.T.; Hasenack, H.; Matsumoto, M.; Ferreira, L.G.; Souza-Filho, P.W.M.; et al. Reconstructing Three Decades of Land Use and Land Cover Changes in Brazilian Biomes with Landsat Archive and Earth Engine. *Remote Sens.* **2020**, *12*, 2735. [CrossRef]
23. Zhang, X.; Liu, L.; Wu, C.; Chen, X.; Gao, Y.; Xie, S.; Zhang, B. Development of a Global 30 m Impervious Surface Map Using Multisource and Multitemporal Remote Sensing Datasets with the Google Earth Engine Platform. *Earth Syst. Sci. Data* **2020**, *12*, 1625–1648. [CrossRef]
24. Chen, D.; Wang, Y.; Shen, Z.; Liao, J.; Chen, J.; Sun, S. Long Time-Series Mapping and Change Detection of Coastal Zone Land Use Based on Google Earth Engine and Multi-Source Data Fusion. *Remote Sens.* **2022**, *14*, 1. [CrossRef]
25. Liu, H.; Gong, P.; Wang, J.; Clinton, N.; Bai, Y.; Liang, S. Annual Dynamics of Global Land Cover and Its Long-Term Changes from 1982 to 2015. *Earth Syst. Sci. Data* **2020**, *12*, 1217–1243. [CrossRef]
26. Liu, C.; Li, W.; Zhu, G.; Zhou, H.; Yan, H.; Xue, P. Land Use/Land Cover Changes and Their Driving Factors in the Northeastern Tibetan Plateau Based on Geographical Detectors and Google Earth Engine: A Case Study in Gannan Prefecture. *Remote Sens.* **2020**, *12*, 3139. [CrossRef]
27. Höhne, N.; Wartmann, S.; Herold, A.; Freibauer, A. The Rules for Land Use, Land Use Change and Forestry under the Kyoto Protocol—Lessons Learned for the Future Climate Negotiations. *Environ. Sci. Policy* **2007**, *10*, 353–369. [CrossRef]
28. Peng, K.; Jiang, W.; Ling, Z.; Hou, P.; Deng, Y. Evaluating the Potential Impacts of Land Use Changes on Ecosystem Service Value under Multiple Scenarios in Support of SDG Reporting: A Case Study of the Wuhan Urban Agglomeration. *J. Clean. Prod.* **2021**, *307*, 127321. [CrossRef]
29. Shen, Z.; Wang, Y.; Su, H.; He, Y.; Li, S. A Bi-Directional Strategy to Detect Land Use Function Change Using Time-Series Landsat Imagery on Google Earth Engine: A Case Study of Huangshui River Basin in China. *Sci. Remote Sens.* **2022**, *5*, 100039. [CrossRef]
30. Fu, Z.H.; Zhao, H.J.; Wang, H.; Lu, W.T.; Wang, J.; Guo, H.C. Integrated Planning for Regional Development Planning and Water Resources Management under Uncertainty: A Case Study of Xining, China. *J. Hydrol.* **2017**, *554*, 623–634. [CrossRef]
31. Li, H.; Wan, W.; Fang, Y.; Zhu, S.; Chen, X.; Liu, B.; Hong, Y. A Google Earth Engine-Enabled Software for Efficiently Generating High-Quality User-Ready Landsat Mosaic Images. *Environ. Model. Softw.* **2019**, *112*, 16–22. [CrossRef]
32. Farr, T.G.; Rosen, P.A.; Caro, E.; Crippen, R.; Duren, R.; Hensley, S.; Kobrick, M.; Paller, M.; Rodriguez, E.; Roth, L.; et al. The Shuttle Radar Topography Mission. *Rev. Geophys.* **2007**, *45*, 361. [CrossRef]
33. McNally, A.; Arsenault, K.; Kumar, S.; Shukla, S.; Peterson, P.; Wang, S.; Funk, C.; Peters-Lidard, C.D.; Verdin, J.P. A Land Data Assimilation System for Sub-Saharan Africa Food and Water Security Applications. *Sci. Data* **2017**, *4*, 170012. [CrossRef] [PubMed]
34. Ye, Y.; Wang, Y.; Liao, J.; Chen, J.; Zou, Y.; Liu, Y.; Feng, C. Spatiotemporal Pattern Analysis of Land Use Functions in Contiguous Coastal Cities Based on Long-Term Time Series Remote Sensing Data: A Case Study of Bohai Sea Region, China. *Remote Sens.* **2022**, *14*, 3518. [CrossRef]
35. Li, C.; Wang, J.; Wang, L.; Hu, L.; Gong, P. Comparison of Classification Algorithms and Training Sample Sizes in Urban Land Classification with Landsat Thematic Mapper Imagery. *Remote Sens.* **2014**, *6*, 964–983. [CrossRef]
36. Haralick, R.M.; Shanmugam, K.; Dinstein, I. Textural Features for Image Classification. *IEEE Trans. Syst. Man Cybern.* **1973**, *SMC-3*, 610–621. [CrossRef]
37. Conners, R.W.; Harlow, C.A. A Theoretical Comparison of Texture Algorithms. *IEEE Trans. Pattern Anal. Mach. Intell.* **1980**, *PAMI-2*, 204–222. [CrossRef]
38. Wang, S.; Liu, F.; Zhou, Q.; Chen, Q.; Liu, F. Simulation and Estimation of Future Ecological Risk on the Qinghai-Tibet Plateau. *Sci. Rep.* **2021**, *11*, 17603. [CrossRef]
39. Caro, C.; Marques, J.C.; Cunha, P.P.; Teixeira, Z. Ecosystem Services as a Resilience Descriptor in Habitat Risk Assessment Using the InVEST Model. *Ecol. Indic.* **2020**, *115*, 106426. [CrossRef]
40. Grizzetti, B.; Lanzanova, D.; Liquete, C.; Reynaud, A.; Cardoso, A.C. Assessing Water Ecosystem Services for Water Resource Management. *Environ. Sci. Policy* **2016**, *61*, 194–203. [CrossRef]
41. Zhao, W.; Liu, Y.; Daryanto, S.; Fu, B.; Wang, S.; Liu, Y. Metacoupling Supply and Demand for Soil Conservation Service. *Curr. Opin. Environ. Sustain.* **2018**, *33*, 136–141. [CrossRef]
42. Myers, N. Environmental Services of Biodiversity. *Proc. Natl. Acad. Sci. USA* **1996**, *93*, 2764–2769. [CrossRef] [PubMed]
43. Niu, L.; Shao, Q.; Ning, J.; Huang, H. Ecological Changes and the Tradeoff and Synergy of Ecosystem Services in Western China. *J. Geogr. Sci.* **2022**, *32*, 1059–1075. [CrossRef]
44. Zhao, G.; Mu, X.; Wen, Z.; Wang, F.; Gao, P. Soil Erosion, Conservation, and Eco-Environment Changes in the Loess Plateau of China. *Land Degrad. Dev.* **2013**, *24*, 499–510. [CrossRef]
45. Ibáñez, J.; Valderrama, J.M.; Puigdefábregas, J. Assessing Desertification Risk Using System Stability Condition Analysis. *Ecol. Model.* **2008**, *213*, 180–190. [CrossRef]
46. Hemati, M.; Hasanlou, M.; Mahdianpari, M.; Mohammadimanesh, F. A Systematic Review of Landsat Data for Change Detection Applications: 50 Years of Monitoring the Earth. *Remote Sens.* **2021**, *13*, 2869. [CrossRef]
47. Immitzer, M.; Vuolo, F.; Atzberger, C. First Experience with Sentinel-2 Data for Crop and Tree Species Classifications in Central Europe. *Remote Sens.* **2016**, *8*, 166. [CrossRef]
48. Belgiu, M.; Drăguţ, L. Random Forest in Remote Sensing: A Review of Applications and Future Directions. *ISPRS J. Photogramm. Remote Sens.* **2016**, *114*, 24–31. [CrossRef]

49. Li, X.; Chen, Y.; Liu, X.; Xu, X.; Chen, G. Experiences and Issues of Using Cellular Automata for Assisting Urban and Regional Planning in China. *Int. J. Geogr. Inf. Sci.* **2017**, *31*, 1606–1629. [CrossRef]
50. Gounaridis, D.; Chorianopoulos, I.; Symeonakis, E.; Koukoulas, S. A Random Forest-Cellular Automata Modelling Approach to Explore Future Land Use/Cover Change in Attica (Greece), under Different Socio-Economic Realities and Scales. *Sci. Total Environ.* **2019**, *646*, 320–335. [CrossRef]
51. Qin, X.; Fu, B. Assessing and Predicting Changes of the Ecosystem Service Values Based on Land Use/Land Cover Changes With a Random Forest-Cellular Automata Model in Qingdao Metropolitan Region, China. *IEEE J. Sel. Top. Appl. Earth Obs. Remote Sens.* **2020**, *13*, 6484–6494. [CrossRef]
52. Jun, M.-J. A Comparison of a Gradient Boosting Decision Tree, Random Forests, and Artificial Neural Networks to Model Urban Land Use Changes: The Case of the Seoul Metropolitan Area. *Int. J. Geogr. Inf. Sci.* **2021**, *35*, 2149–2167. [CrossRef]
53. Li, X.; Liu, X.; Yu, L. A Systematic Sensitivity Analysis of Constrained Cellular Automata Model for Urban Growth Simulation Based on Different Transition Rules. *Int. J. Geogr. Inf. Sci.* **2014**, *28*, 1317–1335. [CrossRef]
54. Pontius, R.G.; Boersma, W.; Castella, J.-C.; Clarke, K.; de Nijs, T.; Dietzel, C.; Duan, Z.; Fotsing, E.; Goldstein, N.; Kok, K.; et al. Comparing the Input, Output, and Validation Maps for Several Models of Land Change. *Ann. Reg. Sci.* **2008**, *42*, 11–37. [CrossRef]
55. Pontius, R.G. Criteria to Confirm Models That Simulate Deforestation and Carbon Disturbance. *Land* **2018**, *7*, 105. [CrossRef]

Article

Inner Dynamic Detection and Prediction of Water Quality Based on CEEMDAN and GA-SVM Models

Zhizhou Yang [1,2], Lei Zou [1,*], Jun Xia [1,3], Yunfeng Qiao [2,4] and Diwen Cai [1,2]

[1] Key Laboratory of Water Cycle & Related Land Surface Processes, Institute of Geographic Sciences and Natural Resources Research, Chinese Academy of Sciences, Beijing 100101, China; yangzz.18b@igsnrr.ac.cn (Z.Y.); xiajun@igsnrr.ac.cn (J.X.); caidw.17b@igsnrr.ac.cn (D.C.)
[2] University of Chinese Academy of Sciences, Beijing 100049, China; qiaoyf@igsnrr.ac.cn
[3] State Key Laboratory of Water Resources and Hydropower Engineering Science, Wuhan University, Wuhan 430072, China
[4] Key Laboratory of Ecosystem Network Observation and Modeling, Institute of Geographic Sciences and Natural Resources Research, Chinese Academy of Sciences, Beijing 100101, China
* Correspondence: zoulei@igsnrr.ac.cn

Citation: Yang, Z.; Zou, L.; Xia, J.; Qiao, Y.; Cai, D. Inner Dynamic Detection and Prediction of Water Quality Based on CEEMDAN and GA-SVM Models. *Remote Sens.* **2022**, *14*, 1714. https://doi.org/10.3390/rs14071714

Academic Editors: Qiusheng Wu, Jun Li, Xinyi Shen and Chengye Zhang

Received: 6 February 2022
Accepted: 30 March 2022
Published: 1 April 2022

Publisher's Note: MDPI stays neutral with regard to jurisdictional claims in published maps and institutional affiliations.

Abstract: Urban water quality is facing strongly adverse degradation in rapidly developing areas. However, there exists a huge challenge to estimating the inner features and predicting the variation of long-term water quality due to the lack of related monitoring data and the complexity of urban water systems. Fortunately, multi-remote sensing data, such as nighttime light and evapotranspiration (ET), provide scientific data support and reasonably reveal the variation mechanisms. Here, we develop an integrated decomposition-reclassification-prediction method for water quality by integrating the CEEMDN method, the RF method mothed, and the genetic algorithm-support vector machine model (GA-SVM). The degression of the long-term water quality was decomposed and reclassified into three different frequency terms, i.e., high-frequency, low-frequency, and trend terms, to reveal the inner mechanism and dynamics in the CEEMDAN method. The RF method was then used to identify the teleconnection and the significance of the selected driving factors. More importantly, the GA-SVM model was designed with two types of model schemes, which were the data-driven model (GA-SVMd) and the integrated CEEMDAN-GA-SVM model (defined as GA-SVMc model), in order to predict urban water quality. Results revealed that the high-frequency terms for NH_3-N and TN had a major contribution to the water quality and were mainly dominated by hydrometeorological factors such as ET, rainfall, and the dynamics of the lake water table. The trend terms revealed that the water quality continuously deteriorated during the study period; the terms were mainly regulated by the land use and land cover (LULC), land metrics, population, and yearly rainfall. The predicting results confirmed that the integrated GA-SVMc model had better performance than single data-driven models (such as the GA-SVM model). Our study supports that the integrated method reveals variation rules in water quality and provides early warning and guidance for reducing the water pollutant concentration.

Keywords: CEEMDAN method; GA-SVM model; decomposition; prediction; water quality

1. Introduction

Recently, urban water quality degradation has become a considerable restricting factor for achieving the goal of the green development in metropolises, and thus has caused worldwide concern [1,2]. Urbanization rates and the urban built-up area confirm that urban area tends to continuously expand [3] and thus change the structure of the water system, causing potential water pollution [4,5]. For instance, 32% of surface water in China was facing water pollution disasters [6,7]. Waterbody quality strongly varies in time due to uneven development of the urban area; the ongoing drastic change of the effective soil water amounts, nutrient levels, and land use and land cover; and point sources of pollution discharged from residential and industrial sources [8,9]. Thus, accurately detecting the inner dynamics and predicting potential water pollution issues caused by the varied driving factors are the key points to preventing and reducing the degree of water pollution and require immediate attention [10]. Particularly in the urban-rural marginal area, urban expansion has a substantial influence on the hydrology and water environment. Moreover, the high disturbance in the urban has caused more complex hydraulic conditions and more sources of pollutants [11,12].

To detect the inner variation features of the water quality, plenty of methods exist and have provided reasonable results [13–21]. However, there also exist some limitations that have restricted the application of these methods. The Mann–Kendall test is mainly used to detect the tendency of time-related data, and thus is widely used for analyzing long-term rainfall, runoff datasets, and water quality [13,15]. However, water quality for urban areas undergoing rapid expansion may not have a long time series of detection data. Moreover, more decomposed features are necessary to analyze the dynamic of water quality. The Fourier transformation (FT) method has also been used to detect the dynamic pattern of time series data; however, the features of stationary and linear processes and priori basis restrict its application for water quality [16]. The wavelet transform (WT) method, which solves the shortage of FT method in the single resolution of short time, is a time-frequency based method and thus is widely used for rainfall, runoff, and water quality transformation [17–20]. The WT method is suitable for non-stationary signals and is extremely dependent on the wavelet basis function. When the signal-to-noise ratio is small or the data is not linear [21], the denoising effect of WT cannot obtain reasonable results. The empirical mode decomposition (EMD) method has been proposed [22], as the EMD method can compose the non-stationary and non-linear into linearizing and stabilizing series, and the EMD method can select the basis function based on the time scale characteristics of the signals themselves. Furthermore, with the development of other improved EMD methods, such as the complete ensemble empirical mode decomposition with adaptive noise (CEEMDAN) method and ensemble empirical mode decomposition (EEMD), EMD, EEMD, and CEEMDAN have been widely used to decompose time series data of the climatic oscillation, runoff, water quality, and landslides [23–28].

Prediction of the variation of water quality also significantly supports improving waterbody deterioration. Several models and information systems have been proposed to predict the variation of water quality and obtain reasonable results. Among these models, physical-based models, i.e., hydrologic-environmental models, have been widely used in urban areas. For example, Joshi et al. [29] used the storm water management model (SWMM) to reduce combined sewer overflows with reasonable cost-effectiveness for sustainable urban drainage systems. The InfoWorks ICM model or the full hydrodynamic (FH) models were widely used for multi-scale catchments in real-time control (RTC) and obtained optimum results [30,31]. The Mike URBAN model contains distributed water systems including combined sewer overflow system and separate stormwater system. More importantly, the Mike URBAN covers two-dimensional overland flow and thus has good performance in urban areas with rapid urbanization and climate change [32]. These models are both supported by rigorous physical theory and are easily acceptable. However, the rigorous physical theory-based models also need high-quality monitoring data to satisfy the accuracy of the model.

However, the rapid expansion of urban areas is always accompanied by drastic changes in the underlying surface, urban pipe networks, hydrological conditions, and water environment conditions. Moreover, all of these changes are not always well monitored or do not have high-quality data available. Therefore, data-driven models, such as machine learning models, have also been used for water quality predictions. Zhi et al. [33] used machine learning models in 236 minimally disturbed watersheds of the US and confirmed that machine learning models can predict results well in data-lacking areas. With the immense and urgent demand for good-quality prediction of water quality variation with the rapid development of urban areas, more machine learning models have been presented and compared. Qiao et al. [34] used 12 machine learning algorithms to evaluate water quality, and both models obtained reasonable results. Compared with the neural network model, Mohammadpour et al. [35] also analyzed the SVM model and artificial neural networks (ANNs), and revealed that the SVM model could obtain better results with limited monitoring data. Recently, a few types of integrated models, which can decompose the data series into more inner sequences and which are then coupled with the machine learning model, were analyzed to evaluate the inner dynamic and provide better modeling performance. For example, the EMD-ANN model and EMD-Auto-Regressive and Moving Average (ARMA) model were integrated to predict runoff, and revealing that the EMD-based integrated model performed better than the single model, i.e., the ANN model and the ARMA model, in the hindcast experiment performed [36]. Yuan et al. [25] integrated the EEMD and Long Short-Term Memory (LSTM) models to forecast daily runoff, and confirmed that the integrated model significantly improved the simulation results compared to the LSTM model. The EEMD and the SVM model also were integrated to predict water quality and landslide displacement, and results revealed that the integrated model increased the prediction accuracy [27,28]. However, some of the EMD-based integrated models were not data-based [27], and some forecast results of the integrated models performed worse than the original models [36].

To evaluate the inner dynamic and achieve better prediction performance of water quality with limited data, this study integrated the CEEMDAN method, the random forest method, and the GA-SVM model. The CEEMDAN method was used to decompose the long-term water quality data; then, the decomposed sequences were reclassified into three sequences according to the variance proportion, i.e., the high-frequency term, the low-frequency term, and the trend term. Furthermore, the RF method was used to identify the importance of the driving data on the water quality series, the high-frequency term, the low-frequency term, and the trend term. More importantly, we then used the GA-SVM model and the identified driving factors of the high-frequency, low-frequency, and trend terms to predict the different terms, which were then coupled to predict the water quality. In contrast, the data-driven model, i.e., the identified driving factors of the water quality series coupled with the GA-SVM model, was set to forecast water quality.

2. Materials and Methods

2.1. The CEEMDAN Method

The complete ensemble empirical mode decomposition with adaptive noise (CEEMDAN) method was developed from empirical mode decomposition (EMD) and ensemble empirical mode decomposition (EEMD) by adding adaptive white noise to suppress the aliasing of the EMD [22,37,38] The CEEMADAN model is an efficient decomposed method for the adaptive decomposition of non-stationary and non-linear data into many intrinsic mode functions (IMF). The main progress is as follows:

Step 1 Define the long-term data $x_i(t)$ as the original input signal.

$$x_i(t) = x(t) + \varepsilon\omega_i(t) \tag{1}$$

where ε represents a noise coefficient and $\omega_i(t)$ indicates white noise sequences.

Step 2 Decompose the IMF_1. The first decomposed IMF averaged by the EMD method:

$$\mathrm{IMF}_1(t) = \frac{1}{N}\sum_{i=1}^{N}\mathrm{IMF}_{i1}(t) \tag{2}$$

The residue is defined as:

$$r_1(t) = x(t) - \mathrm{IMF}_1(t) \tag{3}$$

Step 3 Decompose the IMF_2.

$$IMF_2(t) = \frac{1}{N}\sum_{i=1}^{N} IMF_1(r_1(t) + \varepsilon_1 IMF_1(\omega(t))) \tag{4}$$

Step 4 Decompose the other IMFs unless the extreme points are less than two. Therefore, the final signal sequences $x(t)$ are decomposed as follows:

$$x(t) = \sum_{i=1}^{N} IMF_2(t) + r(t) \tag{5}$$

In the decomposing process, the IMFs and trend term can extract series terms for the high-frequency to low-frequency and trend terms. In this study, the t-test was used to reclassify the IMFs based on fine-to-coarse reclassification [39].

2.2. Driving Factors Selection and the Relative Importance Analysis

Urban water quality was influenced by many factors due to the complexity of the urban water system [12], such as the heavy variation of LULC, land metrics, rainfall, the human control of the lake water table, multi-point sources, complex rainfall-induced runoff, and non-point pollutants. Therefore, identifying the important driving factors under the condition of limited monitoring data and remote sensing data was the key point to achieving more accurate predictions. Before evaluating the importance of the driving factors, the Pearson method was used to analyze the correlation between the selected factors and to exclude the variables with high correlation. The random forest (RF) method split each partition into a random subset to search for the best feature variable, which produces better overall performance and thus has been widely used for identifying the importance of the driving factors for water quality [40]. Therefore, the RF method was used to identify the importance of the driving factors for the water quality series, the high-frequency term, the low-frequency term, and the trend term.

2.3. GA-SVM Model

The support vector machine (SVM) model is a nonlinear regression and is widely used for predicting hydrological issues and water quality issues. In this study, we used the SVM model to predict the water quality; the input data were divided into training data and test data. Furthermore, the GA imitates biological evolution to approach the best solution of the minimum project [41], and thus was used to search for the best matching kernel function and parameters for the SVM model.

The Nash–Sutcliffe efficiency coefficient (NSE) and the root mean squared error ($RMSE$) were used to estimate the model performance.

$$NSE = 1 - \frac{\sum_{i=1}^{n}\left(y_{mod} - y_{obs}\right)^2}{\sum_{i=1}^{n}\left(y_{mod} - \overline{y_{obs}}\right)^2} \tag{6}$$

$$RMSE = \sqrt{\frac{\sum_{i=1}^{n}\left(y_{mod} - y_{obs}\right)^2}{n}} \tag{7}$$

where the y_{mod} and y_{obs} represent the modeled and observed water quality. $\overline{y_{obs}}$ represents the observed mean of water quality and n represents the number of water quality samplings.

2.4. Experimental Schemes Design

In this study, we integrated a framework that realized the decomposition-reclassification-driving factors identification-prediction for the water quality series. We decomposed the water quality sequences and reclassified them to evaluate the inner dynamic of water quality. Additionally, the water quality and the reclassified terms were set as the inputs for the GA-SVM model. We designed and examined two types of GA-SVM models based on the selected 10 driving factors for each corresponding term (the water quality term, the high-frequency term, the low-frequency term, and the trend term). We named the data-driven GA-SVM model for water quality the GA-SVMd model. More importantly, we used the RF method to identify the corresponding important driving factors for the high-frequency term, the low-frequency term, and the trend term. Then, the GA-SVM model was used to predict each term sequence. Finally, the water quality was obtained by the sum of each predicted term sequence. Thus, this model was defined as the GA-SVMc model (Figure 1).

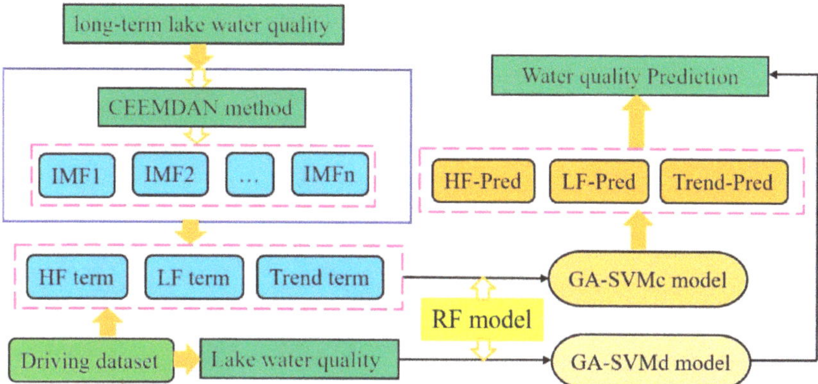

Figure 1. The two designed integrated experimental schemes. HF term and LF term represent the high-frequency and low-frequency terms. HF-Pre, LF-Pre, and Trend-Pre indicate the prediction of the high-frequency, low-frequency, and trend terms.

3. Case Study of Beihu Lake, Wuhan City, China

3.1. Study Area

The Beihu catchment is situated on the eastern expansion edge of the Wuhan City and includes the majority of heavy industrial parks (Figure 2). As a result, the Beihu catchment has a relatively lagging underground pipe network and sewage treatment capacity. Furthermore, sewage water sources, such as industrial, domestic, runoff, and agricultural sources, contribute vastly without reasonable water treatment, and discharge directly in the surface water body. Thus, the multiple sources of sewage have caused the downstream water body of Beihu Lake to be heavily polluted for a long time. Recently, many countermeasures have been performed to control water pollution; however, significant improvement in water quality has not been observed [42].

The Beihu Lake is a semi-natural lake regulated by a pumping station. Furthermore, the Beihu Lake catchment is situated in the rural-urban marginal area. As a result, the complex LULC, urban stormwater network, and multi-source water pollution create a complex urban water system. In the wet season, the high frequency of rainfall events causes a large amount of runoff; moreover, the water level of the outer river is higher than the water table of the Beihu Lake. Therefore, pumping stations are needed to drain the lake water to the outer river. In the dry season, the runoff of the Beihu Lake is quite small, and the water table of the outer river is lower than the water table of the Beihu Lake; therefore, the water of the lake is free to discharge to the outer river. Consequently, the water table of the Beihu Lake level is a key factor, since it is a significant indicator of whether the pumping station needs to drain water from the lake and whether the lake can be discharged into the outer river via free flow. Evapotranspiration (ET) is also a key factor since the lake is an open water body with a large surface.

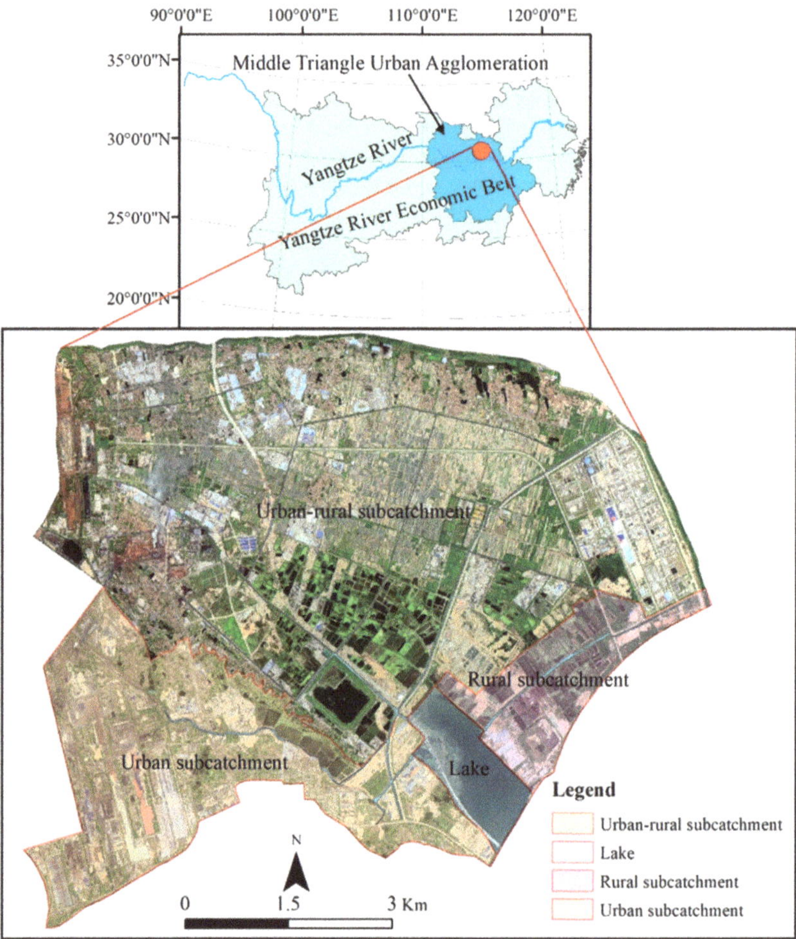

Figure 2. The study area of the Beihu catchment.

3.2. Water Quality and Other Monitored Datasets

In this study, the monthly water quality series from 15 January 2014 to 15 November 2021 were collected partly form environmental measurements by water samples and partly from Wuhan Ecological Environment Bureau [43], and the modeling period was set to the same period as the monitoring period. According to the monitored results, NH_3-N and TN were confirmed to be the main pollutants in the study area (Table 1); therefore, NH_3-N and TN were selected as the main water quality variables in this study. The hourly rainfall data and the water table of the Beihu lake were also monitored. Furthermore, the sums of 5-day, 10-day, 15-day, 20-day, monthly, seasonal, and yearly rainfall were calculated based on the hourly rainfall data. The 5-day, 10-day, 15-day, 20-days, and monthly average water table and accumulated variations of the water table were also calculated based on the hourly water table of the Beihu lake.

Table 1. Statistics of the water quality in the Beihu catchment.

Water Quality Variable	Units	Mean	Standard Deviation
NH_3-N	mg/L	2.59	1.69
TN	mg/L	4.56	2.29

3.3. Remote Sensing-Based Data

Remote sensing-based data have significant contributions to the prediction of long-term water quality. In this study, three types of remote sensing data were used: land use and land cover, the ET dataset, and the nighttime light dataset (NTL) (Table 2). In detail, three periods of the Chinese Gaofen (GF)-1 data (resolution of 2 m) were manually identified to obtain the land use and land cover land metrics dataset for the years of 2014, 2017, 2020. The 8-day ET dataset [44], which ranges from 15 June 2015 to 15 November 2021, was downloaded from the MODIS Land Products (Net Evapotranspiration 8-Day L4 Global 500 m) (https://ladsweb.modaps.eosdis.nasa.gov/search/, accessed on 15 December 2021). Then, the monthly potential ET data were obtained by summing the total potential ET data on the 8-day total ET dataset for four periods of every month. Domestic wastewater discharge is a critical point pollutant source to the lake water quality; therefore, accurately evaluating the population has a significant effect on predicting water quality. The NTL dataset has been proven to be an effective dataset to obtain population data [45]. In this study, the yearly Visible Infrared Imaging Radiometer Suite (VIIRS) Day/Night Band (DNB) dataset was chosen to calculate the population [46].

Table 2. Statistics of the related data details.

No.	Data	Unit	Temporal and Spatial Resolution	Data Sources
1	Rainfall	mm	1 h	Field investigation
2	Lake water table	m	1 h	
3	Annual District Population	people	1 year	Statistical yearbook
4	Land use and land cover	m	2 m	Chinese Gaofen (GF)-1
5	ET	mm	8 days, 500 m	https://ladsweb.modaps.eosdis.nasa.gov/search/, accessed on 15 December 2021
6	Nighttime light	m	1 year, 500 m	https://eogdata.mines.edu/products/vnl/, accessed on 15 December 2021

The yearly VIIRS data were first corrected based on the assumption that the NTL value of the previous year is smaller than that of the next year (Equation (8)) [46].

$$DN_{(n,i)} = \begin{cases} DN_{(n-1,i)} & DN_{(n-1,i)} \geq DN_{(n,i)} \\ DN_{(n,i)} & DN_{(n,i)} > DN_{(n-1,i)} \end{cases} \tag{8}$$

Literature has proven that the NPP-VIIR NTL data can obtain a reasonable estimation of distributed population [46]. The correlation between NPP-VIIR NTL radiance and population follows Equation (9).

$$POP_c = ax^3 + bx^2 + cx + d \tag{9}$$

The precision of the calculated population and the real population was evaluated by Equation (10). If the calculated population had a relatively large error, the power function was then used to recorrect the calculated population until it obtained a reasonable result (Equation (11)).

$$\gamma = \frac{|POP_c - POP_s|}{POP_s} \times 100\% \tag{10}$$

$$\begin{cases} c_n = POP_n / POP_{total} \\ f_n = c_n * DNB_n \end{cases} \tag{11}$$

where γ indicates the relative error. POP_c and POP_s indicate the calculated population by the NTL and statistical population. c_n and f_n indicate the correction and the adjusted DBN. POP_n and POP_{total} represent the nth yearly statistical population and the total statistical population during the calculated period.

4. Results

4.1. The Main Input Data from the Remote Sensing Dataset

4.1.1. Land Use and Land Cover (LULC), Land Metrics

LULC has a remarkable influence on urban water system quality due to different rainfall-runoff response mechanisms and non-point sources pollution generation mechanisms. Especially in urban-rural marginal areas, land use types are significantly altered, changing the effective water amounts, nutrient levels, and surface roughness of the land surface directly, and thus changing the urban hydrological processes and ecological environments. In this study, the years 2014, 2017, 2020 were interpreted for water quality prediction. Results revealed that 11 types of LULC mainly existed in the Beihu catchment, i.e., lake, rivers, roads, grassland, forest land, ponds, paddy fields, bare land, industrial land, and residential land (Figure 3). The area of the ponds and paddy fields slightly declined during the study period, while the area of the industrial land and residential land increased due to the expansion of the urban area (Figure 3a–d). The area of forest/grassland also had a substantial influence on non-point sources, and was chosen as a driving factor in this study. The chosen land metrics were the patch density (PD) and contagion index (CONTAG) due to the high Person's correlations of the other factors, such as the landscape shape index (LSI) and the largest patch index (LPI).

4.1.2. ET and POP Dataset

The average potential ET of the Beihu catchment exhibited strong seasonal variation (Figure 4a). The population calculated by the NPP-VIIR NTL radiance of Wuhan City performed well. The population of the Beihu catchment tended to decrease slowly in the early period and increase rapidly in the later period (Figure 4b). This might be due to the Beihu catchment being located at the edge of the urban area; people tended to migrate to the urban area in the early period, while when the urban gradually expanded, more areas of the catchment became urban areas; therefore, the population showed a trend of rapid growth in the later period. This result was consistent with the statistical data of Qingshan District, Wuhan City [47].

4.2. Decomposition and Reclassification of the Water Quality Series

All water quality sequences, i.e., the NH_3-N and TN monitoring data from 15 January 2014 to 15 November 2021, were decomposed by the CEEMDAN method (Figure 5). Then, the decomposed IMFs terms and trend terms were reclassified based on the *t*-test (Figure 6).

The decomposed IMFs and the residue term by the CEEMDAN for NH_3-N and TN are shown in Figure 5. In this study, 500 trials were implemented and the white noise coefficient was given as 0.2. Results revealed that both NH_3-N and TN had four IMFs. From the high-frequency IMF to low-frequency IMF, the frequencies and amplitudes changed significantly and the amplitudes became smaller (Figure 5). The amplitudes of NH_3-N and TN were 3 for IMF1 and then declined to 1.5–2 for IMF2-IMF3, while the amplitudes increased to 3 for IMF4 for NH_3-N and TN. The residue term for NH3-N and TN increased and had relatively small amplitudes (Figure 5).

The mean period [39], mean values, the variance of each IMF, the percentage of the variance of the IMFs, and the Pearson correlation between each IMF with the water quality series were analyzed in this study (Table 3). Results revealed that IMF1 and IMF2 had more frequent fluctuations and had different mean periods for NH_3-N and TN, while IMF3 and IMF4 had larger and similar mean periods (Table 3). The percentage of the variance of the IMFs confirmed that the IMF1 and IMF4 had the greatest proportion of contribution on the water quality.

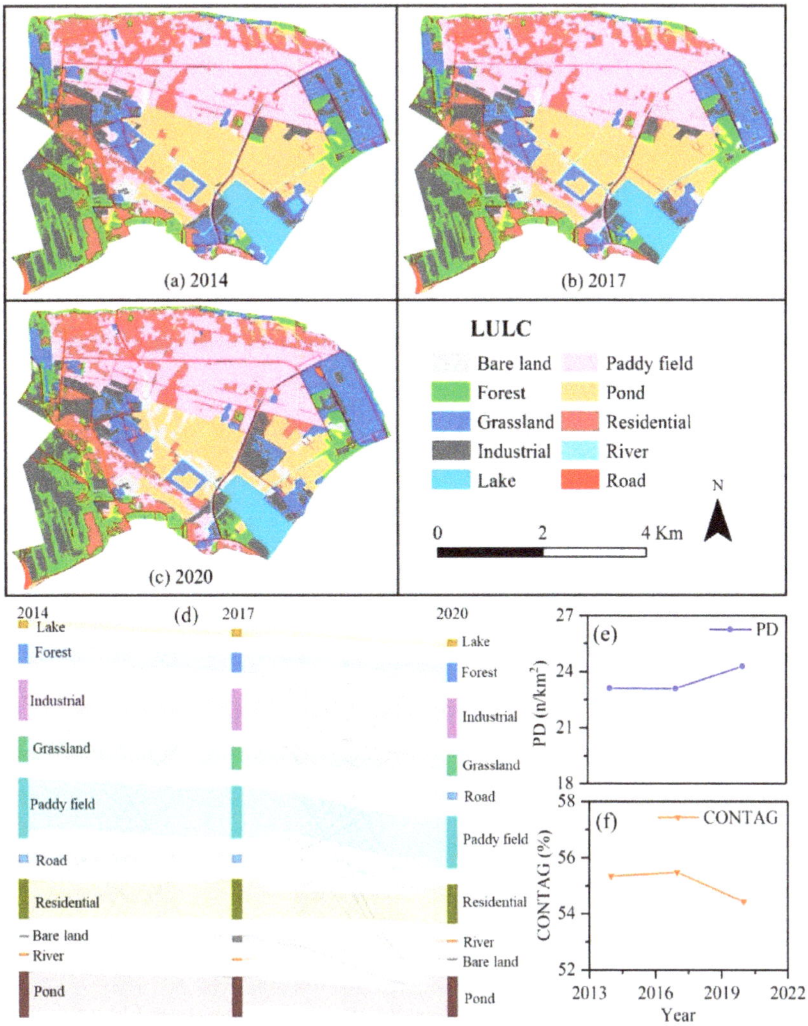

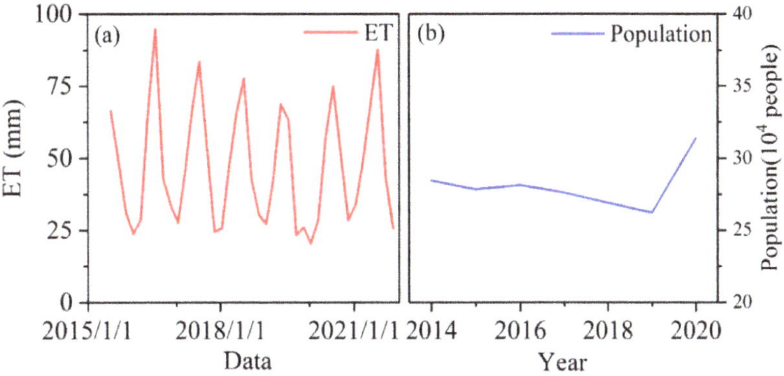

Figure 3. Land use and land change (**a**–**d**); land metrics (**e**,**f**).

Figure 4. The potential monthly ET (**a**); the yearly population of the Beihu catchment (**b**).

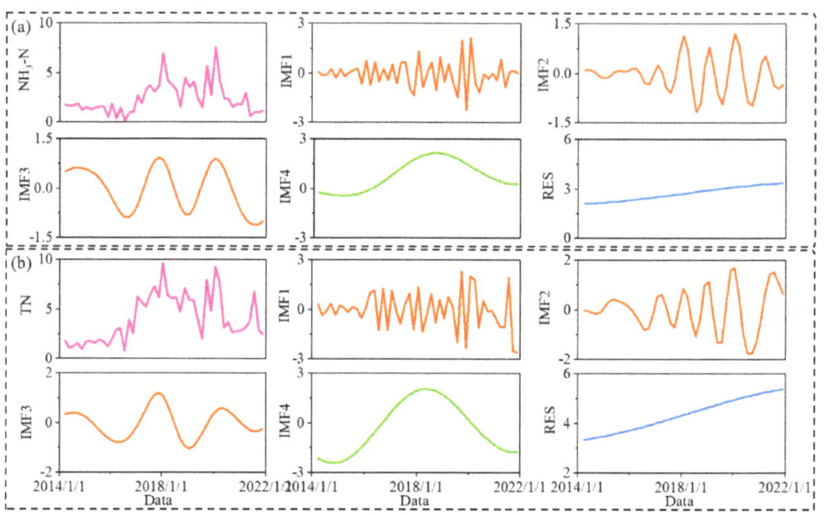

Figure 5. The decomposition of water quality sequences by the CEEMDAN method. (**a**) NH$_3$-N; (**b**) TN.

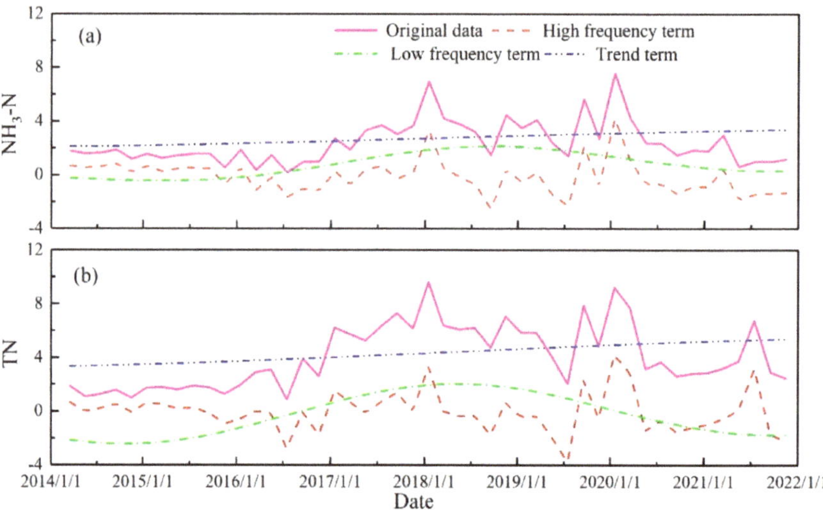

Figure 6. The reclassified terms and the original water quality series. (**a**) NH$_3$-N; (**b**) TN.

We reclassified the decomposed water quality of IMF1 to IMF4 and residual term based on the *t*-test in this study. The residual term was set as the trend term, IMF1 to IMF3 were reclassified as the high-frequency term for both NH$_3$-N and TN due to the significant difference among the IMFs, and for IMF4, both NH$_3$-N and TN were reclassified as the low-frequency term. The original water quality series and the reclassified high-frequency, low-frequency, and trend term are shown in Figure 6. Results confirmed that the high-frequency terms for both NH$_3$-N and TN had stronger fluctuation frequencies, which were similar to the water quality series. The low-frequency terms for both NH$_3$-N and TN showed a tendency of first increasing and then decreasing, and the trend of the low-frequency terms was similar to the water quality trend. The trend terms for NH$_3$-N and TN increased in the whole monitoring period and gradually leveled at the end of the monitoring period.

Table 3. The IMFs and the residue values for the decomposed long-term water quality data.

	Variable	IMF1	IMF2	IMF3	IMF4	Residue
Mean period	NH$_3$-N	1.31	2.94	7.83	23.5	47
(Month)	TN	1.47	3.92	7.83	23.5	47
Mean	NH$_3$-N	−0.054	−0.030	−0.024	0.796	2.742
	TN	−0.052	0.038	−0.018	−0.273	4.348
Variance	NH$_3$-N	0.83	0.52	0.65	0.89	0.41
	TN	1.15	0.83	0.57	1.55	0.65
Variance as % of	NH$_3$-N	25.17	15.86	19.70	26.96	12.31
(ΣIMFs + residual)	TN	24.15	17.56	12.08	32.57	13.64
Pearson correlation	NH$_3$-N	0.50	0.46	0.44	0.65	0.27
	TN	0.46	0.32	0.28	0.74	0.42

4.3. Evaluation of the Importance of Driving Factors

We used the RF method to estimate the importance of driving factors (determined by the relative importance for all driving factors), which was used for both the GA-SVMd model and the GA-SVMc model. As shown in Figure 7, the relative importance of driving factors significantly varied between the water quality variables and the corresponding data series and different frequencies.

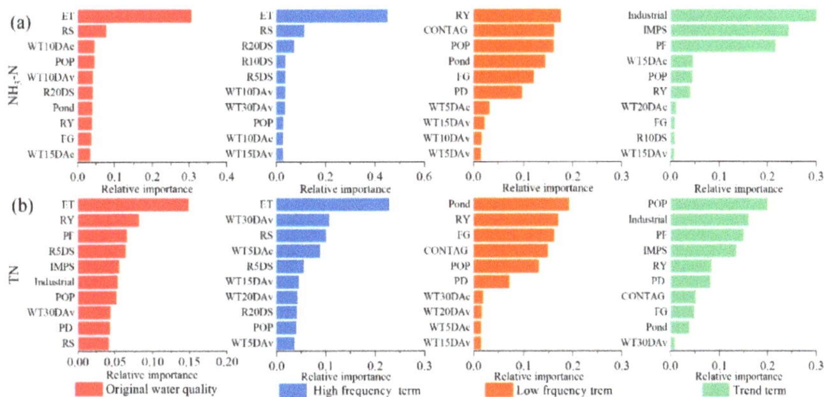

Figure 7. The relative importance of the driving factors. The R5DS, R10DS, R20DS, RS, and RY represent the sum rainfall over 5 days, 10 days, 20 days, seasonally, and yearly. The WT5DAc, WT10DAc, WT15DAc, WT20DAc, and WT30DAc indicate the cumulative magnitude of change in the lake water table over 5 days, 10 days, 15 days, 20 days, and 30 days. The WT5Av, WT10Av, WT15Av, WT20Av, and WT30Av indicate the average lake water table over 5 days, 10 days, 15 days, 20 days, and 30 days. FG represents forest/grassland. POP represents the population. IMPS indicates impervious surface.

Regarding the NH$_3$-N of the Beihu Lake for the GA-SVMd model, the main driving factors were hydro-meteorological factors, i.e., ET, seasonal rainfall, the cumulative magnitude of change in the lake water table over 10 days, the average lake water table over 10 days, and the sum rainfall over 20 days. Furthermore, population, pond land, and forest/grassland were also relatively important influences. When analyzing the importance of the driving factors on the decomposed and reclassified results of the CEEMDAN, the driving factors of the high-frequency term for NH$_3$-N were dominated by the hydrometeorological factors; only the population had a slight effect. The driving factors on the low-frequency term for NH$_3$-N were dominated by yearly rainfall, the population, LUCC (such as the pond, the forest/grassland, and the paddy field), and land metrics (the PD, the CONTAG). The different days of the cumulative magnitude lake water table also had a relatively significant impact on NH_3-N. In detail, field investigation confirmed that the industrial point sources were the main source of NH_3-N; therefore, the industrial land area had the most significant effect on the trend term of NH_3-N. The impervious surface, the paddy field, the population, and the cumulative magnitude of change in the lake water table over 5, 20, and 10 days were also the main driving factors for the trend term of NH_3-N.

Compared to the GA-SVMd model, the driving factors for TN also included many hydro-meteorological factors, i.e., the ET, the yearly rainfall, the sum rainfall over 5 days, and the average lake water table over 30 days. Moreover, paddy field land, impervious surface (defined as the sum of the residential, industrial, and road land), population, and PD also had significant influence on TN (Figure 7). When analyzing the importance of the driving factors on the decomposed and reclassified results of the CEEMDAN, the driving factors also significantly varied from the high-frequency term to trend terms for TN. The driving factors of the high-frequency term for TN were dominated by the hydrometeorological factors; only the population had a slight effect. The driving factors on the low-frequency term for TN were also dominated by yearly rainfall, population, LUCC, and land metrics. The driving factors on the trend term for TN were mainly influenced by the population, the LUCC, land metrics, and yearly rainfall. Residential recharge was proven to be the main source of TN by previous studies (Hwang et al., 2016; Paule et al., 2014), which is related to the most important driving factor (the population) for the trend term of TN (Figure 7).

4.4. Prediction of Water Quality by the GA-SVMd Model and the GA-SVMc Model

The proportion of the calibration period and the validation period was set as 0.7 for both NH_3-N and TN; i.e., the period from 1 July 2015 to 15 January 2020 was set as the calibration period, and the period from 15 January 2015 to 15 November 2021 was set as the validation period. The results modeling with the GA-SVMd model and the GA-SVMc model both showed reasonable performance (Figure 8, Table 4). Apparently, the GA-SVMc model performed better in the prediction of water quality. Furthermore, the GA-SVMc model provided more accurate prediction results on the strong variations of water quality. However, the simulation accuracy of GA-SVMd model and the GA-SVMc were poor when the water quality dramatically changed, which may be due to the lack of measured runoff data in the study area. Non-point source pollution was usually the main pollution source during the rainfall-runoff process [48]. Therefore, it is necessary to strengthen the monitoring of runoff and water quality during rainfall in future research.

Table 4. The modeled accuracy results in the calibration period and the validation period.

Water Quality Variables	Evaluation Function	Calibration Period		Validation Period	
		GA-SVMd	GA-SVMc	GA-SVMd	GA-SVMc
NH_3-N	*NSE*	0.63	0.81	0.51	0.62
	RMSE	0.97	1.28	1.7	1.14
TN	*NSE*	0.57	0.77	0.55	0.61
	RMSE	1.46	1.07	1.57	1.48

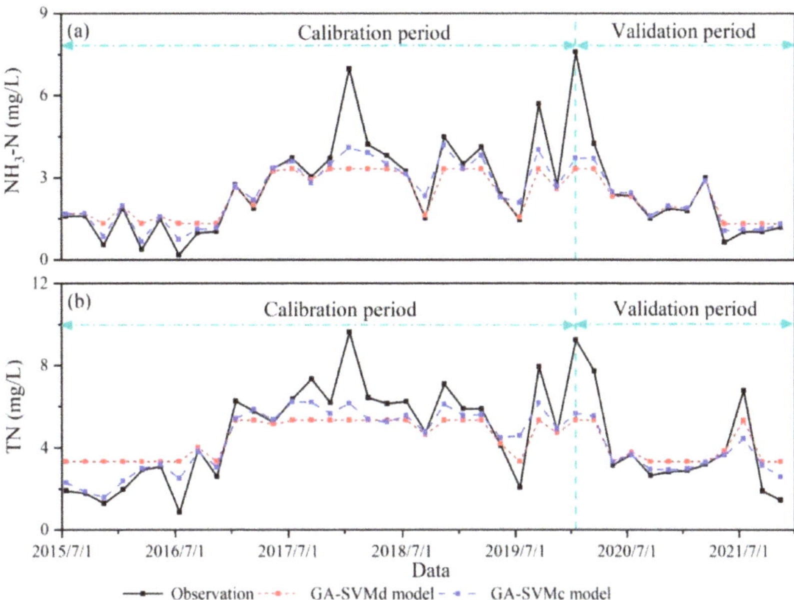

Figure 8. The prediction results of the GA-SVMd model and GA-SVMc model. (**a**) NH$_3$-N; (**b**) TN.

5. Discussion

5.1. Important Factors Dominating Water Pollution and Different Frequency Terms of Water Quality

The water pollution of urban-rural marginal areas is attributed to many factors, such as LULC, land metrics, hydro-meteorological factors, and point sources recharged by the domestic and industrial [4,33,49]. Land surface runoff is usually set as an essential factor for predicting water quality in data-driven models [33]. The runoff was significantly complex due to the multiple inputs and has not been monitored over a long time series. In this study, the lake water table was affected by the recharge of rainfall-runoff sources, domestic sources and industrial sources; controlled by the pumping gate, it could be set as a substitute factor for runoff to predict water quality and obtain reasonable performance. The factors of the LULC and land metrics for the urban-rural marginal area changed significantly due to the rapid expansion of the urban area, and thus had notable impacts on water quality, as has confirmed by many studies [50–52]. In this study, results confirmed that LULC and land metrics had a relatively high impact on the low-frequency and trend term of water quality. Point sources, such as industrial wastewater discharge and domestic wastewater discharge, also had significant impacts on water quality [12]. Our results confirmed that the population was the dominant pollution source of TN, and also had a relative effect on NH$_3$-N. Industrial land had a significant impact on NH$_3$-N and a similar effect on TN. Meteorological conditions, such as rainfall and the ET, also had significant and complex impacts on water quality [51]. For example, the first rainfall was confirmed to have a significant impact on water quality in the urban area, while seasonal rainfall had a greater effect on agricultural land water quality [53,54]. Our results also confirmed that rainfall and the ET had significant effects on water quality, and dominated the high-frequency term of water quality.

5.2. Prediction of the Urban Water Quality by Machine Learning Models

Considering the strongly changing LULC, the complexity of diverse and continuous varied pollution sources and hydro-hydraulic conditions, meteorological conditions with complex dynamic characteristics, and the widespread lack of data in rural-urban marginal areas, developing a prediction model with reasonable performance is still a tremendous challenge [4,33,35,55]. The original data-driven machine learning models seemed to provide a good choice to simulate the urban-rural catchment water quality with complex and data-lacking conditions [33]. In a fact, point sources, such as industrial discharge, have not been well monitored for a long time. Runoff volumes have also not been monitored and the complexity could not easily be modeled. However, the original data-driven machine learning models, i.e., the GA-SVDd model in our study, still performed reasonably due to the substantial data obtained from the remote sensing data and the lake water table (Figure 8).

A successful model could both be used to reveal the inner dynamics and driving mechanisms and provide accurate prediction results. Previous studies have integrated many models to reveal the inner features of the time series data, such as runoff and water quality. The EMD method, the EEMD method, the CEEMDAN method, and the WT method have been widely used to decompose time series data, after which they integrated with machine learning models [25,28,36]. However, not all the integrated models achieve better prediction performance; Zhang et al. [36] confirmed that EMD-based integrated models may perform worse than data-driven models in simulating streamflow. In our study, prediction results from the integrated GA-SVMc model confirmed that the CEEMDAN integrated with the GA-SVM model for water quality can achieve markedly better performance than the original SVM model.

6. Conclusions

Evaluation of the dynamic and influence mechanisms, and the prediction of variations of water quality provide early warning and guidance to reduce water pollution concentration. The limited monitoring data and the complexity of the water system restrict the prediction of long-term water quality. However, the multiple variables derived from remote-sensing data (ET, LULC, etc.) provide scientific data and reasonably reveal the variation mechanism.

In this study, we developed an integrated decomposition-reclassification-prediction method for water quality by integrating the CEEMDAN method, the RF method, and a genetic algorithm-support vector machine model (GA-SVM). The degression of the long-term water quality was decomposed and reclassified into three different frequency terms, i.e., the high-frequency, low-frequency, and trend terms, to reveal the inner mechanisms and dynamics in the CEEMDAN method. The RF method was then used to identify the teleconnection and the significance of the selected driving factors. More importantly, the GA-SVM model was integrated and designed in two types of model schemes, which were the data-driven model (GA-SVMd) and the integrated CEEMDAN-GA-SVM model (defined as GA-SVMc model), in order to predict urban water quality. Results revealed that the high-frequency terms for NH_3-N and TN had a major contribution to the water quality and were mainly dominated by the hydrometeorological factors, such as the ET, rainfall, and dynamics of the lake water table. The low-frequency terms for NH_3-N and TN were both dominated by yearly rainfall, population, LULC, and land metrics. The trend terms revealed that the water quality continuously deteriorated during the study period and was mainly regulated by the LULC and land metrics factor, population, and yearly rainfall. The prediction results confirmed that the integrated GA-SVMc model achieved better performance than a single data-driven model such as GA-SVM.

Author Contributions: Conceptualization, Z.Y. and L.Z.; methodology, Z.Y. and L.Z.; software, Z.Y.; validation, Z.Y., L.Z. and D.C.; formal analysis, Z.Y., L.Z. and J.X.; investigation, Z.Y. and L.Z.; resources, L.Z.; data curation, L.Z. and Y.Q.; writing—original draft preparation, Z.Y.; writing—review and editing, L.Z.; visualization, Z.Y. and L.Z.; supervision, L.Z.; funding acquisition, J.X. and L.Z.; Data curation, D.C.; Investigation, D.C. All authors have read and agreed to the published version of the manuscript.

Funding: This research was funded by the Strategic Priority Research Program of the Chinese Academy of Sciences (Grant No. XDA23040304) and the National Nature Science Foundation of China (No. 41890823).

Data Availability Statement: Not applicable.

Acknowledgments: We thank the anonymous reviewers for their constructive feedback.

Conflicts of Interest: The authors declare no conflict of interest.

References

1. Cheng, F.Y.; Basu, N.B. Biogeochemical hotspots: Role of small water bodies in landscape nutrient processing. *Water Resour. Res.* **2017**, *53*, 5038–5056. [CrossRef]
2. Freni, G.; Mannina, G.; Viviani, G. Assessment of the integrated urban water quality model complexity through identifiability analysis. *Water Res.* **2011**, *45*, 37–50. [CrossRef] [PubMed]
3. Forman, R.; Wu, J. Where to put the next billion people. *Nature* **2016**, *537*, 608–611. [CrossRef]
4. Carey, R.O.; Migliaccio, K.W.; Li, Y.; Schaffer, B.; Kiker, G.A.; Brown, M.T. Land use disturbance indicators and water quality variability in the Biscayne Bay Watershed, Florida. *Ecol. Indic.* **2011**, *11*, 1093–1104. [CrossRef]
5. Mello, K.; de Valente, R.A.; Randhir, T.O.; dos Santos, A.C.A.; Vettorazzi, C.A. Effects of land use and land cover on water quality of low-order streams in Southeastern Brazil: Watershed versus riparian zone. *Catena* **2018**, *167*, 130–138. [CrossRef]
6. Pan, D.; Hong, W.; Kong, F. Efficiency evaluation of urban wastewater treatment: Evidence from 113 cities in the Yangtze River Economic Belt of China. *J. Environ. Manag.* **2020**, *270*, 110940. [CrossRef] [PubMed]
7. Jia, X.; O'Connor, D.; Hou, D.; Jin, Y.; Li, G.; Zheng, C.; Ok, Y.S.; Tsang, D.S.W.; Luo, J. Groundwater depletion and contamination: Spatial distribution of groundwater resources sustainability in China. *Sci. Total Environ.* **2019**, *672*, 551–562. [CrossRef]
8. Dunalska, J.A.; Grochowska, J.; Wiśniewski, G.; Napiórkowska-Krzebietke, A. Can we restore badly degraded urban lakes? *Ecol. Eng.* **2015**, *82*, 432–441. [CrossRef]
9. Freni, G.; Mannina, G. Uncertainty in water quality modelling: The applicability of Variance Decomposition Approach. *J. Hydrol.* **2010**, *394*, 324–333. [CrossRef]
10. Schellart, A.N.A.; Tait, S.J.; Ashley, R.M. Towards quantification of uncertainty in predicting water quality failures in integrated catchment model studies. *Water Res.* **2010**, *44*, 3893–3904. [CrossRef]
11. Dhakal, K.P.; Chevalier, L.R. Urban Stormwater Governance: The Need for a Paradigm Shift. *Environ. Manag.* **2016**, *57*, 1112–1124. [CrossRef] [PubMed]
12. Garnier, J.; Brion, N.; Callens, J.; Passy, P.; Deligne, C.; Billen, G.; Servais, P.; Billen, C. Modeling historical changes in nutrient delivery and water quality of the Zenne River (1790s–2010): The role of land use, waterscape and urban wastewater management. *J. Mar. Syst.* **2013**, *128*, 62–76. [CrossRef]
13. Tong, S.; Li, X.; Zhang, J.; Bao, Y.; Bao, Y.; Na, L.; Si, A. Spatial and temporal variability in extreme temperature and precipitation events in Inner Mongolia (China) during 1960–2017. *Sci. Total Environ.* **2018**, *649*, 75–89. [CrossRef] [PubMed]
14. Ali, R.; Kuriqi, A.; Abubaker, S.; Kisi, O. Long-Term Trends and Seasonality Detection of the Observed Flow in Yangtze River Using Mann-Kendall and Sen's Innovative Trend Method. *Water* **2019**, *11*, 1855. [CrossRef]
15. Yenilmez, F.; Keskin, F.; Aksoy, A. Water quality trend analysis in Eymir Lake, Ankara. *Phys. Chem. Earth Parts A/B/C* **2011**, *36*, 135–140. [CrossRef]
16. Huang, N.E.; Wu, Z. A review on Hilbert-Huang transform: Method and its applications to geophysical studies. *Rev. Geophys.* **2008**, *46*, 228–251. [CrossRef]
17. Chou, C.-M. Wavelet-Based Multi-Scale Entropy Analysis of Complex Rainfall Time Series. *Entropy* **2011**, *13*, 241–253. [CrossRef]
18. Liu, H.-L.; Bao, A.-M.; Chen, X.; Wang, L.; Pan, X. Response analysis of rainfall-runoff processes using wavelet transform: A case study of the alpine meadow belt. *Hydrol. Process.* **2011**, *25*, 2179–2187. [CrossRef]
19. Ruiming, F. Wavelet based relevance vector machine model for monthly runoff prediction. *Water Qual. Res. J.* **2018**, *54*, 134–141. [CrossRef]
20. Kang, S.; Lin, H. Wavelet analysis of hydrological and water quality signals in an agricultural watershed. *J. Hydrol.* **2007**, *338*, 1–14. [CrossRef]
21. Niu, M.; Wang, Y.; Sun, S.; Li, Y. A novel hybrid decomposition-and-ensemble model based on CEEMD and GWO for short-term PM2.5 concentration forecasting. *Atmos. Environ.* **2016**, *134*, 168–180. [CrossRef]
22. Huang, N.E.; Shen, Z.; Long, S.R. The empirical mode decomposition and the Hilbert spectrum for nonlinear and nonstationary time series analysis. *Proc. R. Soc. Lond.* **1998**, *454*, 903–995. [CrossRef]
23. Xiao, X.; He, J.; Yu, Y.; Cazelles, B.; Li, M.; Jiang, Q.; Xu, C. Teleconnection between phytoplankton dynamics in north temperate lakes and global climatic oscillation by time-frequency analysis. *Water Res.* **2019**, *154*, 267–276. [CrossRef] [PubMed]
24. Ouyang, Q.; Lu, W.; Xin, X.; Zhang, Y.; Cheng, W.; Yu, T. Monthly Rainfall Forecasting Using EEMD-SVR Based on Phase-Space Reconstruction. *Water Resour. Manag.* **2016**, *30*, 2311–2325. [CrossRef]
25. Yuan, R.; Cai, S.; Liao, W.; Lei, X.; Zhang, Y.; Yin, Z.; Dingm, G.; Wang, J.; Xu, Y. Daily Runoff Forecasting Using Ensemble Empirical Mode Decomposition and Long Short-Term Memory. *Front. Earth Sci.* **2021**, *9*, 621780. [CrossRef]

26. Fijani, E.; Barzegar, R.; Deo, R.; Tziritis, E.; Konstantinos, S. Design and implementation of a hybrid model based on two-layer decomposition method coupled with extreme learning machines to support real-time environmental monitoring of water quality parameters. *Sci. Total Environ.* **2018**, *154*, 267–276. [CrossRef]

27. Huan, J.; Cao, W.; Qin, Y. Prediction of dissolved oxygen in aquaculture based on EEMD and LSSVM optimized by the Bayesian evidence framework. *Comput. Electron. Agric.* **2018**, *150*, 257–265. [CrossRef]

28. Zhang, J.; Tang, H.; Tannant, D.D.; Lin, C.; Xia, D.; Liu, X.; Zhang, Y.; Ma, J. Combined forecasting model with CEEMD-LCSS reconstruction and the ABC-SVR method for landslide displacement prediction. *J. Clean. Prod.* **2021**, *293*, 126205. [CrossRef]

29. Joshi, P.; Leitão, J.P.; Maurer, M.; Bach, P.M. Not all SuDS are created equal: Impact of different approaches on combined sewer overflows. *Water Res.* **2021**, *191*, 116780. [CrossRef] [PubMed]

30. Van Daal-Rombouts, P.; Sun, S.; Langeveld, J.; Bertrand-Krajewski, J.-L.; Clemens, F. Design and performance evaluation of a simplified dynamic model for combined sewer overflows in pumped sewer systems. *J. Hydrol.* **2016**, *538*, 609–624. [CrossRef]

31. Coutu, S.; Del Giudice, D.; Rossi, L.; Barry, D.A. Parsimonious hydrological modeling of urban sewer and river catchments. *J. Hydrol.* **2012**, *464–465*, 477–484. [CrossRef]

32. Tan, K.M.; Seow, W.K.; Wang, C.L.; Kew, H.J.; Parasuraman, S.B. Evaluation of performance of Active, Beautiful and Clean (ABC) on stormwater runoff management using MIKE URBAN: A case study in a residential estate in Singapore. *Urban Water J.* **2019**, *16*, 156–162. [CrossRef]

33. Zhi, W.; Feng, D.; Tsai, W.-P.; Sterle, G.; Harpold, A.; Shen, C.; Li, L. From Hydrometeorology to River Water Quality: Can a Deep Learning Model Predict Dissolved Oxygen at the Continental Scale? *Environ. Sci. Technol.* **2021**, *55*, 2357–2368. [CrossRef]

34. Qiao, Z.; Sun, S.; Jiang, Q.; Xiao, L.; Wang, Y.; Yan, H. Retrieval of Total Phosphorus Concentration in the SurfaceWater of Miyun Reservoir Based on Remote Sensing Data and Machine Learning Algorithms. *Remote Sens.* **2021**, *13*, 4662. [CrossRef]

35. Mohammadpour, R.; Shaharuddin, S.; Chang, C.K.; Zakaria, N.A.; Ghani, A.A.; Chan, N.W. Prediction of water quality index in constructed wetlands using support vector machine. *Environ. Sci. Pollut. Res.* **2014**, *22*, 6208–6219. [CrossRef] [PubMed]

36. Zhang, X.; Peng, Y.; Zhang, C.; Wang, B. Are hybrid models integrated with data preprocessing techniques suitable for monthly streamflow forecasting? Some experiment evidences. *J. Hydrol.* **2015**, *530*, 137–152. [CrossRef]

37. Wu, Z.; Huang, N.E. Ensemble empirical mode decomposition: A noiseassisted data analysis method. *Adv. Adapt. Data Anal.* **2009**, *1*, 1–41. [CrossRef]

38. Yeh, J.R.; Shieh, J.S.; Huang, N.E. Complementary ensemble empirical mode decomposition: A novel noise enhanced data analysis method. *Adv. Adapt. Data Anal.* **2010**, *2*, 135–156. [CrossRef]

39. Zhang, X.; Lai, K.K.; Wang, S.-Y. A new approach for crude oil price analysis based on Empirical Mode Decomposition. *Energy Econ.* **2008**, *30*, 905–918. [CrossRef]

40. Wu, C.; Fang, C.; Wu, X.; Zhu, G. Health-Risk Assessment of Arsenic and Groundwater Quality Classification Using Random Forest in the Yanchi Region of Northwest China. *Expo. Health* **2020**, *12*, 761–774. [CrossRef]

41. Marler, R.; Arora, J. Survey of multi-objective optimization methods for engineering. *Struct. Multidiscip. Optim.* **2004**, *26*, 369–395. [CrossRef]

42. Wuhan Ecological Environment Bureau. Report of Water Quality of Centralized Drinking Water Sources in Urban and County Level of Wuhan City. 2021. Available online: http://hbj.wuhan.gov.cn/hjsj/ (accessed on 15 December 2021). (In Chinese)

43. Wuhan Ecological Environment Bureau. Hubei Province Pollution Source Environmental Information Release System. 2021. Available online: http://113.57.151.5:4504/EAFMS/Guest.aspx (accessed on 15 December 2021). (In Chinese)

44. Mu, Q.; Zhao, M.; Running, S.W. Improvements to a MODIS global terrestrial evapotranspiration algorithm. *Remote Sens. Environ.* **2011**, *115*, 1781–1800. [CrossRef]

45. Yu, M.; Guo, S.; Guan, Y.; Cai, D.; Zhang, C.; Fraedrich, K.; Liao, Z.; Zhang, X.; Tian, Z. Spatiotemporal Heterogeneity Analysis of Yangtze River Delta Urban Agglomeration: Evidence from Nighttime Light Data (2001–2019). *Remote Sens.* **2021**, *13*, 1235. [CrossRef]

46. Elvidge, C.D.; Baugh, K.; Zhizhin, M.; Hsu, F.C.; Ghosh, T. VIIRS night-time lights. *Int. J. Remote Sens.* **2017**, *38*, 5860–5879. [CrossRef]

47. Wuhan Municipal Bureau of Statistics; NBS Survey Office in Wuhan. *Wuhan Statistical Yearbook*; China Statistics Press Co., Ltd.: Wuhan, China, 2021. Available online: http://tjj.hubei.gov.cn/tjsj/sjkscx/tjnj/gsztj/whs/202201/P020220125601159778648.pdf (accessed on 15 January 2022). (In Chinese)

48. Huang, F.; Wang, X.; Lou, L.; Zhou, Z.; Wu, J. Spatial variation and source apportionment of water pollution in Qiantang River (China) using statistical techniques. *Water Res.* **2010**, *44*, 1562–1572. [CrossRef] [PubMed]

49. Seeboonruang, U. A statistical assessment of the impact of land uses on surface water quality indexes. *J. Environ. Manag.* **2012**, *101*, 134–142. [CrossRef] [PubMed]

50. Yang, K.; Luo, Y.; Chen, K.; Yang, Y.; Shang, C.; Yu, Z.; Xu, J.; Zhao, Y. Spatial–Temporal variations in urbanization in Kunming and their impact on urban lake water quality. *Land Degrad. Dev.* **2020**, *31*, 1392–1407. [CrossRef]

51. Mello, K.; de Taniwaki, R.H.; Paula FR de Valente, R.A.; Randhir, T.O.; Macedo, D.R.; Leal, C.G.; Rodrigues, C.B.; Hughes, R.M. Multiscale land use impacts on water quality: Assessment, planning, and future perspectives in Brazil. *J. Environ. Manag.* **2020**, *270*, 110879. [CrossRef] [PubMed]

52. Zhang, J.; Li, S.; Jiang, C. Effects of land use on water quality in a River Basin (Daning) of the Three Gorges Reservoir Area, China: Watershed versus riparian zone. *Ecol. Indic.* **2020**, *113*, 106226. [CrossRef]

53. Lewis, W.M. Physical and Chemical Features of Tropical Flowing Waters. *Trop. Stream Ecol.* **2008**, 1–21. [CrossRef]
54. Shi, P.; Zhang, Y.; Li, Z.; Li, P.; Xu, G. Influence of land use and land cover patterns on seasonal water quality at multi-spatial scales. *Catena* **2017**, *151*, 182–190. [CrossRef]
55. Jia, Q.M.; Li, Y.P.; Liu, Y.R. Modeling urban eco-environmental sustainability under uncertainty: Interval double-sided chance-constrained programming with spatial analysis. *Ecol. Indic.* **2020**, *115*, 106438. [CrossRef]

 *remote sensing*

Article

Mapping Large-Scale Forest Disturbance Types with Multi-Temporal CNN Framework

Xi Chen [1,2,3,†], Wenzhi Zhao [1,2,†], Jiage Chen [4,*], Yang Qu [5], Dinghui Wu [3,6] and Xuehong Chen [1,2]

[1] State Key Laboratory of Remote Sensing Science, Faculty of Geographical Science, Institute of Remote Sensing Science and Engineering, Beijing Normal University, Beijing 100875, China; 211804020027@home.hpu.edu.cn (X.C.); wenzhi.zhao@bnu.edu.cn (W.Z.); chenxuehong@bnu.edu.cn (X.C.)

[2] Beijing Engineering Research Center for Global Land Remote Sensing Products, Faculty of Geographical Science, Institute of Remote Sensing Science and Engineering, Beijing Normal University, Beijing 100875, China

[3] State Key Laboratory of Information Engineering in Surveying, Mapping and Remote Sensing, Wuhan University, Wuhan 430079, China; 2019165110@stu.sdjzu.edu.cn

[4] National Geomatics Center of China, Beijing 100830, China

[5] School of Remote Sensing and Information Engineering, Wuhan University, Wuhan 430079, China; 2021102130005@whu.edu.cn

[6] School of Surveying and Geo-Informatics, Shandong Jianzhu University, Jinan 250101, China

* Correspondence: jiagechen@ngcc.cn

† Both authors contributed equally to this work and should be considered co-first authors.

Citation: Chen, X.; Zhao, W.; Chen, J.; Qu, Y.; Wu, D.; Chen, X. Mapping Large-Scale Forest Disturbance Types with Multi-Temporal CNN Framework. *Remote Sens.* **2021**, *13*, 5177. https://doi.org/10.3390/rs13245177

Academic Editors: Qiusheng Wu, Jun Li, Xinyi Shen and Chengye Zhang

Received: 24 October 2021
Accepted: 16 December 2021
Published: 20 December 2021

Publisher's Note: MDPI stays neutral with regard to jurisdictional claims in published maps and institutional affiliations.

Abstract: Forests play a vital role in combating gradual developmental deficiencies and balancing regional ecosystems, yet they are constantly disturbed by man-made or natural events. Therefore, developing a timely and accurate forest disturbance detection strategy is urgently needed. The accuracy of traditional detection algorithms depends on the selection of thresholds or the formulation of complete rules, which inevitably reduces the accuracy and automation level of detection. In this paper, we propose a new multitemporal convolutional network framework (MT-CNN). It is an integrated method that can realize long-term, large-scale forest interference detection and distinguish the types (forest fire and harvest/deforestation) of disturbances without human intervention. Firstly, it uses the sliding window technique to calculate an adaptive threshold to identify potential interference points, and then a multitemporal CNN network is designed to render the disturbance types with various disturbance duration periods. To illustrate the detection accuracy of MT-CNN, we conducted experiments in a large-scale forest area (about 990 km^2) on the west coast of the United States (including northwest California and west Oregon) with long time-series Landsat data from 1986 to 2020. Based on the manually annotated labels, the evaluation results show that the overall accuracies of disturbance point detection and disturbance type recognition reach 90%. Also, this method is able to detect multiple disturbances that continuously occurred in the same pixel. Moreover, we found that forest disturbances that caused forest fire repeatedly appear without a significant coupling effect with annual temporal and precipitation variations. Potentially, our method is able to provide large-scale forest disturbance mapping with detailed disturbance information to support forest inventory management and sustainable development.

Keywords: forest disturbance mapping; multitemporal CNN; large-scale long time-series; disturbance type

1. Introduction

Amid radical global environmental degradation, forests with worldwide coverage play a vital role in promoting human sustainable development. In terms of balancing the regional ecology system, forests are also able to regulate regional climate as well as carbon and water cycles [1–3]. Nevertheless, most existing forest ecosystems are continuously disturbed by natural and human-made events, such as fires, pests, and deforestation, which seriously harms the local ecosystem and wildlife population structure [4,5]. It is urgently needed to develop appropriate land management strategies using accurate forest disturbance maps through long time-series archived remote sensing data. Conventional forest ecology methods often require intensive field investigation and consume abundant resources, but it is difficult to achieve satisfactory forest disturbance detection results on large-scales [6]. Therefore, developing a timely and accurate forest disturbance detection strategy is urgently needed.

Earth observation satellites are designed to acquire land surface images constantly with a long-time span and large-scale coverage, which was long considered an ideal solution to comprehensively investigate forest disturbances [7]. Since the free opening of the U.S. Geological Service (USGS) Landsat data archive in 2008, it provides large amounts of medium-resolution optical imagery dating back to 1972 [8,9]. Densely formed time-series remote sensing data can dynamically identify the annual forest disturbance and characterize the magnitude and duration of the disturbance [10]. Aiming to detect forest disturbances, intensive studies emerged for the sake of abundant time-series remote sensing imagery, which significantly promoted the relevant research. However, direct identifying forest disturbances almost impossible due to substantial random noises produced by atmospheric effects, the sunlit angle variation, and sensor degradation. Therefore, the challenge of detecting large-scale forest disturbance is how to identify disturbance signals from a large amount of noisy background.

The conventional forest disturbance identification strategy in remote sensing society is to use the vegetation index (such as the normalized difference vegetation index (NDVI) and normalized burn ratio (NBR)) quantifying the degree of changes compared to the standard unchanged targets' profile [11,12]. For instance, The VCT algorithm designed the integrated forest Z-score (IFZ) index to represent the possibility of forest pixels and developed a set of rules to determine forest disturbances [13,14]. Such methods usually require predefined standard time-series curves before iterative comparison for disturbance determination [15]. However, due to the influence of random noise, even the same type of forest may present great differences in time-series vegetation indices. For this reason, the conventional difference measurements (such as Euclidean distance or Mahalanobis distance) are often failed to calculate inconsistency between the target time-series and standard ones. To alleviate the above problem, Landtrendr method uses straight-line segments to fit the representative features of the Landsat time-series signals. It simplifies the complex time-series into straight-line feature comparison, which reduces random noise interference to a certain degree [16]. Following a different strategy, dynamic time warping (DTW) constructs the optimal nonlinear alignment between two time-series, which can overcome the time deviation effectively [17,18]. Meanwhile, The Breaks For Additive Season and Trend Monitor (BFAST) algorithm decomposes Landsat time-series pixels into the trend, season, and noise components and determines forest disturbance by continually comparing the disturbed time-series profile with the historically stable forest profile [19,20]. Moreover, the CCDC algorithm constructs a physical prediction model and detects disturbance points by comparing the prediction with the real-time series. The model can adapt to noise when given enough coefficients [21,22]. Currently, CCDC was integrated on the Google Earth Engine (GEE) platform and plays as the benchmark for forest disturbance validation [23–25]. Still, these methods are built on prior knowledge or assumptions, and the process of disturbances detection must go through a series of iterative curve filtering or parameter optimization. Therefore, the accuracy of detection solely depends on the selection of threshold parameters and complete rule formulation, which inevitably deteriorate the detection accuracies and automation levels [26]. Thus, it is necessary to develop self-adaptative and autonomous detection methods for the improvement of forest disturbance identification.

Deep learning strategies, which extract high-level features through hierarchy struc-tures, were widely used in time-series information extraction tasks [27]. Compared with traditional methods based on hand-crafted rules or thresholds, the high-level features learned by deep learning are more representative [28]. At present, deep networks such as CNN and LSTM were intensively used for large-scale agricultural and forestry monitoring, such as forest coverage prediction and crop phenology detection [29,30]. Specifically, Kong et al. (2018) [31] proposed a strategy to learn stable forest time-series with LSTM network from historical data, which then allows for the forest disturbance detection to be identified by comparing the predicted data and real data. Ban et al. (2020) [32] achieved near real-time localization of forest fires with the assistance of a multilayer CNN framework. Thus, the mentioned studies are designated to detect forest disturbances given the abnormality. Still, the types of disturbances (i.e., fire, deforestation) is often needed before a reasonable decision can be made for forest protection.

For a typical forest disturbance time-series, it can be decomposed into two subprofiles that are the disturbance process (shown as a rapid decline in the values of vegetation index) and a duration process (shown as a slow recovery of the vegetation values), respectively [33]. For example, after a forest fire event, the ecological forest infrastructure is fully destroyed, and it will take 10 years or more to restore to the original level, while for most deforestations, it can be quickly restored within few years (i.e., 3–5) [12,34,35]. But, most of the existing forest disturbance detection methods focus on the fixed-scale for disturbance process recognition and neglecting the recovery process measurement. Therefore, the relationship between disturbance types and their scale-variate duration process is often overlooked [36]. Moreover, different disturbances may constantly occur through a specific temporal range. How to iteratively detect forest disturbances also remains unexploited.

To solve the above problems, we propose a multitemporal convolutional network framework (MT-CNN) to recognize different types of forest disturbances. Specifically, we first detect potential forest disturbance points through a set of sliding windows techniques, which inevitably may contain a certain degree of noise. After that, the rough detection results are further being refined with a well-trained multitemporal CNN network. In addition, during the process of refinement, the disturbance types of the forest also to be determined. The main contributions of this paper are:

(1) The sliding window scheme is being integrated with multiscale temporal CNN for forest disturbance type recognition with various duration periods.
(2) The proposed MT-CNN can simultaneously achieve long-term and large-scale multiple forest disturbance detection without human intervention.
(3) The prediction accuracy of forest disturbances is above 90% on the USA west coast region with the past 35 years of Landsat time-series data.

The remainder of this paper is organized as follows: the research data and the details of the MT-CNN method are described in Section 2, and Section 3, respectively. Section 4 introduces the experiment of forest disturbance type detection and shows the experimental results. Discussion and conclusion are presented in Sections 5 and 6, respectively.

2. Study Area and Data Preparation

2.1. Study Area

To fully demonstrate the performance of our proposed forest disturbance detection method, we conducted tests on a large-scale area at the west coast of USA. As shown in Figure 1), it is mainly located in northwestern California and western Oregon. The region mainly has a Mediterranean climate, with dry summers and rainy winters. Due to its unique geographical location, the region has a variety of forest types. Among them, the coastal area is mainly wet temperate rain forest, and the inland area has a large area of coniferous forest. Affected by the vast forest area and human activities, forest fires and anthropogenic deforestation constantly occurred.

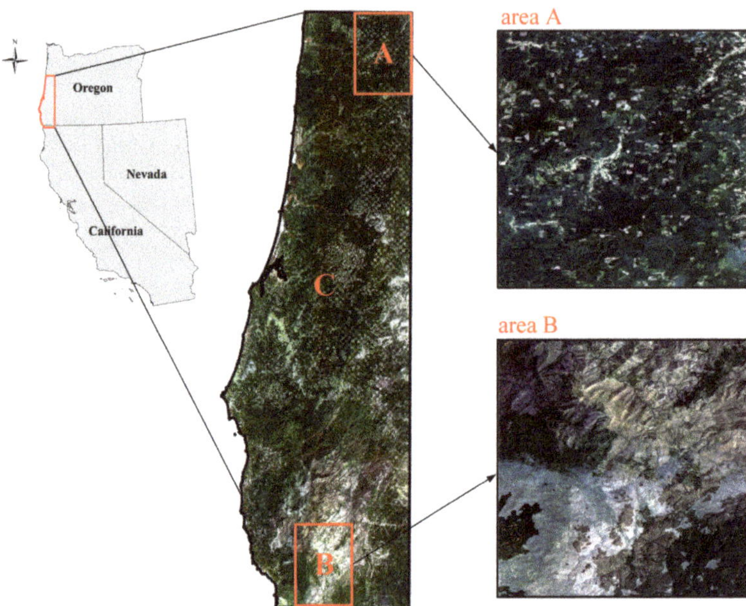

Figure 1. Study on west coast area of USA. Area A is dominated by forest decline, while Area B is dominated by wildfires.

2.2. Time-Series Data Acquisitions

2.2.1. Landsat Data

We constructed long time-series data from 1986 to 2020 for the study area, mainly consisting of annual composite images of Landsat in June, July, and August for each year. Among them, Landsat5 TM and Landsat8 OLI are the main data sources, while Landsat7 ETM+ is the auxiliary data to make up for the lack of data observation. To reduce the influence of clouds and cloud shadows, we do cloud masking for each image, and use the median of all available images each year as the annual composite pixel value. For forest disturbance detection, deforestation is manifested by the decrease of near-infrared (NIR) reflectance and forest fire has an obviously increase in short-wave infrared($SWIR_2$) reflectance [37,38]. Therefore, we preprocessed long time-series data to normalized burn ratio (NBR), as it can effectively distinguish different types of forest disturbances such as forest fire and deforestation. The NBR is formulated as $NBR = \frac{(NIR-SWIR_2)}{(NIR+SWIR_2)}$.

We drew some typical NBR curves in this area. As shown in Figure 2, wildfires and deforestation showed different characteristics in NBR time-series curves. For example, harvest/deforestation activities are usually only lasting for a short period, while a quick recovery can be observed. Differently, the forest fires usually with a long recovery period compared to human intervened deforestation.

Moreover, Since the long time-series data was constructed between three different sensors, TM and ETM+ were harmonized into OLI processing to maintain consistency [39]. All of the above processes are done on the GEE platform.

2.2.2. Disturbance Reference Data

We learned that in 2002, 2018, and 2020, several catastrophic wildfires occurred in Medford and Eugene in southern Oregon through related articles and fire monitoring websites (see area B in Figure 1). Meanwhile, the ever-increasing demand for the wood industry also caused the constant decline in forest inventory.We found a large number of harvest/deforestation areas from google earth platform (see area A in Figure 1). Therefore,we refer to two types of disturbances, that are harvest/deforestation and forest fires, as our forest disturbance detection objectives [40].

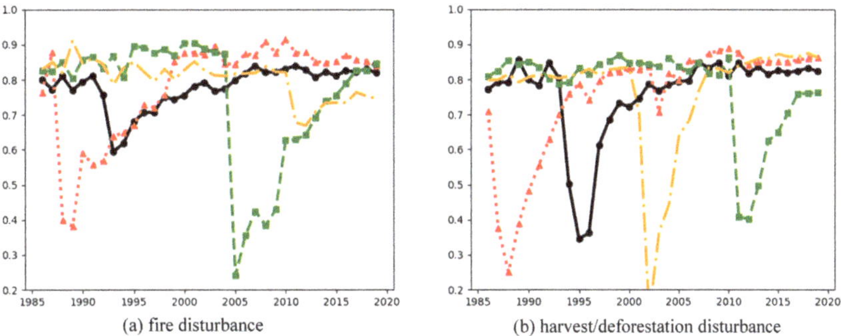

Figure 2. The typical time-series curves for different types of forest disturbances. (**a**) represents fire disturbance while (**b**) represents harvest/deforestation.

To validate the forest disturbance results by the proposed MT-CNN method, we obtained the high-precision forest disturbance samples by interactively combining the forest disturbance events and TimeSync interpretation tool [41]. We first convert the Landsat time series data into the standard format required by Timesync, and then visually distinguished fire and harvest/deforestation samples from nonchange ones combined with spatial information and NBR index.To ensure the uniform distribution of the samples, we randomly select a certain number of samples in area A and area B. In the end, we get 6000 samples in area A and area B, respectively (12,000 in total), of which 2000 stable forest samples were in each of the two regions, as shown in Figure 3.

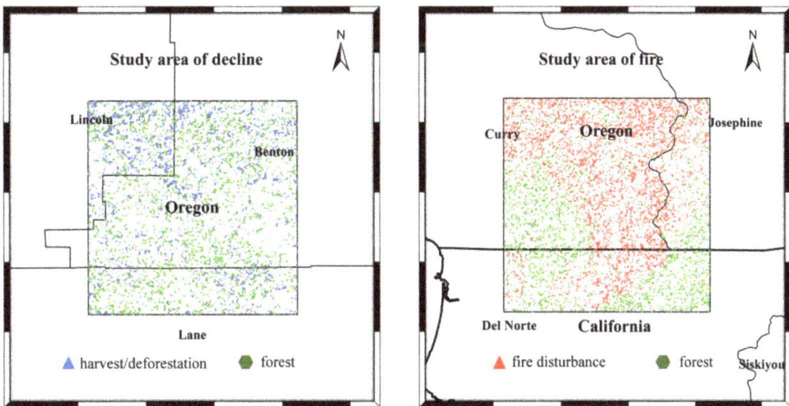

Figure 3. Two annotated areas with different types of forest disturbances. Harvest/deforestation disturbance samples are shown in **left**, while fire disturbance samples are shown in **right**.

3. Methodology

The proposed MT-CNN mainly consists of three modular units, that is, the data preparation module, the time-series breakpoint detection module, and the multiscale CNN disturbance type identification module. Specifically, we first construct the Landsat time-series stack and compute the NBR index to amplify forest disturbance signals. Then, a sliding window technique is applied to filter time-series signals which aim to detect possible breakpoints for forest disturbances. In this process, the thresholds for potential forest disturbances are automatically calculated by a self-adaptive scheme without human intervention. As a result, both real and pseudo disturbances may be detected. Finally, the rough detection results are input into the multitemporal CNN model to determine the type of forest disturbance and eliminate false detections. Based on the two-step detection process, various forest disturbances and their types can be determined through long time series data. The flowchart of the proposed MT-CNN method is shown in Figure 4.

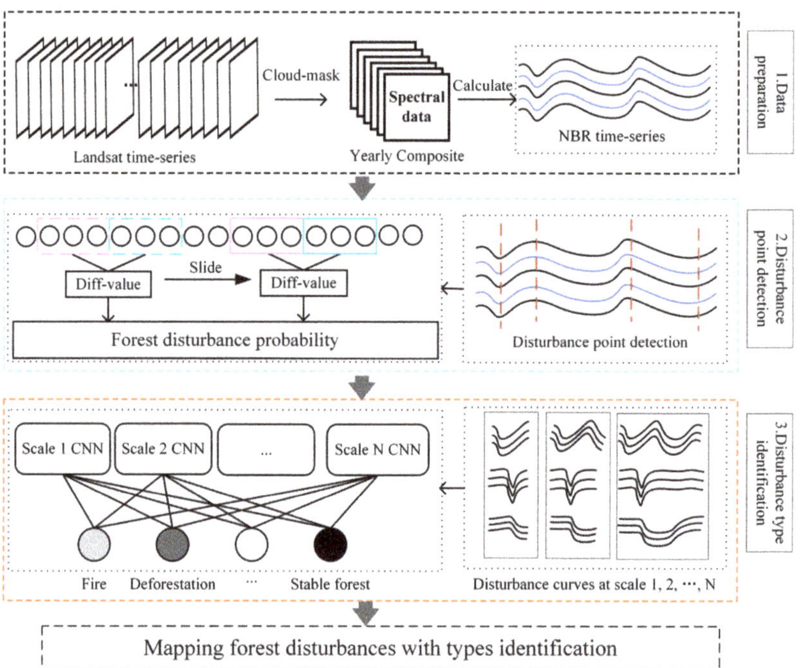

Figure 4. Flowchart of proposed integrated MT-CNN. (**1**) yearly composite data are prepared for analysis; (**2**) disturbance point is detected with sliding window technique; (**3**) disturbance type recognition with multitemporal time-series fragments.

3.1. Disturbance Point Detection

For most forest disturbance scenarios, the disturbance point is defined as the timing of a sudden decrease in the NBR value. Therefore, the sharp decline of NBR detection is the key to determine the forest disturbance time. For this purpose, we use a self-adaptive sliding window strategy to achieve disturbance point detection. For a given time series $t = t_1, t_2, \ldots, t_n$, the corresponding NBR value is denoted by V_t. We use two sliding windows side by side along with the time series to calculate the index difference between windows for disturbance points detection. Suppose that T_i represents the time interval that in range of $[t_i, t_{i+b}]$ with window size b, $T_i = [V_i, V_{i+1}, \ldots, V_{i+b}]$. To catch the overall trend information, we compute the 2-norm for each window to represent the statistical characteristics of the signal. The discrepancy measurement for each sliding window can be expressed as:

$$N(T_i) = \|T_i\|_2 = \sqrt{V_{t_i}^2 + V_{t_{i+1}}^2 + \cdots + V_{t_{i+b-1}}^2} \tag{1}$$

where $N(T_i)$ represents the L2-norm of the given time-series signals. We predict the disturbance point by the self-adaptive dynamic threshold technique. Specifically, the possible disturbance points are automatically evaluated by a quartile statistical method. Firstly, we arrange the window trend signals obtained in the previous step from small to large and get the values q_1 and q_3 of their 1/4 and 3/4 positions according to the linear rule. Then the interquartile range (r) of the signal can be expressed as:

$$r = (q_3 - q_1) \tag{2}$$

We take the self-adaptive factor σ on the quarterback difference as the abnormal threshold, then the range of normal data can be expressed as:

$$\text{range} = [q_1 - \sigma \cdot r, q_3 + \sigma \cdot r] \tag{3}$$

Since, the forest disturbance scene usually shows that the NBR value suddenly drops, so we only take the lower bound of this range. The range formula of the disturbance point is $N(T_i) \leq q_1 - \sigma \cdot r$.

3.2. MT-CNN for Disturbance Type Recognition

The flowchart of our proposed multitemporal CNN framework is shown in Figure 5. To achieve the purpose of disturbance type recognition, we developed a multitemporal scheme to extract rapid decline and recovery process time-series fragments through forest regions at different scales. Meanwhile, the convolutional neural network is applied to automatically detect the disturbance types of the extracted time-series fragments. Specifically, we extracted the multitemporal fragments with various lengths $l = [3, 4, 5, \ldots, L]$ begin with the disturbance points detected in Section 3.1. For the part with insufficient length, the last point of the time series is replicated to the extent of desired lengths. For the extracted time-series fragments at different scales $[S_1, S_2, S_3, \ldots, S_L]$, we use multitemporal convolutional neural networks to extract cross-scale disturbance features.

$$\begin{bmatrix} d_1 \\ d_2 \\ d_3 \\ \vdots \\ d_L \end{bmatrix} = \begin{bmatrix} w_1^n & 0 & 0 & 0 & 0 \\ 0 & w_2^n & 0 & 0 & 0 \\ 0 & 0 & w_3^n & 0 & 0 \\ 0 & 0 & 0 & \ddots & 0 \\ 0 & 0 & 0 & 0 & w_L^n \end{bmatrix} \begin{bmatrix} s_1 \\ s_2 \\ s_3 \\ \vdots \\ s_L \end{bmatrix} + \begin{bmatrix} b_1^n \\ b_2^n \\ b_3^n \\ \vdots \\ b_L^n \end{bmatrix} \tag{4}$$

where d_L represents the time-series fragments at different temporal scales, w_L^n and b_L^n represent learnable weights and bias for the multitemporal CNN framework. With increasing length of time series fragments, the number of layers n of convolutional layer and subsampling layer can be automatically adjusted. To accurately measure the general disturbance characteristics for each temporal scale, we proposed an overall multitemporal attention mechanism to the CNN framework. This process aims to learn the relative features of each time-series fragment relative to the whole time-series, so as to increase its discrimination. For the attention mechanism, we have

$$a_l^m = \sum_{m=1}^{M} \frac{\exp\left((q_S)^T d_l^m\right)}{\sum_{j=1}^{N} \exp\left((q_S)^T d_l^m\right)} \tag{5}$$

where q_S represents the whole time-series, a_l^m represents the attention indicator of the deep features extracted from multitemporal CNN for the specific scale l. M represents the dimension of extracted deep features. During the training process, the multitemporal attention term gradually learns the relationship between the disturbance types and multitemporal deep features. Since the weights of the cross-scale network are determined, the probability of disturbance type for each scale also can be deduced. This process can be represented as:

$$\begin{bmatrix} p_1 \\ p_2 \\ p_3 \\ \vdots \\ p_L \end{bmatrix} = \sigma \left(\begin{bmatrix} d_1 \\ d_2 \\ d_3 \\ \vdots \\ d_L \end{bmatrix} + \begin{bmatrix} d_1 a_1^m \\ d_2 a_2^m \\ d_3 a_3^m \\ \vdots \\ d_L a_L^m \end{bmatrix} \right) \tag{6}$$

Finally, the output probabilities of cross-scale CNNs are connected through the fully connected layer, allowing the model to be trained by optimizing a multiscale loss function.

In this process, we get the final prediction for each sample across all-scale time-series fragments on forest disturbance types. Specifically, to strengthen the cross-scale ability of MT-CNN, we integrated all probabilities on various time-series fragments to get the final prediction. Therefore, the final forest disturbance prediction can be formulated as

$$y(l,c) = \max\left(\sum_{l=1}^{L}\sum_{c=1}^{C} WP_l^c\right), \forall c \in [1,2,\dots,C] \tag{7}$$

where C describes the number of forest disturbance types. Thus, in this process, CNN predictions at different scales are integrated into the final predict disturbance types. In addition, this modular is also able to correct the false detection in Section 3.1.

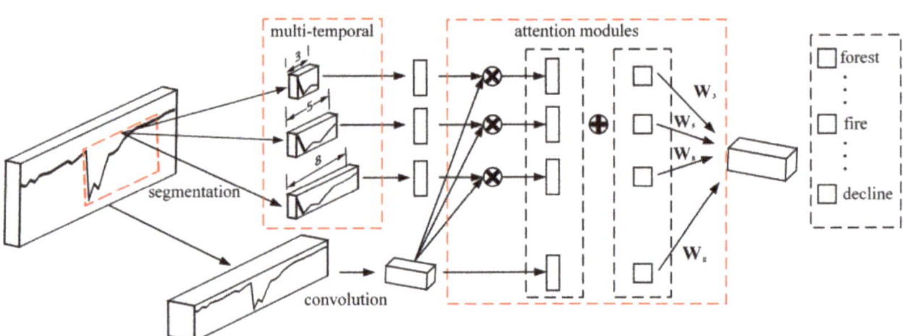

Figure 5. Flowchart of the multitemporal CNN framework.

4. Experiments and Results

In this section, we apply the MT-CNN algorithm to two areas on the west coast of the United States that contain specific types of disturbances for qualitative evaluation (area A and area B in Figure 1). To illustrate the capability of the proposed method, we randomly select 50% of the manually labeled samples as the training set and the rest as the test ones. Accuracy metrics such as users' accuracy, producers' accuracy, confusion matrix, and precision rate are used to evaluate the detection accuracy of forest disturbance. In addition, we also analyzed the evaluation results into three levels: disturbance type, disturbance time, and the number of occurrences for each time-series signal. In the following sections, we conduct a qualitative and quantitative evaluation of the disturbance detection in the study area.

4.1. Large-Scale Forest Disturbance Mapping

To demonstrate the robustness of the proposed method, we used the MT-CNN method to predict forest disturbances over a large-scale area of the USA west coast. With the proposed modular, the forest disturbances with different types are accurately identified. The forest disturbance results are illustrated in Figure 6.

Firstly, for the disturbance point detection from Figure 6a, we can conclude that most disturbances occurred in 2002 and 2019 as there were two devastating wildfires broke out. From 1986 to 2020, the yearly area of forest disturbances has witnessed a continuous increase, especially since the year 2001. From a regional perspective, small forest disturbances can be observed in the north region of the study area, while the large disturbance areas are mainly located in the south region. Before the year 2000, forest disturbances are sparsely distributed in the study area and the majority of them are near the fringe area of the deep forest. While, after the year 2000, large-scale forest disturbances are frequently observed, especially for the mountainous and deep forest regions.

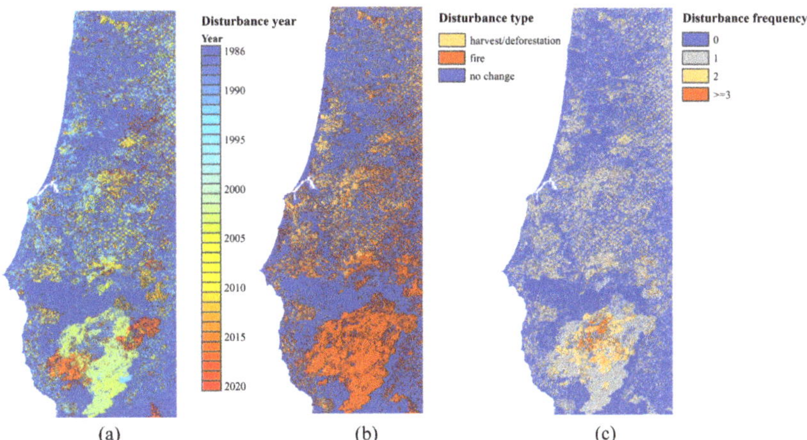

Figure 6. (**a–c**) Disturbance mapping results on disturbance point, type, and frequency.

Secondly, for the disturbance type identification from Figure 6b, the majority disturbance type of our study area is fire instead of harvest/deforestation. In general, there are 10,894,868 pixels of detected fire disturbances and 4,361,100 pixels of harvest/deforestation disturbances. Similarly, to the disturbance year, a large area of fire disturbances occurred in the south region of the study area, while the north part of the study area is mainly with harvest/deforestation. With the spatial integration of disturbance point and disturbance type, we can conclude that sparsely distributed forest disturbances in the north region are mainly harvest/deforestation. The reason for this phenomenon probably is human intervened urbanization. While for the south region of the study area, forest fire is the prominent impact factor for forest sustainability.

Thirdly, from the perspective of disturbance frequency, we can identify the most unstable area that suffers from continuous disturbances, as shown in Figure 6c. For the detected forest disturbances, there is 69.1% (9,704,841) pixels had occurred one-time disturbance over the last 3 decades. Another 27.4% detected pixels had encountered two-time disturbances, such as fire-fire or harvest/deforestation-fire. Combining the mapping results of disturbance point and type, we can deduce that high-frequency forest disturbance areas are located in the deep forest region. Meanwhile, the disturbance type is mainly forest fire with the first appearance in the year 1990. Over the last three decades, this region is continuously suffered from forest fire disturbance until 2020.

In general, the study area has mainly two types of forest disturbances, i.e., fire and harvest/deforestation. Most disturbances that occurred in this region is caused by wildfire outbreak. Compared to that of the existing forest disturbance detection algorithm, the MT-CNN is able to detect repeatedly occurred disturbances with types over the given time-series signal. Therefore, we can deduce useful information for forest risks analysis and management. To further evaluate the performance of MT-CNN on forest disturbance detection, we performed a quantitative evaluation on two smaller study areas (area A and area B in Figure 7) with exhausted annotation labels. According to the disturbance point, type, and frequency identification, we compared the prediction results of the MT-CNN with human labeled data.

4.2. Quantitative Evaluation on Forest Disturbance Detection

For area A, the major disturbance type is harvest/deforestation caused by human logging.To validate the harvest/deforestation samples, we visually inspected the disturbance area by comparing historical high-resolution satellite images from Google Earth. Two sub-regions (c and d) had occurred deforestation event for the year of 1990 and 2009, respectively. While, for area B, there are two massive forest fire that occurred in sub-regions a and b (as shown in Figure 7). With the accurately labeled data, we further validate the disturbance detection results through the perspective of disturbance point, type, and frequency.

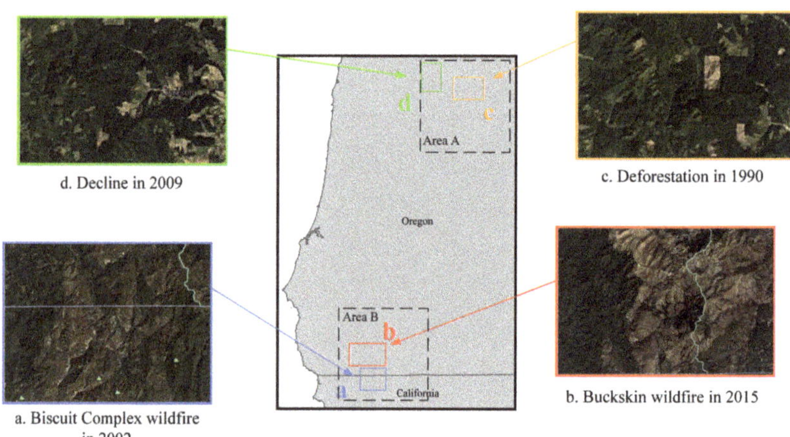

Figure 7. Finer-scale illustration of forest disturbance events with specific year.

4.2.1. Disturbance Point Accuracy Analysis

The disturbance point represents the beginning of forest disturbances from the time-series signal. Therefore, the accurate identification of disturbance points is a prerequisite for forest disturbance type understanding. To identify the disturbance point within the time-series data, the sliding window technique is applied to progressively detect suspected forest disturbances. With the automatic weights learning mechanism, the probability of being a forest disturbance can be determined.

We mapped the disturbance point of areas A and B consecutively with the help of MT-CNN. To better illustrate the prediction accuracy of the proposed method. We created three accumulated forest disturbance maps for area A in the year 1994, 2002, and 2015, and for area B in the year 1990, 2000, and 2009 (as shown in Figure 8), respectively. For area A, we can see that fire disturbance (the Biscuit Complex wildfire (2002)) have occurred in the year 2002 for the black square window. It continuously expanded to larger areas, a specific disturbance region inside of red box have appeared in the year 2015 (the Buckskin wildfire (2015)). For area B, there are several deforestation events that happened inside of this area. At the beginning, a few number of disturbance pixels are identified in the year 1990. Then, a few disturbances are identified in 2000 with rectangular-like shapes, which mainly produced by timber harvest, as shown in the green and yellow box. In 2009, massive disturbances can be observed around the outskirts area of previous logging sites. These maps are consistent with the time of the disturbances observed by Google Earth. The mapping results indicate that the proposed MT-CNN is able to continuously detect the disturbance point regardless of the types.

To quantitatively measure the accuracy of disturbance point detection, we compared the annotated label with the predicted output. For the annotated label, we extracted the staring magnitude as the standard disturbance point. Also, for the circumstance that disturbance happened more than once given a time-series signal, we only extracted the first disturbance point for comparison. The time span of our study area is more than 30 years; therefore, we re-grouped the detection results with 2-year bins for an easy plot. We used another 50% annotated samples to test the accuracy of the proposed MT-CNN in disturbance point detection. Given the 2-year bin as a statistic unit, the confusion matrix in disturbance point detection is generated, as shown in Figure 9. The overall accuracy in terms of disturbance point detection reaches 91%, and the kappa coefficient is 0.88. For the confusion matrix, the majority of detection accuracies are above 85%. For instance, the disturbance point detection accuracy reaches 99% around the year 2010–2011. Low-confident detection rates appeared at the beginning and end of time- series, where prior information is limited and coupled with significant noises. For the no-change (NC) class, the stable forests also demonstrated great variations in the time-series NBR index that inevitably lead to false predictions.

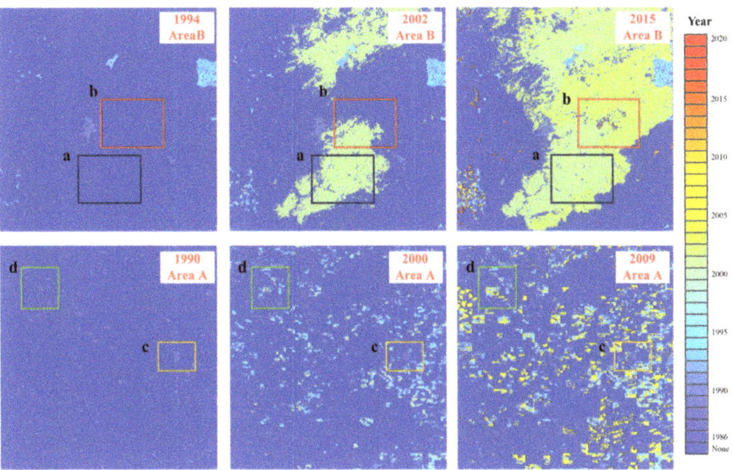

Figure 8. Disturbance point detection results on area A and B. For area A, maps with forest fire in years of 1994–2002–2015 are illustrated. For area B, maps with harvest/deforestation in years of 1990–2000–2009 are illustrated.

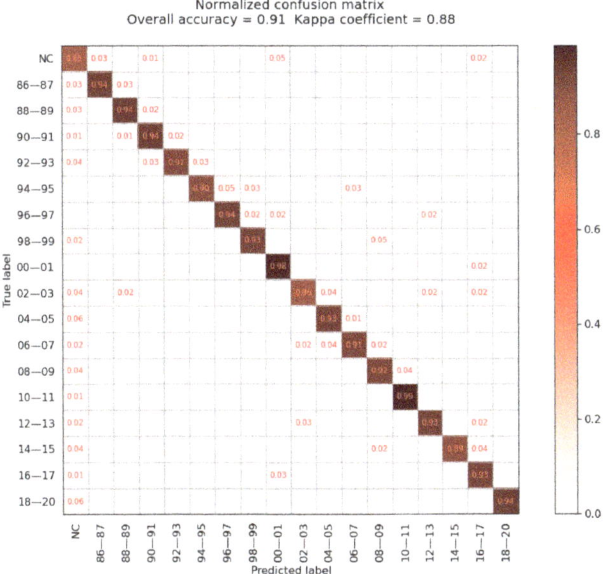

Figure 9. Confusion matrix of disturbance point detection from 1986 to 2020.

4.2.2. Disturbance Type Accuracy Analysis

Disturbance type identification is one of the most challenging tasks in time-series analysis. The proposed MT-CNN developed a multitemporal scheme to capture change patterns from various time-series fragments. To feed MT-CNN, the time-series fragments were extracted based on the disturbance point identified from the previous step. The accumulative disturbance maps on two areas (area A and area B in Figure 7) are demonstrated with two types of disturbance, that are, fire and harvest/deforestation, as shown in Figure 10. We visually inspected the disturbance type maps and found clear boundaries between stable forest and disturbances. For the left area B, the dominant disturbance type is fire and it is mainly covered by the mountain area, while with the right area A, the dominant disturbance type is deforestation caused by human intervention, as square-shaped logging sites can be clearly observed.

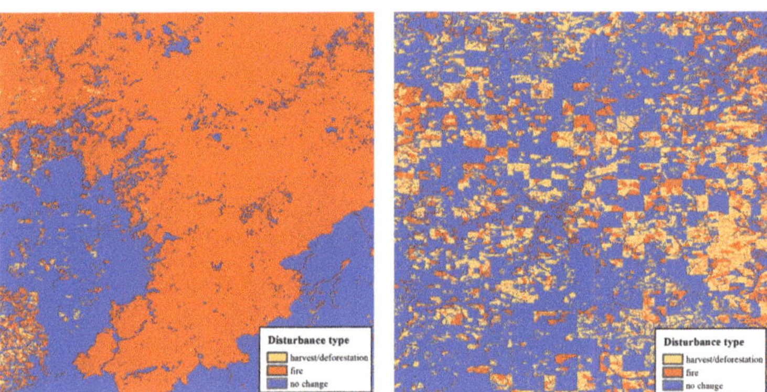

Figure 10. Disturbance type identification on area A and area B. Left area B is dominated by forest fire, while right area A is dominated by harvest/deforestation.

To quantitatively validate the disturbance type detection results, we used 5795 available test samples to calculate disturbance accuracy. For the classes of no change (stable forest), fire, and harvest/deforestation, the user accuracy is 94%, 85%, 93%, and the producer accuracy is 88%, 98%, 84%. The overall accuracy of disturbance type identification as high as 90%, and the kappa coefficient is 0.85, as shown in Table 1. The results of disturbance type detection further proved the robustness of the MT-CNN regardless of the complex background.

Table 1. Accuracies of disturbance type identification.

	Manual Annotation			**Users Accuracy**
	No Change	**Fire**	**Harvest/Deforestation**	
Algorithm				
No change	1749	4	114	0.94
Fire	177	1952	177	0.85
harvest/deforestation	71	44	1507	0.93
Producers accuracy	0.88	0.98	0.84	
Overall users accuracy	0.90	Overall accuracy	0.90	
Overall producers accuracy	0.90	Kappa coefficient	0.85	

4.2.3. Disturbance Frequency Accuracy Analysis

Compare to the disturbance point and type identification, the continuity of detection is another important indicator to evaluate the detection ability of the proposed method. To illustrate the continuity detection capability, we mapped the disturbance frequency for every single pixel on areas A and B, as shown in Figure 11. In this figure, we counted the disturbance times with one, two, and multiple (MT > 2) for each pixel,the statistical results are shown in the Figure 12. For the left area A, most disturbed areas have a one-time fire event, while inside the mountain area, 2-time disturbances also can be found. It indicates that this area suffers from continuous fire disturbances that significantly reduce the stability of the forest. Meanwhile, for the right area B, a large proportion of detected areas encountered more than 2-time disturbances. The reason for this phenomenon is probably the regular harvest of planted trees. From a quantitative point of view, we randomly selected 5795 test samples from areas A and B, and there were 3317 and 607 detected disturbances that occurred once or twice over the past 35 years. In general, the proposed method illustrated great robustness in multiple-disturbance detection.

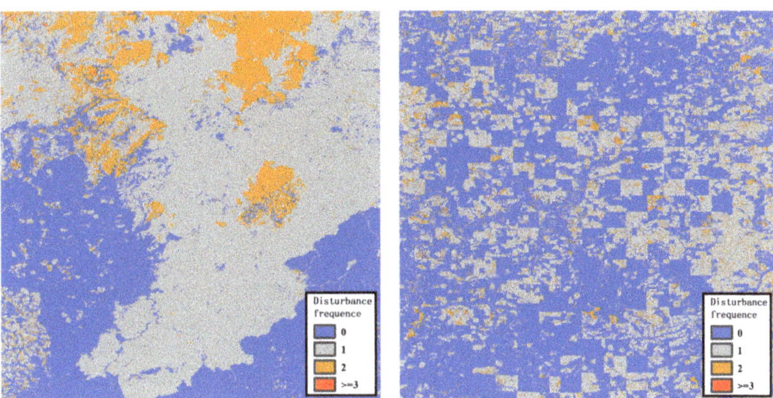

Figure 11. Mapping results of forest disturbance frequencies. Left area B is dominated by 1 or 2 times disturbances while right area A dominated with more than 3 times number of forest disturbances within 35 years.

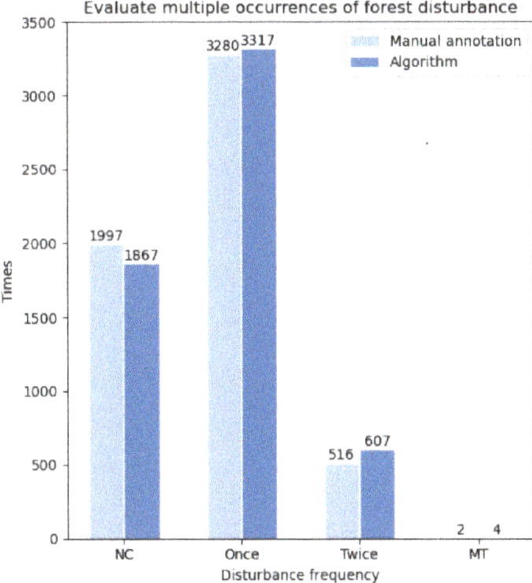

Figure 12. Statistical results on multiple disturbance with 5795 selected test samples.

5. Discussion

The proposed MT-CNN uses an integrated strategy to identify forest disturbances in terms of disturbance point, type, and frequency. Compared with the most existing forest disturbance detection strategies, the proposed method focuses on identifying the type of time-series fragments which include both decline and duration process. Since the disturbance point is determined, the time-series fragments with various lengths/scales can be fed into the MT-CNN for further type understanding. This multitemporal based forest disturbance detection strategy proved to be efficient at frequent, multiple-type disturbance identification. Still, the proposed method requires multitemporal inputs, which left an open question on temporal scale selection. In other words, how to choose the best temporal scale for disturbance type identification is remains unexploited. In addition, from the disturbance results of the USA west coast, we can observe the significant differences between the north and south region over the past 35 years in terms of forest disturbances. Also, the continuous fire disturbances happened inside of mountain areas, and the reason for this phenomenon is worth noticing. Therefore, in this section, we discussed two important factors that could impact the disturbance detection results.

5.1. Contribution of Multitemporal Scheme

To quantify the impact factor of a multitemporal scheme for disturbance detection. We set up an ablation experiment with different lengths of time-series fragments as multitemporal inputs. For a specific length of the fragment, the MT-CNN is able to automatically adjusted to fit the dimension of input data. To train and test the accuracy of disturbance types, we split all available samples into 50% and 50%. To determine both the final estimate of the expected accuracy and the standard deviation, we ran this train-test process 10 times on every single temporal scale. The quantitative evaluation of prediction results is plotted in a boxplot as shown in Figure 13. In this figure, boxes show the median and interquartile range of commission and omission errors for fire or harvest/deforestation disturbance type given a specific temporal scale, and whiskers show full data range. In general, the commission error for harvest/deforestation is relatively lower than that of fire disturbances. For omission error, it shows the opposite trend. With the increasing size of the temporal fragment, the commission error of harvest/deforestation and the omission error of forest fire witnessed a slow decline trend, and the values stay around 0.1 without significant variations. However, for the commission error of the forest fire and the omission error of the harvest/deforestation, there are three distinct valleys on the temporal scale of 3–4, 6–7, and 9–10. Consequently, different scales temporal-series fragments have different abilities to distinguish different disturbances types.

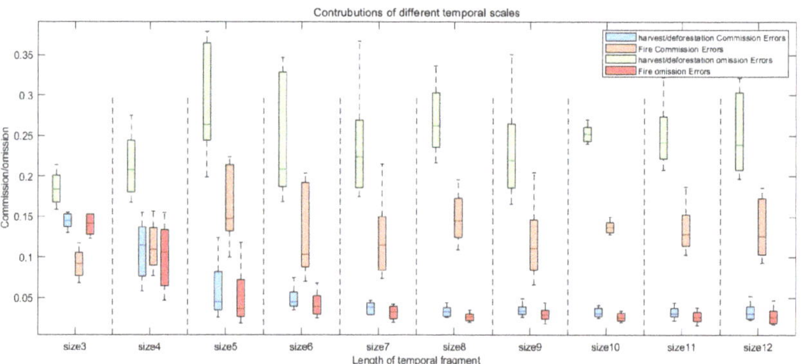

Figure 13. Overall accuracy regarding disturbance types with different lengths of temporal fragment.

Therefore, the multitemporal scheme is effective in terms of disturbance type identification. Specifically, for the fire disturbance, the predicted accuracies increase with the larger input temporal scale. Meanwhile, the decline disturbance is not sensitive to the temporal scale as it with smaller duration temporal window. The conclusion will further guide us to select proper temporal scales for various forest disturbances.

5.2. Influencing Factors of Forest Disturbance

The study area is located on the west coast of the USA, which is dominated by a Mediterranean climate. With the increasing crisis of global warming, this area is continuously impacted by heatwaves and wildfires. In the northern region of the study area, the low-altitude forests with intense human management are much more protected against wildfires, but regular tree harvesting is a major disturbance. The northwest region is the most stable region, and it greatly benefits from cooler temperatures and sufficient precipitation. By contrast, for the southern region, mountains with solid rocks and high-altitude drive plants tend to be extremely dry and flammable. From the previous disturbance maps, we can conclude that forests in the southern region continuously suffer from fire disturbances. The extremely large-scale wildfires broke out in the years 2002 and 2015. The frequency of fire disturbances in the southern mountain area is relatively high, especially for deep forest regions. Comparatively, for the northern region, most disturbances are forest decline caused by human activities. These harvest/deforestation regions are near to cities or towns, to where timber can be easily transported. However, the relationship between terrain features (such as DEM), weather indicators (precipitation and temperature), and disturbance frequency is needed to be deduced.

We obtained DEM, temperature and precipitation data on the GEE platform through two data sets named "PRISM Monthly Spatial Climate Dataset AN81m" and "NASA-DEM".To determine the relationship between DEM and disturbance frequency, we have selected a representative "south-north" profile for quantitative analysis. To better represent the statistical results, we set the statistical bin to 50 pixels. That is, the DEM is the average value of all the 50 pixels. And, the number of disturbances was summarized from 50 pixels for better understanding. Began with the south region, we can see that a sharp increase in DEM as it is the mountain region. In the meantime, a sharp increase also can be observed in the number of disturbances inside of this region. The highest disturbance number reaches 125 times for a statistical unit over the past 35 years. Then, the number of disturbances almost reaches 0 as DEM drops to 0. It means that the valley area with sufficient water and cooler temperature will prevent wildfire expansion. For DEM in the range of 0–500 m, the disturbances occurred without significant patterns. Statistical units that far than 160 km demonstrate a much more stable curve than the southern regions. While the disturbance type of this region is mainly harvest/deforestation with human intervention. In general, we can conclude that the number of disturbances roughly coincided with the distribution of DEM. That is, the higher DEM may suffer frequent disturbances than the plain area.

As the study area is mainly with the Mediterranean climate, scarce rains in hot summer usually coexistence with severe fire disturbances. We collected the annual precipitation rates over the last 35 years, and plotted it with the occurrence of observed disturbances, as shown in Figure 14. The highest precipitation rate reaches almost 300 mm in 1996 that coupled with the lowest frequency of fire disturbance. On the contrary, the minimum annual precipitation is about 50 mm which appeared in 2014 within this study area. The medium precipitation is around 175 mm per year, and five large-scale fire disturbances appeared in the year of 1988, 2002, 2003, 2018, and 2019, respectively. Comparatively, the harvest/deforestation is relatively stable that varies from the range of 0–100 cases that happened in the 300 km profile over the last 35 years. After 2015, the deforestation caused by human activities has significantly dropped, while fire disturbance continuously increases. From the perspective of temperature variation, we also plotted the mean annual temperature over the selected profile. The overall trend of the plotted curve demonstrates small variations between 10 °C to 12 °C. And 2015, 2014, and 1992 are the top three hottest years, while 2011, 2008, and 1990 are the top three coolest years. The highest temperature almost reaches 12.5 °C in 2015 and the coolest temperature near to 10 °C in 2011. Coupled with the precipitation information, the frequency of fire disturbance significantly increases when little rain with high temperature both appeared within a year. Although the scorching heat did not immediately cause heat fire in 2015, the dryer soil and dead trees become one of the most important factors for wildfire outbroke in 2018 and 2019. Therefore, both DEM, precipitation, and temperature potentially impact the frequency of forest disturbance in the west coast region of the USA. The reason behind this phenomenon is worth being quantitatively evaluated in future studies.

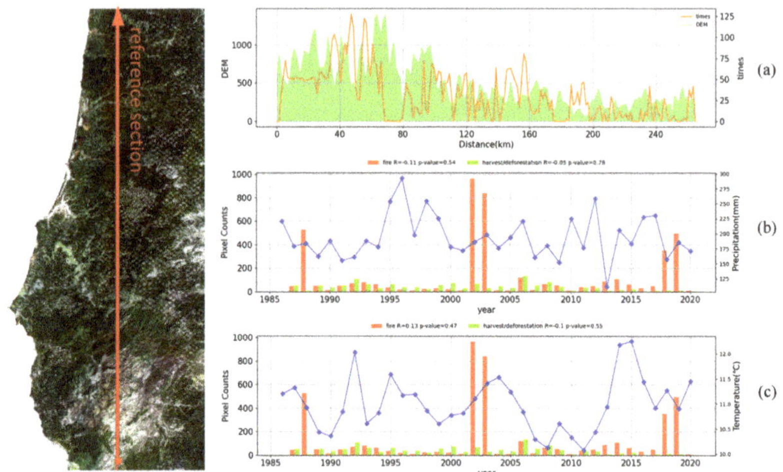

Figure 14. (**a–c**) Influences of terrain and weather factors on different types of forest disturbances.

6. Conclusions

Forest is one of the fundamental elements for human survival on Earth. With ever-increasing global warming and land detrition, timely and accurate identification of forest disturbances in terms of their location, type, and frequency are urgently needed. Aiming to detect forest disturbances from long time-series satellite data, we developed a multitemporal convolutional neural network (MT-CNN) to automatically identify forest disturbances with the feature of the beginning point, type, and frequency. The main contribution of this method is threshold-free multiscale identification of forest disturbances, regardless of duration periods variations of different events. To validate the performance of the proposed method, we selected a large-scale study area of the west coast area of the USA. For the 35-year time-series data, we identified forest disturbances with respect to their occurrence time, type, and frequency for each pixel. The overall accuracy reaches 90% compared to the manually labeled data. From the forest disturbance map, we also concluded that forest disturbances coupled with several indicators including DEM, precipitation, and temperature.

Author Contributions: Conceptualization, X.C. (Xi Chen) and W.Z.; methodology, X.C. (Xi Chen) and W.Z.; data analyses, J.C., Y.Q. and D.W.; writing—original draft preparation, X.C. (Xi Chen) and W.Z.; writing—review and editing, X.C. (Xi Chen), W.Z. and J.C.; visualization, X.C. (Xi Chen), W.Z., Y.Q., D.W. and J.C.; supervision, X.C. (Xuehong Chen), W.Z. and J.C.; project administration, X.C. (Xuehong Chen), W.Z. and J.C. All authors have read and agreed to the published version of the manuscript.

Funding: This research project was funded by National Natural Science Foundation of China (No. 41871224) , National Key R&D Program of China (2018YFC1508903), and Beijing Municipal Natural Science Foundation (No. 4214065).

Conflicts of Interest: The authors declare no conflict of interest.

References

1. Kurz, W.; Dymond, C.; White, T.; Stinson, G.; Shaw, C.; Rampley, G.; Smyth, C.; Simpson, B.; Neilson, E.; Trofymow, J. CBM-CFS3: A model of carbon-dynamics in forestry and land-use change implementing IPCC standards. *Ecol. Model.* **2009**, *220*, 480–504. [CrossRef]
2. Li, R.; Buongiorno, J.; Turner, J.A.; Zhu, S.; Prestemon, J. Long-term effects of eliminating illegal logging on the world forest industries, trade, and inventory. *For. Policy Econ.* **2008**, *10*, 480–490. [CrossRef]
3. Cohen, W.B.; Healey, S.P.; Yang, Z.; Zhu, Z.; Gorelick, N. Diversity of algorithm and spectral band inputs improves Landsat monitoring of forest disturbance. *Remote Sens.* **2020**, *12*, 1673. [CrossRef]

4. Mirabel, A.; Hérault, B.; Marcon, E. Diverging taxonomic and functional trajectories following disturbance in a Neotropical forest. *Sci. Total Environ.* **2020**, *720*, 137397. [CrossRef] [PubMed]

5. Zhu, F.; Wang, H.; Li, M.; Diao, J.; Shen, W.; Zhang, Y.; Wu, H. Characterizing the effects of climate change on short-term post-disturbance forest recovery in southern China from Landsat time-series observations (1988–2016). *Front. Earth Sci.* **2020**, *14*, 816–827. [CrossRef]

6. Senf, C.; Pflugmacher, D.; Hostert, P.; Seidl, R. Using Landsat time series for characterizing forest disturbance dynamics in the coupled human and natural systems of Central Europe. *ISPRS J. Photogramm. Remote Sens.* **2017**, *130*, 453–463. [CrossRef] [PubMed]

7. Cohen, W.B.; Goward, S.N. Landsat's role in ecological applications of remote sensing. *Bioscience* **2004**, *54*, 535–545. [CrossRef]

8. Hansen, M.C.; Potapov, P.V.; Moore, R.; Hancher, M.; Turubanova, S.A.; Tyukavina, A.; Thau, D.; Stehman, S.; Goetz, S.J.; Loveland, T.R. High-resolution global maps of 21st-century forest cover change. *Science* **2013**, *342*, 850–853. [CrossRef]

9. Wulder, M.A.; White, J.C.; Goward, S.N.; Masek, J.G.; Irons, J.R.; Herold, M.; Cohen, W.B.; Loveland, T.R.; Woodcock, C.E. Landsat continuity: Issues and opportunities for land cover monitoring. *Remote Sens. Environ.* **2008**, *112*, 955–969. [CrossRef]

10. Kennedy, R.E.; Andréfouët, S.; Cohen, W.B.; Gómez, C.; Griffiths, P.; Hais, M.; Healey, S.P.; Helmer, E.H.; Hostert, P.; Lyons, M.B. Bringing an ecological view of change to Landsat-based remote sensing. *Front. Ecol. Environ.* **2014**, *12*, 339–346. [CrossRef]

11. Cohen, W.B.; Yang, Z.; Healey, S.P.; Kennedy, R.E.; Gorelick, N. A LandTrendr multispectral ensemble for forest disturbance detection. *Remote Sens. Environ.* **2018**, *205*, 131–140. [CrossRef]

12. Meng, Y.; Liu, X.; Ding, C.; Xu, B.; Zhou, G.; Zhu, L. Analysis of ecological resilience to evaluate the inherent maintenance capacity of a forest ecosystem using a dense Landsat time series. *Ecol. Inform.* **2020**, *57*, 101064. [CrossRef]

13. Huang, C.; Goward, S.N.; Schleeweis, K.; Thomas, N.; Masek, J.G.; Zhu, Z. Dynamics of national forests assessed using the Landsat record: Case studies in eastern United States. *Remote Sens. Environ.* **2009**, *113*, 1430–1442. [CrossRef]

14. Myroniuk, V.; Bilous, A.; Khan, Y.; Terentiev, A.; Kravets, P.; Kovalevskyi, S.; See, L. Tracking Rates of Forest Disturbance and Associated Carbon Loss in Areas of Illegal Amber Mining in Ukraine Using Landsat Time Series. *Remote Sens.* **2020**, *12*, 2235. [CrossRef]

15. Kennedy, R.E.; Cohen, W.B.; Schroeder, T.A. Trajectory-based change detection for automated characterization of forest disturbance dynamics. *Remote Sens. Environ.* **2007**, *110*, 370–386. [CrossRef]

16. Kennedy, R.E.; Yang, Z.; Cohen, W.B. Detecting trends in forest disturbance and recovery using yearly Landsat time series: 1. LandTrendr—Temporal segmentation algorithms. *Remote Sens. Environ.* **2010**, *114*, 2897–2910. [CrossRef]

17. Maus, V.; Câmara, G.; Cartaxo, R.; Sanchez, A.; Ramos, F.M.; De Queiroz, G.R. A time-weighted dynamic time warping method for land-use and land-cover mapping. *IEEE J. Sel. Top. Appl. Earth Obs. Remote Sens.* **2016**, *9*, 3729–3739. [CrossRef]

18. Yan, J.; Wang, L.; Song, W.; Chen, Y.; Chen, X.; Deng, Z. A time-series classification approach based on change detection for rapid land cover mapping. *ISPRS J. Photogramm. Remote Sens.* **2019**, *158*, 249–262. [CrossRef]

19. Verbesselt, J.; Hyndman, R.; Zeileis, A.; Culvenor, D. Phenological change detection while accounting for abrupt and gradual trends in satellite image time series. *Remote Sens. Environ.* **2010**, *114*, 2970–2980. [CrossRef]

20. Verbesselt, J.; Zeileis, A.; Herold, M. Near real-time disturbance detection using satellite image time series. *Remote Sens. Environ.* **2012**, *123*, 98–108. [CrossRef]

21. Zhu, Z.; Woodcock, C.E. Continuous change detection and classification of land cover using all available Landsat data. *Remote Sens. Environ.* **2014**, *144*, 152–171. [CrossRef]

22. Zhu, Z.; Woodcock, C.E.; Holden, C.; Yang, Z. Generating synthetic Landsat images based on all available Landsat data: Predicting Landsat surface reflectance at any given time. *Remote Sens. Environ.* **2015**, *162*, 67–83. [CrossRef]

23. Ye, S.; Rogan, J.; Zhu, Z.; Eastman, J.R. A near-real-time approach for monitoring forest disturbance using Landsat time series: Stochastic continuous change detection. *Remote Sens. Environ.* **2021**, *252*, 112167. [CrossRef]

24. Zhu, Z.; Fu, Y.; Woodcock, C.E.; Olofsson, P.; Vogelmann, J.E.; Holden, C.; Wang, M.; Dai, S.; Yu, Y. Including land cover change in analysis of greenness trends using all available Landsat 5, 7, and 8 images: A case study from Guangzhou, China (2000–2014). *Remote Sens. Environ.* **2016**, *185*, 243–257. [CrossRef]

25. Zhu, Z.; Zhang, J.; Yang, Z.; Aljaddani, A.H.; Cohen, W.B.; Qiu, S.; Zhou, C. Continuous monitoring of land disturbance based on Landsat time series. *Remote Sens. Environ.* **2020**, *238*, 111116. [CrossRef]

26. Shimizu, K.; Ota, T.; Mizoue, N.; Yoshida, S. A comprehensive evaluation of disturbance agent classification approaches: Strengths of ensemble classification, multiple indices, spatio-temporal variables, and direct prediction. *ISPRS J. Photogramm. Remote Sens.* **2019**, *158*, 99–112. [CrossRef]

27. Zhan, Y.; Fu, K.; Yan, M.; Sun, X.; Wang, H.; Qiu, X. Change detection based on deep siamese convolutional network for optical aerial images. *IEEE Geosci. Remote Sens. Lett.* **2017**, *14*, 1845–1849. [CrossRef]

28. Kislov, D.E.; Korznikov, K.A.; Altman, J.; Vozmishcheva, A.S.; Krestov, P.V. Extending deep learning approaches for forest disturbance segmentation on very high-resolution satellite images. *Remote Sens. Ecol. Conserv.* **2021**, *7*, 355–368. [CrossRef]

29. Yang, Q.; Shi, L.; Han, J.; Yu, J.; Huang, K. A near real-time deep learning approach for detecting rice phenology based on UAV images. *Agric. For. Meteorol.* **2020**, *287*, 107938. [CrossRef]

30. Ye, L.; Gao, L.; Marcos-Martinez, R.; Mallants, D.; Bryan, B.A. Projecting Australia's forest cover dynamics and exploring influential factors using deep learning. *Environ. Model. Softw.* **2019**, *119*, 407–417. [CrossRef]

31. Kong, Y.-L.; Huang, Q.; Wang, C.; Chen, J.; Chen, J.; He, D. Long short-term memory neural networks for online disturbance detection in satellite image time series. *Remote Sens.* **2018**, *10*, 452. [CrossRef]
32. Ban, Y.; Zhang, P.; Nascetti, A.; Bevington, A.R.; Wulder, M.A. Near real-time wildfire progression monitoring with Sentinel-1 SAR time series and deep learning. *Sci. Rep.* **2020**, *10*, 1322. [CrossRef] [PubMed]
33. Kennedy, R.E.; Yang, Z.; Cohen, W.B.; Pfaff, E.; Braaten, J.; Nelson, P. Spatial and temporal patterns of forest disturbance and regrowth within the area of the Northwest Forest Plan. *Remote Sens. Environ.* **2012**, *122*, 117–133. [CrossRef]
34. Bright, B.C.; Hudak, A.T.; Kennedy, R.E.; Braaten, J.D.; Khalyani, A.H. Examining post-fire vegetation recovery with Landsat time series analysis in three western North American forest types. *Fire Ecol.* **2019**, *15*, 8. [CrossRef]
35. Turner, M.G. Disturbance and landscape dynamics in a changing world. *Ecology* **2010**, *91*, 2833–2849. [CrossRef] [PubMed]
36. Rodman, K.C.; Andrus, R.A.; Veblen, T.T.; Hart, S. Disturbance detection in landsat time series is influenced by tree mortality agent and severity, not by prior disturbance. *Remote Sens. Environ.* **2021**, *254*, 112244. [CrossRef]
37. Chen, D.; Fu, C.; Hall, J.V.; Hoy, E.E.; Loboda, T.V. Spatio-temporal patterns of optimal Landsat data for burn severity index calculations: Implications for high northern latitudes wildfire research. *Remote Sens. Environ.* **2021**, *258*, 112393. [CrossRef]
38. Gao, Y.; Solórzano, J.V.; Quevedo, A.; Loya-Carrillo, J.O. How BFAST Trend and Seasonal Model Components Affect Disturbance Detection in Tropical Dry Forest and Temperate Forest. *Remote Sens.* **2021**, *13*, 2033. [CrossRef]
39. Roy, D.P.; Kovalskyy, V.; Zhang, H.; Vermote, E.F.; Yan, L.; Kumar, S.; Egorov, A. Characterization of Landsat-7 to Landsat-8 reflective wavelength and normalized difference vegetation index continuity. *Remote Sens. Environ.* **2016**, *185*, 57–70. [CrossRef]
40. DeVries, B.; Decuyper, M.; Verbesselt, J.; Zeileis, A.; Herold, M.; Joseph, S. Tracking disturbance-regrowth dynamics in tropical forests using structural change detection and Landsat time series. *Remote Sens. Environ.* **2015**, *169*, 320–334. [CrossRef]
41. Cohen, W.B.; Yang, Z.; Kennedy, R.Detecting trends in forest disturbance and recovery using yearly Landsat time series: 2. TimeSync—Tools for calibration and validation. *Remote Sens. Environ.* **2010**, *114*, 2911–2924. [CrossRef]

 *remote sensing*

Article

Multi-Scale Response Analysis and Displacement Prediction of Landslides Using Deep Learning with JTFA: A Case Study in the Three Gorges Reservoir, China

Yanan Jiang [1,2], **Lu Liao** [3,4,*], **Huiyuan Luo** [2], **Xing Zhu** [2] and **Zhong Lu** [5]

[1] School of Earth Sciences, Chengdu University of Technology, Chengdu 610059, China; jiangyanan@cdut.edu.cn

[2] State Key Laboratory of Geological Hazard Prevention and Geological Environment Protection, Chengdu University of Technology, Chengdu 610059, China; huiyuanluo@cdut.edu.cn (H.L.); zhuxing15@cdut.edu.cn (X.Z.)

[3] Technology Service Center of Surveying and Mapping, Sichuan Bureau of Surveying, Mapping and Geoinformation, Chengdu 610081, China

[4] School of Resources and Environment, University of Electronic Science and Technology of China, Chengdu 610054, China

[5] Huffington Department of Earth Sciences, Southern Methodist University, Dallas, TX 75275, USA; zhonglu@smu.edu

[*] Correspondence: liaolusc@uestc.edu.cn

Citation: Jiang, Y.; Liao, L.; Luo, H.; Zhu, X.; Lu, Z. Multi-Scale Response Analysis and Displacement Prediction of Landslides Using Deep Learning with JTFA: A Case Study in the Three Gorges Reservoir, China. *Remote Sens.* **2023**, *15*, 3995. https://doi.org/10.3390/rs15163995

Academic Editors: Qiusheng Wu, Xinyi Shen, Jun Li and Chengye Zhang

Received: 30 June 2023
Revised: 30 July 2023
Accepted: 8 August 2023
Published: 11 August 2023

Abstract: Reservoir water and rainfall, leading to fluctuations groundwater levels, are the main triggering factors that induce landslides in the Three Gorges Reservoir area. This study investigates the response mechanism of landslide deformation under reservoir water and rainfall variations through long-time on-site observations. To address the non-stationary characteristics of the time-series records, joint time-frequency analysis (JTFA) is first introduced into our landslide prediction model. This model employs optimal variational mode decomposition (VMD) to obtain specific signal components with clear physical meaning, such as trend component and periodic components. Then, multi-scale response analysis between the displacement and external factors three wavelet methods was conducted. The analysis results show a 1 year primary cycle of the time series associated with the landslide evolution. The reservoir water level and rainfall show anti-phase fluctuations. The periodic displacement correlates significantly with rainfall, lagging by about two months. The reservoir water is anti-phase with the landslide displacement, preceding it by approximately three months ($-51 \pm 8°$ phase difference). For landslide displacement prediction, the gated recurrent units (GRU) neural network model is integrated into the deep learning forecasting architecture. The model takes into account the correlation and hysteresis effect of input variables. Through six experiments, we investigate the effect of data volume on model predictions to determine the optimal model. The results demonstrate that our proposed model ensures high performance in landslide prediction. Moreover, a comparison with six other intelligent algorithms shows the advantages of our model in terms of time-effectiveness and long-sequence forecasting.

Keywords: joint time-frequency analysis (JTFA); multi-scale response analysis; hysteresis effect; deep learning forecasting model

1. Introduction

Since the Three Gorges Reservoir (TGR) was put into operation in June 2003, the stability of slopes along the TGR bank has changed dramatically in response to the periodic fluctuation of the rainfall intensity and the reservoir storage [1–3], resulting in thousands of reactivated landslides. The geomechanical and hydrologic characteristics of these landslides are continually affected by the consistent water levels and rainfall changes, leading to catastrophic events [4]. For example, the July 2003 Qianjiangping landslide developed during the initial reservoir impoundment, causing 24 fatalities and significant economic losses [5]. Since then, research on long-term monitoring and early warning of these landslides has steadily expanded, leading to a deeper understanding of the inherent patterns and the response mechanism of reservoir landslides.

Studies have revealed that the evolution of the reservoir landslide involves a complex, non-stationary and nonlinear dynamic process [6,7]. Consequently, the long-term observations collected on the landslide's behavior exhibit non-stationary time-varying characteristics with specific trends and/or seasonality. Although the deformation evolution and response analysis of reservoir landslides in TGR have been extensively studied and discussed in the time domain [8–10], their characteristics in the frequency domain are relatively unexplored, concealing valuable spectral information. Furthermore, non-stationary time series show time-frequency characteristics [6,11,12], making time-frequency representation of the signal essential for identifying the dominant fluctuation modes/spectra of different signals and understanding how those modes vary over time.

The empirical mode decomposition (EMD), developed by Huang et al. [13], is well-known for its time-frequency analysis, which is suitable for processing non-stationary time series. This method decomposes the data into different band-limited intrinsic mode functions (IMFs), but is sensitive to noise and sampling [14]. The variational modal decomposition (VMD), a more recent signal decomposition technique, enables fully intrinsic and adaptive decomposition [14]. In addition, Isham et al. also pointed out the importance of determining the number of VMD modes to obtain specific signal components with explicitly physical meaning [15]. To achieve efficient decomposition, the Shannon entropy, reflecting the uniformity of probability distribution, can be used as a fitness function to derive optimal parameters through an optimization method, such as the harmony search (HS) algorithm [16] and the wolf optimization algorithm (GWO) [17] employed in this study. GWO is easily implemented and exhibits advantages in strong convergence with few parameters.

Moreover, it is crucial to explore the response mechanisms of landslide deformation [18] to unveil the changing patterns of specific signal components at different scales. In 1998, Torrence et al. proposed wavelet analysis (WA) to study non-stationary time series in signal processing [19]. Presently, wavelet decomposition and denoising are the most popular WA methods in time series analysis widely used in Geophysics [19], Geography [20] and Meteorology [21]. However, WA methods, such as continuous wavelet transform (CWT), cross-wavelet transform (XWT) and wavelet coherency (WTC), are more suitable for analyzing non-stationary time series at different scales [7,22–24]. Unfortunately, they are rarely used to analyze long-term non-stationary observations associated with landslide behavior. As a result, this study will extend WA methods and optimized VMD into a deep learning framework to conduct a joint time-frequency analysis (JTFA) before establishing a landslide forecasting model.

So far, various approaches have been developed for landslide hazards research, ranging from physics-based to statistics-based models [9,25,26]. Over the past decades, the integration of machine learning (ML) methods into landslide research has led to significant advancements [2,8,9]. In recent years, deep-learning approaches, like recurrent neural networks (RNN) and their variants, designed to address the gradient explosion problem, have shown impressive success in predicting landslide displacements based on GNSS time series data [4,10,27,28]. Among these variants, the gated recurrent unit (GRU) proposed by Chung et al. [29] stands out, surpassing other RNN variants, such as the LSTM, in terms of training time, parameter update and generalization ability. Its outstanding performance extends to diverse fields, including economics [30], flood control [31] and disaster reduction [10], among others.

This paper presents a deep learning neural network model for predicting landslide displacement, incorporating a joint time-frequency analysis (JTFA) module. The JTFA employs optimal variational mode decomposition (VMD) to obtain specific signal components with clear physical meaning (e.g., trend component and periodic component).

Subsequently, multi-scale response analysis between the displacement and external factors is carried out by wavelet transform (CWT), cross-wavelet transform (XWT) and wavelet coherency (WTC). The enhanced GRU model based on deep learning architecture is employed to forecast landslide displacement (Figure 1e–g). Additionally, the model considers the correlation and hysteresis effect of input variables, such as impact factors, during the modeling process.

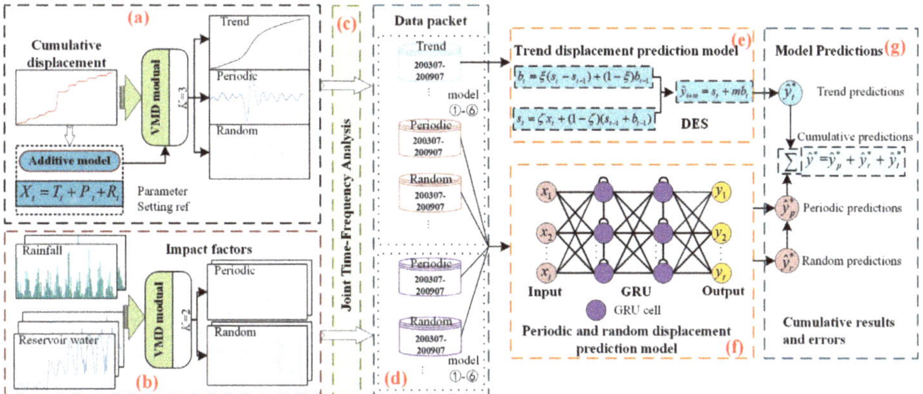

Figure 1. Architecture of the forecasting model.

2. Methodology

As can be seen in Figure 1, the forecasting model consists of seven parts: Figure 1 (a) signal decomposition of landslide displacement, Figure 1 (b) signal decomposition of the impacting factors, Figure 1 (c) multi-scale response analysis of landslide deformation by JTFA, Figure 1 (d) data packet preparation for the deep-learning neural network model, Figure 1 (e) trend displacement prediction by DES, Figure 1 (f) periodic and random displacement prediction by GRU, and Figure 1 (g) cumulative displacements prediction by the optimal model.

First, the on-site measured landslide displacements and the pre-selected impact factor sequences are decomposed into specific IMF components by an optimal VMD (Figure 1a,b). The number and parameters of the VMD modes with explicit physical meanings are determined by employing the Shannon entropy as a fitness function and the GWO optimization method. Then, a multi-scale response analysis between the displacement and external factors is carried out by three wavelet analysis methods to investigate landslide deformation response mechanisms and to reveal the changing patterns of specific signal components at different scales (Figure 1c). Finally, the enhanced GRU and DES model based on deep learning architecture is used to predict the landslide displacement (Figure 1e–g).

2.1. Joint Time-Frequency Analysis (JTFA)

The collected long-term observations associated with the landslide's behavior are non-stationary time series with specific trends or/and seasonality. The landslide displacement is a time-series function controlled by geological conditions, reflecting long-term deformation trends. While subject to external influences, the landslide displacement behaves as a periodic variation with near-white noise associated with random factors. Therefore, the landslide displacement time series can be expressed as follows:

$$Y_t = T_t + P_t + R_t \tag{1}$$

where Y_t is the landslide displacement at time t; while T_t, P_t and R_t are the trend, periodic and random terms, respectively.

While studies on the deformation characteristics of landslides have been extensively explored in the time domain, the significance of time-frequency analysis of the signal should not be overlooked. Time-frequency analysis reveals the time-frequency features, and it provides access to the implicit spectral information of displacement Y_t with different spectra, enabling a time-frequency perspective of information mining. Having a time-frequency representation is indispensable to investigate the dominant fluctuation spectra of deformation signals and understand how those spectra vary over time. This involves one-dimensional time series converting into a diffuse two-dimensional time-frequency image, allowing for a comprehensive examination of the signal's temporal and frequency properties.

2.1.1. Grey Wolf Optimized Variational Mode Decomposition (GWO-VMD)

The VMD enables fully intrinsic and adaptive signal decomposition [14], by which the input displacement signal could be divided into several intrinsic mode functions (IMFs) with specific characteristics determined in advance. To ensure that (1) the central frequency bandwidth of each IMF is limited, (2) the sum of the estimated bandwidths is minimized, and (3) the sum of each IMF equals the original signal f so that the constrained variational equation can be expressed as follows:

$$\min_{\{u_k\},\{w_k\}} \left\{ \sum_k \|\partial_t[(\delta(t) + j/\pi t) * u_k(t)]e^{-jw_k t}\|_2^2 \right\}$$
$$s.t. \quad \sum_k^K u_k = f \tag{2}$$

where u_k is the kth decomposed IMF with a center frequency w_k; and $\delta(t)$ and $*$ indicate the Dirac function and the convolution operator, respectively.

Then, the Lagrange multipliers λ and decomposition parameter ε are introduced to transform the constrained problem into an unconstrained variational one, yielding the augmented Lagrange expression:

$$L(\{u_k\}, \{\omega_k\}, \lambda) =$$
$$\varepsilon\sum_k \|\partial_t[(\delta(t) + j/\pi t) * u_k(t)]e^{-j\omega_k t}\|_2^2 + $$
$$\|f(t) - \sum_k u_k(t)\|_2^2 + \left\langle \lambda(t), f(t) - \sum_k u_k(t) \right\rangle \tag{3}$$

Here, the alternating direction multiplier method (ADMM) is used to search the saddle points of the unconstrained model to obtain the optimal IMFs and the center frequency of the constrained model. This process is defined as a function f_{VMD} in the Algorithm 1.

The VMD is used to decompose complex digital signals, and the empirical decomposition parameter ε is usually $K > 3$. This value will not be physically meaningful when applied to the landslide displacement decomposition. Thus, obtaining suitable VMD modes is crucial to getting specific signal components with explicitly physical meaning. At this point, the Shannon entropy (Equation (4)) is introduced to obtain the optimal IMFs with sparse features: the larger the value is, the better the uniformity of the probability distribution is [32].

$$H(u_k) = -\sum_{i=1}^{n} u_k^i \log u_k^i \tag{4}$$

As for the landslide's periodic displacement, the probability distribution is relatively uniform and sparse. Thus, the value of ε is optimized using the grey wolf optimizer (GWO) with $H(u_k)$ as the fitness function [26] to obtain the periodic displacement with a uniform probability distribution and strong sparsity along with the trend and random terms.

The landslide displacement Y_t is used as input. The population size P of the grey wolf, the upper bound X_{up} and lower bound X_{low} of the individual grey wolves are initialized. Then, the algorithm iterates and updates N times to obtain the optimal value of ε (see the Algorithm 1). In the Algorithm 1, both A and C are coefficient vectors with $A = 2a \cdot r_1 - a$

and $C = 2 \cdot r_2$. r_1 and r_2 are random values between 0 and 1. The convergence factor, $a = 2 - 2n/N$, decreases linearly from 2 to 0 during the iteration.

Algorithm 1 The GWO-VMD

Initialize $\varepsilon_\alpha, \varepsilon_\beta, \varepsilon_\delta, n \leftarrow 0$

$\quad \varepsilon_p^1 \leftarrow X_{low} + (X_{up} - X_{low}) \times rand(0,1)$ $\qquad\qquad\qquad\qquad\qquad$ (5)

repeat

$n \leftarrow n+1$

$\left\{ u_k^p \right\} \leftarrow f_{VMD}\left(Y_t, \varepsilon_p^n \right)$ $\qquad\qquad\qquad\qquad\qquad\qquad\qquad\qquad$ (6)

Update $\varepsilon_\alpha, \varepsilon_\beta, \varepsilon_\delta$:

$\quad \varepsilon_\alpha, \varepsilon_\beta, \varepsilon_\delta \leftarrow \max_{1,2,3} H\left(\left\{ u_{k=2}^p \right\} \right)$ $\qquad\qquad\qquad\qquad\qquad$ (7)

for $p = 1 : P$ do

$\qquad$ Update ε_p:

$X_1 \leftarrow \varepsilon_\alpha - A_1 \left| C_1 \varepsilon_\alpha - \varepsilon_p^n \right|$ $\qquad\qquad\qquad\qquad\qquad\qquad$ (8)

$X_2 \leftarrow \varepsilon_\beta - A_2 \left| C_2 \varepsilon_\beta - \varepsilon_p^n \right|$ $\qquad\qquad\qquad\qquad\qquad\qquad$ (9)

$X_3 \leftarrow \varepsilon_\delta - A_3 \left| C_3 \varepsilon_\delta - \varepsilon_p^n \right|$ $\qquad\qquad\qquad\qquad\qquad$ (10)

$\varepsilon_p^{n+1} \leftarrow mean(X_1, X_2, X_3)$ $\qquad\qquad\qquad\qquad\qquad\qquad$ (11)

end for

until $n = N$.

2.1.2. Wavelet Analysis (WA)

Torrence and Compo [22] proposed WA to study non-stationary time series in signal processing. The most popular WA methods in time series analysis include wavelet decomposition and denoising. Additionally, there are other WA methods, such as the continuous wavelet transform (CWT), the cross-wavelet transform (XWT) and the wavelet coherency (WTC), which excel at non-stationary time series analysis at various scales. However, despite their effectiveness, these WA methods are still rarely used in non-stationary time series analysis associated with the behavior of landslides.

For a discrete time-series x_n ($n = 1, \ldots, N$) with equal time spacing δt, the CWT is used for mapping the changing properties of non-stationary signals. CWT is defined as the convolution of x_n with a scaled and translated version of the wavelet basis function:

$$W_n(s) = \sum_{n'=1}^{N-1} x_{n'} \psi * \left[\frac{(n' - n)\delta t}{s} \right] \qquad (12)$$

where * indicates the complex conjugate, s represents the wavelet scale, and n is the localized time index. The non-orthogonal Morlet wavelets are chosen as the basis function: $\psi_0(\eta) = \pi^{-1/4} e^{iw_0 \eta} e^{-\eta^2/2}$, where η and w_0 are non-dimensional parameters of time and frequency, respectively. When $w_0 = 6$, the wavelet scale s equals the Fourier period. The wavelet basis functions ψ_0 are then normalized to have unit energy, and the CWT is conducted in the Fourier space to improve efficiency.

In Equation (12), $W_n(s)$ is a complex value, and thus, $|W_n(s)|^2$ denotes the wavelet power spectrum that clearly discriminates the periodic fluctuation and intensity in the time series. As the input data is assumed to be cyclic, edge effects appear in the power spectrum when dealing with finite-length time series. Hence, the cone of influence (COI) is defined, representing the corresponding edge effects within the region of the power spectrum. In addition, a Fourier power spectrum is constructed by Monte Carlo to represent the red noise background spectrum and later performs a χ^2 test with a specific confidence level, e.g., 95%, to calculate each scale and construct the confidence contour with a 95% confidence level.

The XWT of two time-series x_n and y_n is defined as $W^{XY} = W^X W^{Y*}$, which measures the similarity between two waveforms. The $arg(W^{XY})$ is the relative local phase between x_n and y_n in the time-frequency space. The mean and confidence interval must be estimated to analyze the phase difference. The phase relation is quantified using the periodic mean of phases over a region with a significance level of 5% outside the COI. The periodic mean of a set of phase angles is defined as follows:

$$a_m = \arg(X, Y) \text{ with } X = \sum_{i=1}^{n} \cos(\alpha_i) \text{ and } Y = \sum_{i=1}^{n} \sin(\alpha_i) \tag{13}$$

Because the phase angles are not independent, it is difficult to calculate the confidence interval of the mean angle. Thus, the solution is obtained by calculating the scatter around the mean using the circular standard deviation $er = \sqrt{-2\ln\left(\sqrt{X^2 + Y^2}/n\right)}$.

The power spectrum of XWT only shows the high common power part, while the WTC measures the coherence of the XWT in the time-frequency space. The WTC of two-time series is defined as follows:

$$R_n^2(s) = \frac{\left|S\left(s^{-1}W_n^{XY}(s)\right)\right|^2}{S\left(s^{-1}|W_n^X(s)|^2\right) \cdot S\left(s^{-1}|W_n^Y(s)|^2\right)} \tag{14}$$

where S is the smooth operator, and $S(W) = S_{scale}(S_{time}(W_n(S)))$. S_{scale} and S_{time} indicate smoothing along the wavelet scale axis and over time, separately.

2.2. Deep Learning Forecasting Model

2.2.1. Gated Recurrent Unit (GRU)

The gated recurrent unit (GRU) is a variant of recurrent neural networks (RNN), which outperforms other RNN variants, e.g., the LSTM [29], in terms of training time, and has fewer parameters. The GRU enables each recurrent unit to capture dependencies adaptively at different time scales. As a refinement of the general RNN structure, GRU is more straightforward, owing to the absence of a separate storage unit (see Figure 2) regulating the internal information flow via a gated unit. Moreover, it is more efficient in training time, parameter update and generalization ability than other RNNs [29].

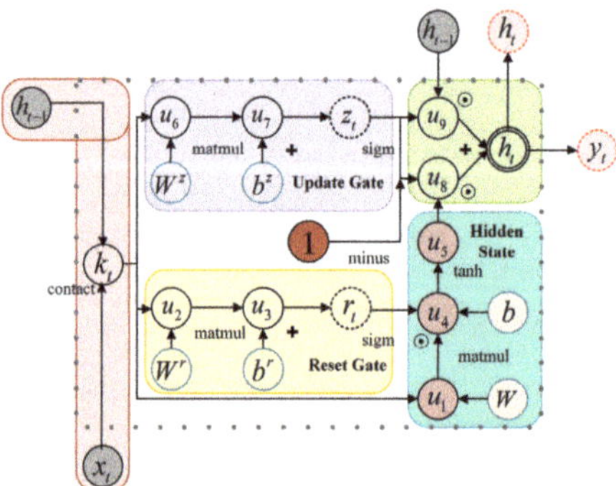

Figure 2. The network structure of GRU.

The update gate z_t is responsible for determining how much of the past information is to be neglected, while the reset gate r_t is responsible for deciding how much past knowledge needs to be passed along into the future. Assume that the current input at t is x_t, then one forward calculation of GRU is as follows:

$$z_t = \sigma(W_z \cdot [h_{t-1}, x_t] + b_z) \tag{15}$$

$$r_t = \sigma(W_r \cdot [h_{t-1}, x_t] + b_r) \tag{16}$$

$$\widetilde{h}_t = \tanh\left(W_{\widetilde{h}} \cdot [r_t * h_{t-1}, x_t] + b_{\widetilde{h}}\right) \tag{17}$$

where σ indicates the sigmoid function, and W and b are the weight and bias parameters, respectively.

The hidden state at the present moment h_t ht of the network is determined by the update gate, the candidate state of the previous time step h_{t-1} and the current moment $\widetilde{h}_t$. The output of the loop $\hat{y}_t$ can be calculated by

$$h_t = (1 - z_t) * h_{t-1} + z_t * \widetilde{h}_t \tag{18}$$

$$\hat{y}_t = \sigma(W_o \cdot h_t) + b_o \tag{19}$$

During the training process, the mean-square error (MSE) is used to measure the errors, and the loss function is defined as $loss = \sum_{t=1}^{T}(\hat{y}_t - x_t)^2 / T$, which is optimized by the adaptive moment estimation (Adam) [33] to obtain the minimization loss, and update W and b iteratively.

2.2.2. Double Exponential Smoothing (DES)

Double exponential smoothing (DES) is a weighted moving average method suitable for predicting time series with certain trends. This method operates by adopting a weighted combination of the past observations and recent observations, with relatively higher weights assigned to the more recent data points. The slope component is updated through exponential smoothing [34], making it well-suited for time-sensitive landslide displacement prediction. Thus, DES is employed to predict the trend component of the landslide displacement.

Given a displacement time series with a specific trend x_t, $t = 1, \ldots, N$. s_t represents the exponential smoothing value, b_t represents the optimal trend estimate. The model's output $\hat{y}_{t+m}$ is an estimate of x_{t+m} at $t + m$ (where $m \geq 1$) and can be written as:

$$s_t = \zeta x_t + (1 - \zeta)(s_{t-1} + b_{t-1}) \tag{20}$$

$$b_t = \xi(s_t - s_{t-1}) + (1 - \xi)b_{t-1} \tag{21}$$

$$\hat{y}_{t+m} = s_t + mb_t \tag{22}$$

where $s_1 = x_1$ and $b_1 = x_1 - x_0$. The ζ and ξ are smoothing factors for the data and the trend, respectively, that range from 0 to 1 and are usually set as $\zeta = 0.98$ and $\xi = 0.99$.

2.2.3. Evaluation Indicators

Two evaluation indicators, namely the coefficient of determination (R^2) and the root mean square error (RMSE), are introduced to evaluate the performance of the deep learning architecture forecasting mode. These indicators are widely used in deep learning regression tasks and are defined as follows:

$$R^2 = 1 - \frac{\sum_{t=1}^{N}(y_t - \hat{y}_t)^2}{\sum_{t=1}^{N}(y_t - \bar{y})^2} \tag{23}$$

$$RMSE = \sqrt{\frac{1}{N}\sum_{t=1}^{N}(y_t - \hat{y}_t)^2} \tag{24}$$

where y_t and $\hat{y}_t$ are the observations and the predicted value at time t, respectively, and $\bar{y}$ represents the mean of the N observed measures.

3. Multi-Scale Response Analysis with JTFA

3.1. Pre-Processing of the Collected Data Sequence

Baishuihe landslide is a typical loose mound landslide in the TGR, about 56 km west of the Three Gorges Dam in Zigui County, Hubei Province (see Figure 3a). The landslide slope is about 30° with an average thickness of about 30 m, and the volume is about 1260×10^4 m^3. The sliding surface is a contact zone between residual slope deposit and bedrock that is about 0.9–3.1 m thick. Its bedrock lithology is a medium-thick sandstone layer with a thin mudstone layer, yielding $15° \angle 36°$, with joints and fissures developed in the rock layer. The slide material consists mainly of residual Quaternary slope deposits of gravelly soil with a gravel content of 20% to 40% [35].

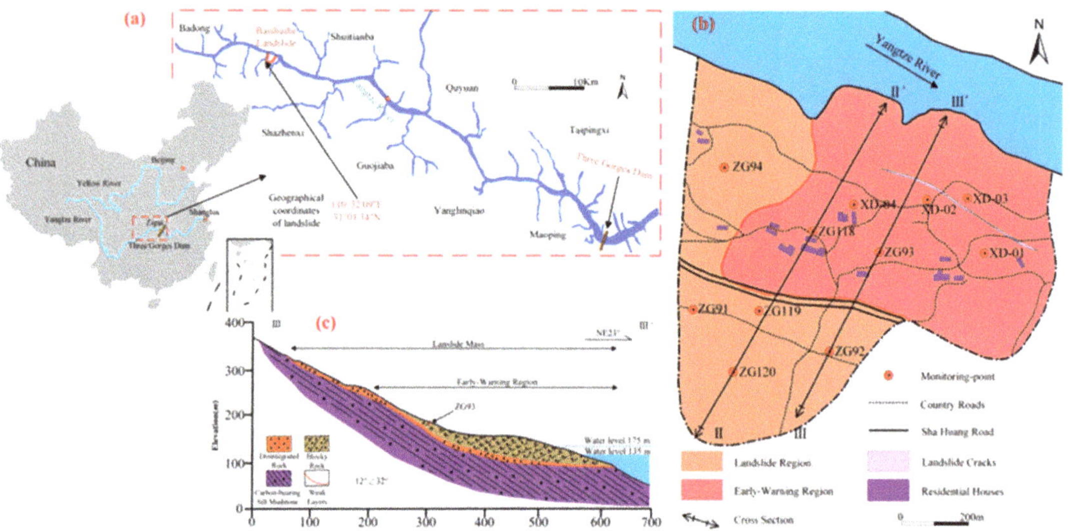

Figure 3. (**a**) Location of the landslide. (**b**) monitoring layout diagram. (**c**) geological profile III′ through point ZG93.

As shown in Figure 3b, 11 GNSS monitoring stations (with a plane accuracy of 5 ± 1 ppm) were deployed on the Baishuihe landslide; six stations, namely the ZG93, ZG118, XD-01, XD-02, XD-03 and XD-04, are located in the warning zone. The observed displacements and the corresponding rainfall and reservoir levels are illustrated in Figure 4. The TGR is in the wet season from May to September when the reservoir begins to play a regulatory role of releasing flood water in moderation per annum. At this time of year, a clear pattern of a step increase in the cumulative displacement appears, indicating a combined effect of the changing reservoir water levels and rainfall on the landslide evolution. However, monitoring stations in the non-warning area give a relatively small displacement variation with an average annual amount of around 35 mm. In the warning zone, the ZG93 station is located on profile III. As shown in Figure 3c, the mid-slip zone runs the entire slope length; the leading edge of the landslide has been submerged in water for years, making the stability of the slope within this area vulnerable to the reservoir water level.

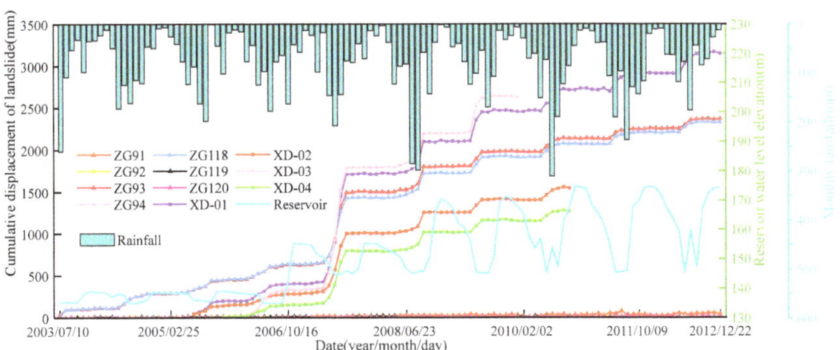

Figure 4. Monitoring data in time series of the Baishuihe landslide.

In the subsequent analysis, the site monitoring data collected by ZG93 is selected as the data source of displacement. Other synchronous on-site data include the monthly reservoir water level data and the local rainfall data in the TGR between July 2003 and December 2012. As shown in Figure 4, the time-series displacement of the landslide shows specific trends, seasonality and noise-induced fluctuations, indicating the need to obtain IFMs based on VMD. Setting $K = 3$ in VMD allow us to acquire three IMFs in an increasing order of frequency. The influence factors, e.g., the reservoir level and rainfall, mainly contribute to the periodic terms, showing cyclical fluctuations. At this point, the Shannon entropy is introduced as the fitness function to obtain the optimal IMFs by searching for the optimum ε using the GWO. The population size of the GWO is set to 20, the number of iterations to 100, and the upper-lower bounds to [0.01,100]. The optimal ε of ZG93 was determined to be 0.55; it is then utilized in the VMD to obtain the trend, the periodic and the random components of the displacement (Figure 5).

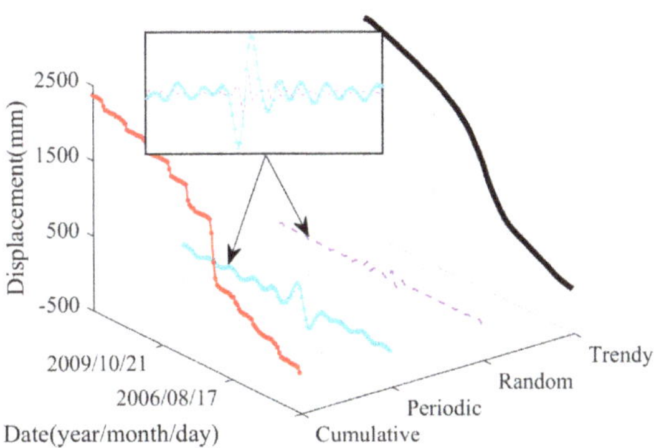

Figure 5. Decomposition result of displacement at ZG93.

3.2. Pre-Selection of the Impact Factors

Surface water infiltration is an essential factor affecting slope stability, especially concerning the matrix suction of the unsaturated zone. When the slope experiences rainfall infiltration, the bulk density increases, causing the matrix suction to decrease, which in turn raises the sliding forces and reduces the shear strength [36]. Moreover, rainfall infiltration into fractures enhances the split effect and raises the groundwater table, resulting in an increase in porewater pressure [37], thus affecting the stability of the landslide. As shown in Figure 6, the effect of rainfall on landslide displacement has an evident hysteresis. For example, the wet season in the TGR area typically begins in May, but the response of landslide deformation does not manifest until July or August. This delayed response highlights the complex interactions between rainfall, infiltration and subsequent landslide movements.

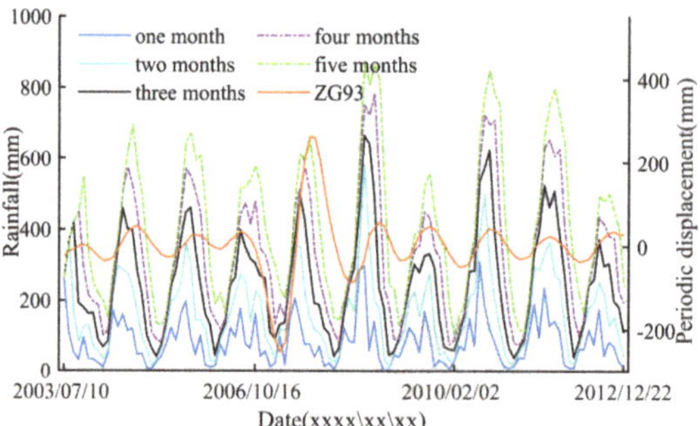

Figure 6. Rainfall effects on landslide displacement.

Studies have shown that the periodic deformation highly correlates with the cumulative rainfall of the current month and the preceding two months [2]. In this study, grey correlation analysis gives a correlation degree of 0.79 and 0.78, further confirming their close relationship with the displacement time series. Thus, the monthly rainfall V_2 and the two-month cumulative rainfall V_3 are selected to characterize the rainfall effects on landslide displacement, thereby avoiding transitional redundancy.

The periodical variation of reservoir level alters the distribution of the seepage field and the stress state of the rock mass, directly influencing the stability of the landslide. The rapid decline in reservoir water results in a higher hydraulic gradient inside and outside the slope [38]. The seepage force along the slope greatly affects the landslide's stability, especially during the late reservoir water decline when rainfall concentration occurs. The landslide deformation and failure process are further accelerated [39]. As shown in Figure 7, the changes in the periodical deformation are consistent with the reservoir level fluctuation.

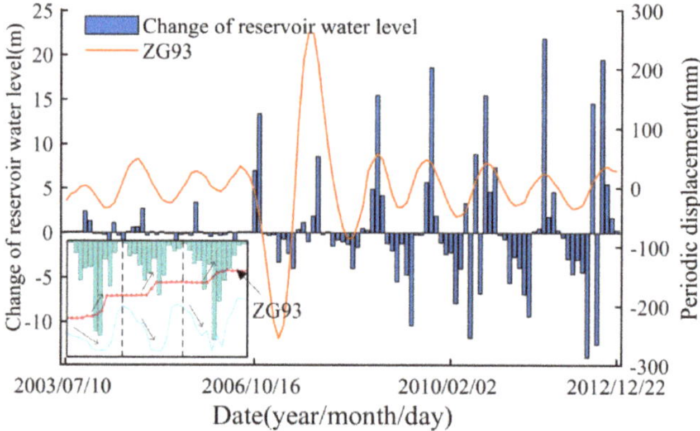

Figure 7. Relations between the periodical deformation and the reservoir level fluctuation.

To characterize the reservoir level effects on landslide displacement, the monthly mean water level V_1, the monthly variation V_4, the monthly change rate V_5 and the bimonthly variation V_6 of the reservoir water level are selected. These variables exhibit a grey correlation degree of 0.78, 0.81, 0.8 and 0.76, respectively, signifying their significant relationship with the displacement of the landslide.

During the decomposition of the influence factors, we set $K = 2$ because they mainly show cyclical fluctuations. The other VMD settings are the same as those used for the landslide displacement. The Shannon entropy of the periodic term of the factors serves as the fitness function to obtain the optimal ε. The six obtained values of ε are 0.23, 0.99, 0.13, 0.98, 0.11 and 0.95, respectively. The results are given in Figure 8, where subscript *p* indicates the periodic term and subscript *s* indicates the random term.

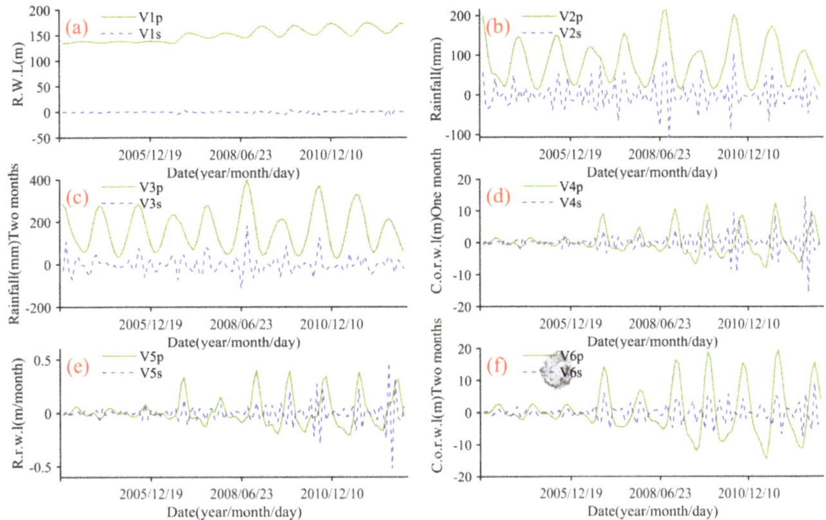

Figure 8. Decomposition results of the six influence factors (V_1–V_6), with the subscript *p* indicates the periodic term and subscript *s* indicates the random term.

3.3. Multi-Scale Response Analysis

Investigating the deformation characteristics and the response analysis at different scales is crucial to revealing the landslide mechanism. To achieve this, the JTFA is introduced to conduct a multi-scale response analysis between the displacement and external factors by three WA methods to investigate the deformation response mechanisms and reveal the changing patterns of specific signal components at different scales. The WA methods employed in this study include the continuous wavelet transform (CWT), the cross-wavelet transform (XWT) and the wavelet coherency (WTC).

3.3.1. Analysis of CWT

The CWT extends time series analysis into a time-frequency domain to intuitively map the changing properties of non-stationary signals. Through CWT, we have observed an apparent yearly periodicity (1 year cycle) in the time series data associated with the Baishuihe landslide. This periodicity is evident in the landslide displacement, rainfall intensity and reservoir levels fluctuation, all coinciding with the primary cycle.

As critical external factors influencing landslides displacement, the variation of the rainfall and reservoir water level (R.w.l) also exhibits specific patterns. With a subtropical monsoon climate, the precipitation in the TGR shows seasonal features characterized by a total rainfall exceeding 1000 mm per year. The rainfall is concentrated mainly in summer and winter [40], which aligns with the 1 year cycle (12 months) of monsoon-related patterns obtained by the CWT analysis (see Figure 9a).

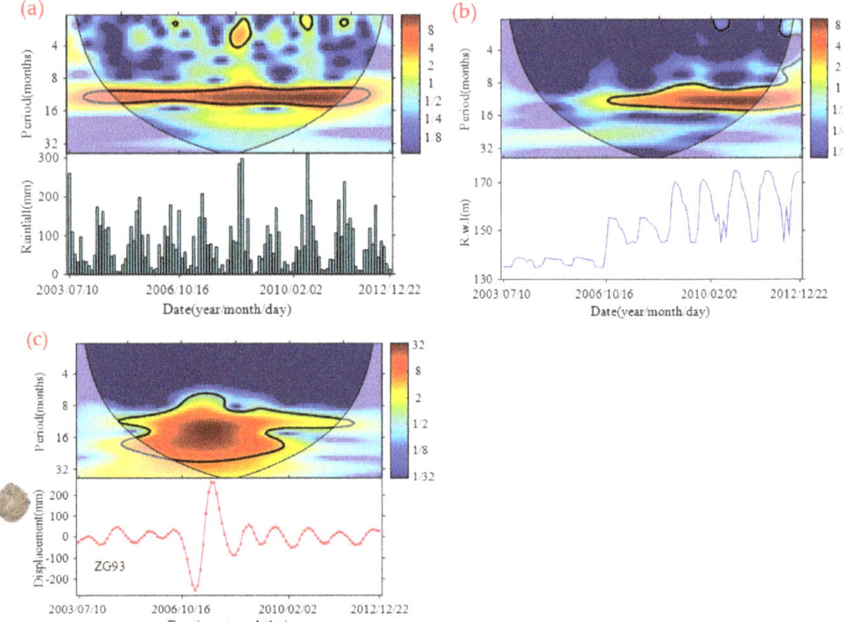

Figure 9. The CWT power spectrum of the rainfall (**a**), the reservoir water level (R.w.l) (**b**) and the displacement at ZG93 (**c**). Regions enclosed by the heavy solid black line show a greater than 95% confidence level of the red noise standard spectrum. The cross-hatched regions on either end indicate the "cone of influence", where edge effects become important. Color changes mean different magnitudes of power.

The storage of the TGR can be categorized into three stages: Stage I (June 2003–August 2006) with an average storage level of 135 m; Stage II (August 2006–August 2008) with the highest storage level raised to 156 m; Stage III starting from November 2008 with the highest storage level regulated to 175 m. Since then, the reservoir water level has fluctuated between 145 m and 175 m, exhibiting a contrary periodic fluctuation to the precipitation intensity due to artificial flood control. As shown in Figure 9b, a clear 1 year cycle (12 months) of R.w.l has been observed since 2006.

Generally, the displacement dominated by geological conditions is approximately monotonic over time, indicating the long-term trend. The displacement affected by external triggering factors (e.g., rainfall and R.w.l variation) can be expressed as a proximate periodic function, leading to different deformation features. Thus, the periodical displacement is the optimal option to illustrate the multi-scale response relationship with the impact factors.

Figure 9c gives the CWT power spectrum of the displacement at ZG93, showing a 1 year cyclic period (12 months), indicating a combined effect of changing reservoir water levels and rainfall on the landslide displacement. In addition, a 2 years cyclic period (24 months) is observed during extreme changes in the landslide displacement. This phenomenon seems related to the deformation responding to the overall rise of the R.w.l, but further research is still needed to confirm the underlying causes. Thus, cross-analysis between the displacement and the influence factors are performed via XWT and WTC later to further study the effect of rainfall and R.w.l variation on the kinematics of the Baishuihe landslide.

3.3.2. Analysis of XWT and WTC

XWT is a measure of similarity between two waveforms, showing the presence of high common power part. On the other hand, the WTC, which combines wavelet transform and coherence analysis, measures the coherence of the XWT in the time-frequency space. It reproduces the consistency and correlation of the sequence in different periods and scales through time-frequency analysis of the time series data [41]. Thus, employing XWT and WTC with dual time series will contribute to further elucidating the characteristics of landslide displacements and impact factors in terms of local coherence and detailed variation.

To explore the underlying correlations and causes in multiple time periods and scales (e.g., monthly or yearly time scales), we construct, the XWT power and the WTC coherence spectrum of the periodic displacement and the six pre-selected impact factors. The red noise standard spectrum is used to verify the reliability of the results, with a 95% confidence level obtained through the standard red noise test enclosed by the heavy solid black line. The thin solid black line enclosed area represents the cone of influence (COI) of the wavelet analysis, where edge effects are significant; thus, the response relationships will not be analyzed outside the COI.

As can be seen from Figure 10, there is a high common power area shared by the R.w.l and the rainfall time series on a time scale of 1 year (12 months) throughout the study period (2003–2012). The two time series are anti-phase relationship within the whole high common power area, with a mean phase difference of about $-178 \pm 3°$, confirming a fluctuation pattern of the R.w.l opposite to the precipitation conditions and thereby guaranteeing the safe operation of the TGR and preventing flooding disasters.

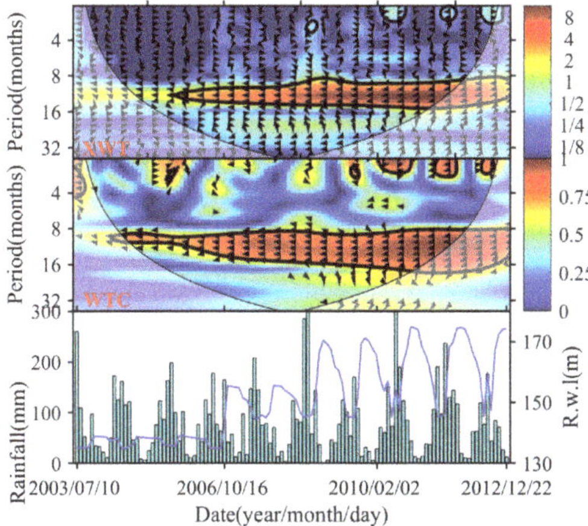

Figure 10. The XWT power and WTC coherence spectrum of R.w.l and rainfall. The arrow direction reflects the phase correlation between the two time series. The arrow points from left to right indicate the in-phase relation, and the opposite suggests the anti-phase. In XWT, the color change indicates the magnitude of power. In WTC, the color differences suggest different magnitudes of coherence.

As shown in Figure 11, the periodic displacement and the associated impact factors have a significant power area throughout the study period. Rainfall shows a natural cyclic fluctuation pattern. Figure 11a,b shows that the periodic displacement shares a continuous significant power sector with the rainfall at a time scale of 1 year (12 months) and is highly coherent. The two time series are in-phase throughout the study period, with a phase difference of about $34 \pm 12°$, indicating about a two-month lag behind the displacement than the rainfall. In addition, a 2 years cyclic period (24 months) coincides with the extreme change in the landslide displacement between 2006 and 2009, showing an anti-phase with the R.w.l. This phenomenon seems related to the deformation response to the rapid rise of the R.w.l; since it is the first time the water level R.w.l has reached its maximum, further research is still needed to confirm the underlying causes of this behavior.

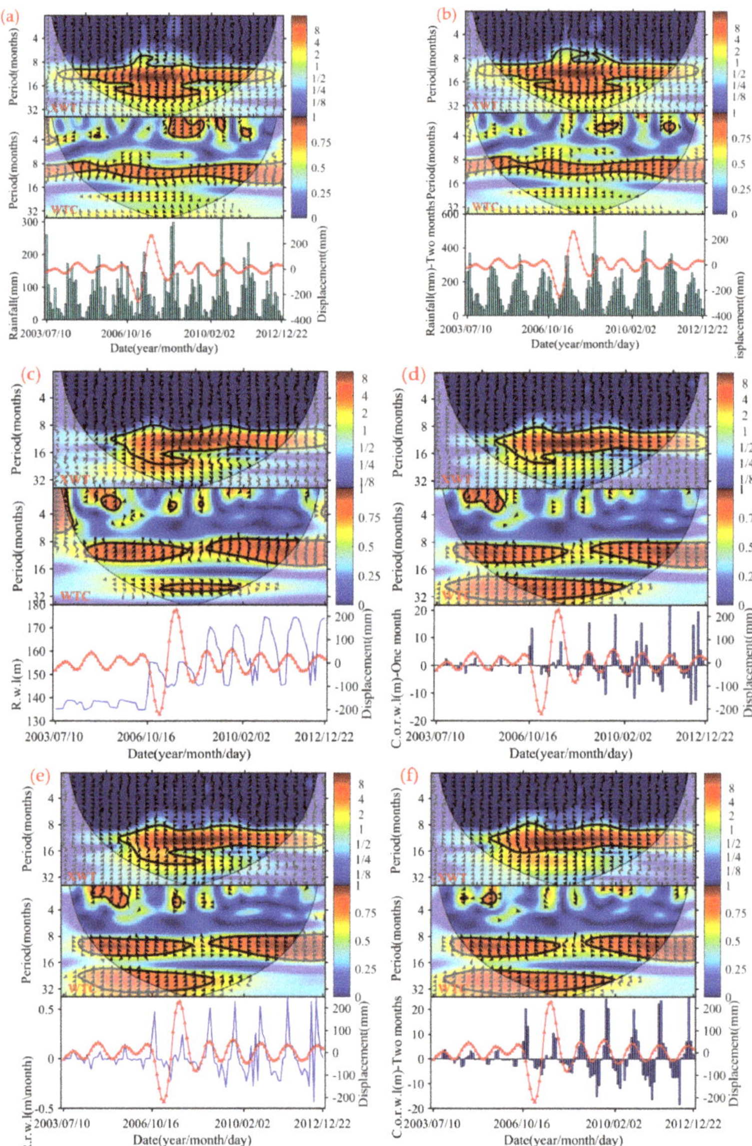

Figure 11. The XWT power and WTC coherence spectrum of the periodic term and six impact factors. (**a**,**b**) indicate rainfall-type factors V_2 and V_3. The remainder indicates reservoir-type factors, where (**c**–**f**) indicate monthly mean level V_1, monthly variation V_4, single-month rate of change V_5 and bi-monthly variation V_6.

As previously stated, the fluctuation of the R.w.l shows a cyclical pattern opposite to the precipitation conditions. As shown in Figure 11c–f, the periodic displacement and the associated four R.w.l factors also have a significant power sector with a 1 year (12 months) cyclic period. However, there is a notable difference with the monthly mean water level V_1 during the rapid impounding of the TGR (2006–2009), as it shows an inverse phase with the displacement. These four factors all contain a significant power sector with a 2 years (24 months) cyclic period, showing anti-phase relations, which may share a similar cause as that seen in the power spectrum of displacement and rainfall. This finding suggests that the rise of the reservoir level is closely related to the extraordinary change in landslide displacement, providing valuable insights into the influence of the reservoir water level on landslide behavior during specific periods.

As shown in Figure 11c, there is a phase difference of $-51 \pm 8°$ between the landslide displacement and the monthly mean level of the reservoir water, indicating the displacement change occurs ahead of the change in reservoir water level by about three months. This phenomenon is likely due to the fact that rising reservoir levels contribute to landslide stability, and the displacement increase often occurs before the reservoir level rises or after it falls. In-phase relations are observed between the displacement and the monthly variation V4, the monthly change rate V5 and the bi-monthly variation V6 of the R.w.l, with a mean phase difference of $-13 \pm 9°$, indicating a consistent variation pattern. All pre-selected impact factors show strong coherence over a 1 year cyclic period (12 months), except during the period from 2006 to 2009.

The analysis above confirms the preselected six impact factors have a strong correlation with the variation trends of landslide displacements. The periodic variation of the rainfall intensity and the reservoir storage leads to regularly varying underground water levels and the seepage field distribution, which adversely affects the Baishuihe landslide's stability. These regular variations alter the equilibrium of the slope and cause periodic changes in landslide displacements time series. The JFTA helps illustrate the periodically changing patterns and the response between the displacement and external factors at multi-scales to reveal the landslide mechanism behind the landslide behavior. By employing JTFA, we gain valuable insights into how the landslide responds to the dynamic interplay of various factors, providing a deeper understanding of the landslide's complex behavior and contributing to the advancement of landslide research and prediction.

4. Model Forecast and Discussion

4.1. Training Dataset and Parameter Setting

In the experiments, monthly GNSS-measured displacements at ZG93 were acquired from July 2003 to March 2013. Additionally, the corresponding monthly reservoir water level and precipitation data are used in the subsequent analysis and modeling. For the training process, datasets collected from July 2003 to June 2009, comprising 72 sets of data, are utilized as model inputs, while 25 sets of data collected from July 2009 to July 2010 are used as forecast datasets, and the reserved 17 sets of data from August 2011 to December 2012 are used for model validation.

Six training sets with different data volumes are obtained by dividing the dataset based on natural years (see Table 1). Using these datasets, six training models are constructed, and the optimal data volume for the prediction model is determined by comparing the model performance. Throughout this process, a sequential prediction strategy is implemented, considering the timeliness of monitoring GNSS displacement.

Table 1. The designed training dataset.

Model Number	Data Packet Time						Date Volume
	200307–200407	200407–200507	200507–200607	200607–200707	200707–200807	200807–200907	
1						√	12
2					√	√	24
3				√	√	√	36
4			√	√	√	√	48
5		√	√	√	√	√	60
6	√	√	√	√	√	√	72

In the experiment, the deep learning architecture of the forecasting model in this paper is built using the Deep Learning Toolbox of Matlab2020. We empirically set the learning rate to 0.01 and the training number to 250. However, the number of hidden neurons is a crucial factor that can affect the prediction precision, and therefore, it should be determined through carefully designed comparison experiments. The results of the seven comparison experiments are shown in Table 2. According to Table 2, the model performance is proportional to its number of neurons. As the number of neurons increases, the RMSE of the model first decreases and then increases, with the optimal RMSE achieved when the number of neurons is 100. However, the computation time keeps growing as the number of neurons increases. Consequently, the number of hidden neurons is set to 100 in the following experiments.

Table 2. The performance of the GRU using different numbers of neurons.

Number of Neurons	*RMSE* (mm)	Time (s)
20	28.581	55.295
40	19.602	56.638
60	17.988	57.699
80	12.603	60.307
100	10.175	60.649
120	10.894	61.635
140	11.131	62.281

4.2. Prediction Results and Analyses

4.2.1. Displacement Components Prediction

The DES is employed to predict the trend component of the landslide displacement by using a weighted combination of past observations with recent observations given relatively higher weights than the older ones. Thus, despite the data volume difference between the designed six training datasets, the predicted trend component displacement is almost identical. Here, the prediction of Model 6 is taken as the final trend result. The DES results and the absolute error are shown in Figure 12a, with $R^2 = 0.998$ and $RMSE = 2.741$ mm.

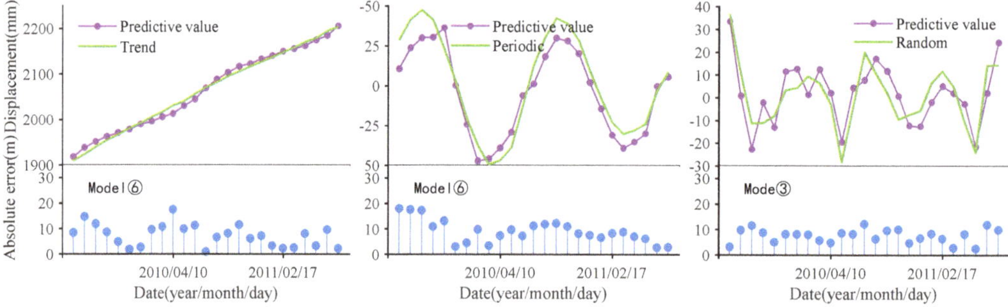

Figure 12. Predictions of each displacement component.

The GRU could adaptively capture dependencies at different time scales, especially suitable for handling non-linear problems, and is thus used to forecast landslide displacements for periodic and random terms. The performance of the GRU in predicting the periodic component of the landslide displacement is shown in Table 3. According to Table 3, the GRU performance increases as the training data volume increases; thus, the optimal result is achieved with Model 6. However, the model performance of the random component displacement first increases and then decreases; the optimal results come from Model 3.

Table 3. The performance of the GRU in predicting periodic and random displacement.

Model Number	ZG93			
	Periodic		Random	
	RMSE (mm)	R^2	*RMSE* (mm)	R^2
1	21.562	0.503	9.647	0.507
2	20.722	0.541	8.729	0.597
3	18.741	0.624	8.127	0.651
4	14.042	0.789	9.107	0.569
5	11.171	0.867	9.196	0.552
6	10.175	0.889	9.468	0.514

The above results indicate that better periodic component prediction needs more trained data; of the six designed models, the larger the amount of the training data, the better. However, it does not apply to random components; for random component forecasts, an appropriate amount of datasets works best.

4.2.2. Cumulative Displacement Prediction

The predicted cumulative displacements can be derived by combining the trend, the periodic and the random predicted components. The optimal prediction of each component is illustrated in Figure 12. According to Figure 12, the absolute forecast error of the trend component remains small (<20 mm) and does not vary much, despite the deformation being around 2 m. In contrast, the absolute forecast error of the periodic component follows a certain regularity and shows a tendency to decrease, and that of the random component shows randomness.

The evaluation indicators, as listed in Table 4, assess the performance of the six models. According to Table 4, both indicators experience an increase and then a decrease as the volume of the training dataset increases. A maximum *RMSE* of 26.043 mm and a minimum R^2 of 0.904 comes from Model 2 with a data volume of two natural years. The turning point found at Model 4 with four years of data volume shows the best results among the six models, with R^2 of 0.952 and *RMSE* of 18.582 mm.

Table 4. Predictions of the cumulative landslide displacement.

Model Number	ZG93	
	RMSE (mm)	R^2
1	24.165	0.917
2	26.043	0.904
3	20.367	0.941
4	18.582	0.952
5	20.106	0.943
6	18.742	0.951
Optimal model	12.301	0.979

According to Section 4.2.1, Model 6 is the best model for predicting the periodic component, while Model 3 gives the best performance for predicting the random part. Consequently, based on this, an optimal combination prediction model for landslide displacement is constructed. The evaluation indicators of the optimal model, shown in Table 4, outperform the other six models in terms of both evaluation indicators, with an *RMSE* of 12.301 mm and R^2 of 0.979. Compared with Model 4, the best forecast model among the six, the RMSE decreases by 6.281 mm and R^2 increases by 0.027 for the optimal model.

4.2.3. Comparative Experiments and Analyses

In Section 3.2, we conducted a pre-selection of the impact factors, and later, through the multi-scale response analysis in Section 3.3, we gained valuable insights into how the landslide responds to the dynamic interplay of the selected factors. The analysis revealed an apparent 1 year primary cycle of the time series associated with the landslide evolution. Of particular interest were the delayed response results, which highlight the complex interactions between the reservoir water level (R.w.l) and rainfall with the subsequent landslide movements. This provides a deeper understanding of the landslide's complex behavior and guides the selection of the landslide prediction model.

GRU and other recurrent neural networks (RNNs) excel at handling the non-linear dynamic characteristics present in complex time series. The GRU model, in particular, has the ability to capture dependencies at different time scales, making it well-suited for time series landslide displacement prediction. Considering the outstanding performance of GRU in time-series forecasting, we conducted comparative experiments to analyze the performance differences of the optimal model screened in Section 4.2.2, both when considering impact factors and when not considering them.

Figure 13 displays the predicted cumulative displacements and absolute errors. Notably, when the impact factors are ignored, the model exhibits more significant deviations, indicating a decline in prediction performance. The calculated $RMSE$ and R^2 are 20.218 mm and 0.942, respectively. However, when considering the impact factors, the model produces an improved performance, yielding an $RMSE$ of 12.301 mm and R^2 of 0.979. This results in a reduction in the RMSE by 7.917 mm and an increase in R2 by 0.037, respectively.

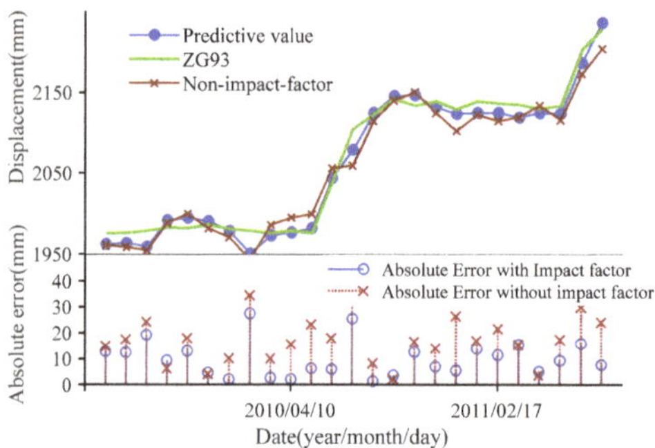

Figure 13. Predictions of the cumulative landslide displacement. The red marker indicates the absolute error without considering influencing factors, while the blue circle indicates the absolute error of our proposed method.

As a result, the impact factors selected in Section 3.2 have been proven to be indispensable for multi-scale response analysis and the prediction model training process. Despite the exceptional performance of GRU in time-series forecasting, it is evident that the influence factors of landslides displacement have a significant impact on the predictions. Hence, the analysis and selection of the impact factors during the modeling process are crucial, underscoring the comprehensiveness of the proposed model.

At this stage, the 17 reserved sets of data from August 2011 to December 2012 are utilized for model validation to further verify the model's performance. Figure 14 displays the cumulative landslide displacement prediction (17 sets) and the absolute errors of the proposed model, resulting in a calculated $RMSE$ of 9.715 mm and R^2 of 0.967. The results demonstrate that the optimal combination prediction model exhibits a reliable capacity for predicting landslide displacement.

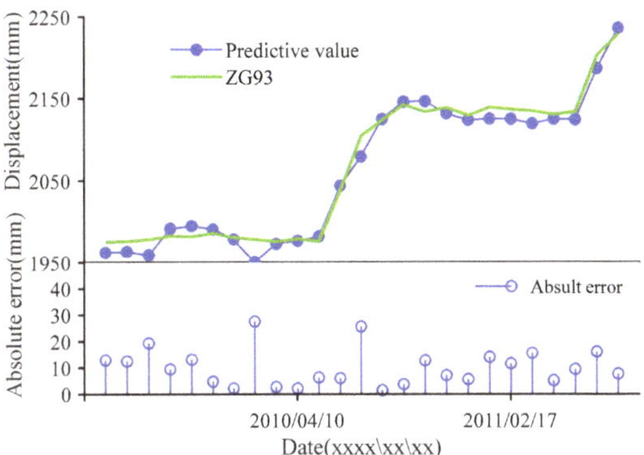

Figure 14. Predictions of the cumulative landslide displacement. The blue circles stands for the absolute errors of the proposed model.

To further illustrate the superiority of the proposed method compared to existing state-of-the-art methods, such as the GWO-MIC-SVR [42], the V/S-LSTM [43], the Chaotic DWT-ELM [44] and the Multi-Chaotic ELM [45], we conduct a comparative analysis of the forecasts generated by these models. The performance of the forecast models is shown in Table 5.

Table 5. Comparison of model prediction results.

Model Name	Forecast Duration (month)	*RMSE* (mm)
The method proposed	17	9.715
PSO-SVR	12	20.770
GWO-MIC-SVR	18	14.024
M-EEMD-ELM	15	-
V/S-LSTM	8	8.950
Chaotic DWT-ELM	15	23.330
Multi-Chaotic ELM	20	23.710

According to Table 5, the proposed model achieves a competitive high-accuracy result in terms of RMSE, ranking in the top two, with an *RMSE* of 9.715 mm, closely following the V/S-LSTM with an *RMSE* of 8.950 mm. However, what makes the proposed model stands out is its ability to makes 17 sets of consecutive forecasts, while V/S-LSTM only forecasts 8 sets.

Considering that forecast errors can accumulate over time, the proposed model's RMSE of 0.765 mm for 17 sets of consecutive forecasts demonstrates its capability for precise and reliable long-term predictions. Thus, from a comprehensive perspective, the proposed model exhibits both time-effectiveness and long-sequence forecasting advantages over the other six intelligent algorithms.

5. Conclusions

This paper presents a novel deep learning architecture specifically designed for predicting reservoir landslide displacement. The evolution of reservoir landslides involves highly complicated and nonlinear dynamics, characterized by specific time-frequency features. To address the complexities, the joint time-frequency analysis (JTFA) module is designed. This module integrates the GWO-optimized VMD and WA methods, facilitating the extraction of essential signal components with clear physical implications. Additionally, the module conducts multi-scale response analysis, thereby revealing deformation variation patterns in the underlying mechanisms governing the landslide's response behavior. For the actual displacement prediction, The GRU is integrated, which possesses the capability of adaptively capturing dependencies at multiple time scales. Moreover, during the modeling process, the correlation and hysteresis effect of the impact factors are also considered, further enhancing the accuracy and reliability of the predictions. The model experiments show that as the amount of training data increases, the periodic component prediction improves significantly. For the random component forecast, an appropriate amount of datasets yields the best results, while the trend component remains almost identical regardless of the data size. This insight led to the construction of an optimal combination prediction model, surpassing the performance of the other six designed models in cumulative landslide displacement predictions. This model achieved impressive results, with a minimum RMSE of 12.301 mm and a maximum R2 of 0.979. Moreover, the proposed architecture's superiority in time-effectiveness and its ability to accurately predict long-sequence landslide displacement have been firmly established through comparative experiments and analyses, in which we evaluated its performance against six other state-of-the-art intelligent methods. The favorable outcomes and impressive forecasting capabilities of our proposed architecture solidify its position as an efficient and accurate tool for landslide displacement prediction.

Author Contributions: Conceptualization, L.L. and Y.J.; methodology and validation, Y.J. and H.L.; formal analysis and investigation, X.Z; resources, Y.J.; writing—original draft preparation, Y.J.; writing—review and editing, Y.J. and Z.L.; visualization, H.L.; funding acquisition X.Z. and Y.J. All authors have read and agreed to the published version of the manuscript.

Funding: This research is financially supported by the National Natural Science Foundation of China (Grant No. 41977255) and the Key Research and Development Program of Sichuan Province (2023YFS0439).

Data Availability Statement: The raw data used in this study can be downloaded for scientific research after approval at http://www.crensed.ac.cn/, accessed on 14 February 2022.

Acknowledgments: The authors would like to thank the National Service Center's Specialty Environmental Observation Stations for providing the in situ monitoring dataset.

Conflicts of Interest: The authors declare no conflict of interest.

References

1. Yin, Y.; Huang, B.; Wang, W.; Wei, Y.; Ma, X.; Ma, F.; Zhao, C. Reservoir-Induced Landslides and Risk Control in Three Gorges Project on Yangtze River, China. *J. Rock Mech. Geotech. Eng.* **2016**, *8*, 577–595. [CrossRef]
2. Zhou, C.; Yin, K.; Cao, Y.; Ahmed, B. Application of Time Series Analysis and PSO–SVM Model in Predicting the Bazimen Landslide in the Three Gorges Reservoir, China. *Eng. Geol.* **2016**, *204*, 108–120. [CrossRef]
3. Tang, H.; Wasowski, J.; Juang, C.H. Geohazards in the Three Gorges Reservoir Area, China–Lessons Learned from Decades of Research. *Eng. Geol.* **2019**, *261*, 105267. [CrossRef]
4. Wang, F.; Li, T. (Eds.) *Landslide Disaster Mitigation in Three Gorges Reservoir, China*; Environmental Science and Engineering; Springer: Berlin/Heidelberg, Germany, 2009; ISBN 978-3-642-00131-4.
5. Wang, F.-W.; Zhang, Y.-M.; Huo, Z.-T.; Matsumoto, T.; Huang, B.-L. The July 14, 2003 Qianjiangping Landslide, Three Gorges Reservoir, China. *Landslides* **2004**, *1*, 157–162. [CrossRef]
6. Schulte, J.A. Wavelet Analysis for Non-Stationary, Nonlinear Time Series. *Nonlin. Process. Geophys.* **2016**, *23*, 257–267. [CrossRef]
7. Rhif, M.; Ben Abbes, A.; Farah, I.R.; Martínez, B.; Sang, Y. Wavelet Transform Application for/in Non-Stationary Time-Series Analysis: A Review. *Appl. Sci.* **2019**, *9*, 1345. [CrossRef]
8. Huang, F.; Huang, J.; Jiang, S.; Zhou, C. Landslide Displacement Prediction Based on Multivariate Chaotic Model and Extreme Learning Machine. *Eng. Geol.* **2017**, *218*, 173–186. [CrossRef]
9. Jiang, Y.; Xu, Q.; Lu, Z.; Luo, H.; Liao, L.; Dong, X. Modelling and Predicting Landslide Displacements and Uncertainties by Multiple Machine-Learning Algorithms: Application to Baishuihe Landslide in Three Gorges Reservoir, China. *Geomat. Nat. Hazards Risk* **2021**, *12*, 741–762. [CrossRef]
10. Zhang, W.; Li, H.; Tang, L.; Gu, X.; Wang, L.; Wang, L. Displacement Prediction of Jiuxianping Landslide Using Gated Recurrent Unit (GRU) Networks. *Acta Geotech.* **2022**, *17*, 1367–1382. [CrossRef]
11. He, B.; Chang, J.; Wang, Y.; Wang, Y.; Zhou, S.; Chen, C. Spatio-Temporal Evolution and Non-Stationary Characteristics of Meteorological Drought in Inland Arid Areas. *Ecol. Indic.* **2021**, *126*, 107644. [CrossRef]
12. Su, L.; Miao, C.; Duan, Q.; Lei, X.; Li, H. Multiple-Wavelet Coherence of World's Large Rivers with Meteorological Factors and Ocean Signals. *J. Geophys. Res. Atmos.* **2019**, *124*, 4932–4954. [CrossRef]

13. Huang, N.E.; Shen, Z.; Long, S.R.; Wu, M.C.; Shih, H.H.; Zheng, Q.; Yen, N.-C.; Tung, C.C.; Liu, H.H. The Empirical Mode Decomposition and the Hilbert Spectrum for Nonlinear and Non-Stationary Time Series Analysis. *Proc. R. Soc. Lond. A* **1998**, *454*, 903–995. [CrossRef]

14. Dragomiretskiy, K.; Zosso, D. Variational Mode Decomposition. *IEEE Trans. Signal Process.* **2014**, *62*, 531–544. [CrossRef]

15. Isham, M.F.; Leong, M.S.; Lim, M.H.; Ahmad, Z.A. Variational Mode Decomposition: Mode Determination Method for Rotating Machinery Diagnosis. *J. Vibroengineering* **2018**, *20*, 2604–2621. [CrossRef]

16. Haghshenas, S.S.; Haghshenas, S.S.; Geem, Z.W.; Kim, T.-H.; Mikaeil, R.; Pugliese, L.; Troncone, A. Application of Harmony Search Algorithm to Slope Stability Analysis. *Land* **2021**, *10*, 1250. [CrossRef]

17. Mirjalili, S.; Mirjalili, S.M.; Lewis, A. Grey Wolf Optimizer. *Adv. Eng. Softw.* **2014**, *69*, 46–61. [CrossRef]

18. Fang, K.; Tang, H.; Li, C.; Su, X.; An, P.; Sun, S. Centrifuge Modelling of Landslides and Landslide Hazard Mitigation: A Review. *Geosci. Front.* **2023**, *14*, 101493. [CrossRef]

19. Olsen, J.; Anderson, N.J.; Knudsen, M.F. Variability of the North Atlantic Oscillation over the Past 5200 Years. *Nat. Geosci* **2012**, *5*, 808–812. [CrossRef]

20. Gan, T.Y.; Gobena, A.K.; Wang, Q. Precipitation of Southwestern Canada: Wavelet, Scaling, Multifractal Analysis, and Teleconnection to Climate Anomalies. *J. Geophys. Res. Atmos.* **2007**, *112*, D10110. [CrossRef]

21. Torrence, C.; Webster, P.J. Interdecadal Changes in the ENSO–Monsoon System. *J. Clim.* **1999**, *12*, 2679–2690. [CrossRef]

22. Torrence, C.; Compo, G.P. A Practical Guide to Wavelet Analysis. *Bull. Amer. Meteor. Soc.* **1998**, *79*, 61–78. [CrossRef]

23. Grinsted, A.; Moore, J.C.; Jevrejeva, S. Application of the Cross Wavelet Transform and Wavelet Coherence to Geophysical Time Series. *Nonlin. Process. Geophys.* **2004**, *11*, 561–566. [CrossRef]

24. Tomás, R.; Li, Z.; Lopez-Sanchez, J.M.; Liu, P.; Singleton, A. Using Wavelet Tools to Analyse Seasonal Variations from InSAR Time-Series Data: A Case Study of the Huangtupo Landslide. *Landslides* **2016**, *13*, 437–450. [CrossRef]

25. Hu, X.; Wu, S.; Zhang, G.; Zheng, W.; Liu, C.; He, C.; Liu, Z.; Guo, X.; Zhang, H. Landslide Displacement Prediction Using Kinematics-Based Random Forests Method: A Case Study in Jinping Reservoir Area, China. *Eng. Geol.* **2021**, *283*, 105975. [CrossRef]

26. Fang, K.; Dong, A.; Tang, H.; An, P.; Zhang, B.; Miao, M.; Ding, B.; Hu, X. Comprehensive Assessment of the Performance of a Multismartphone Measurement System for Landslide Model Test. *Landslides* **2023**, *20*, 845–864. [CrossRef]

27. Jiang, Y.; Luo, H.; Xu, Q.; Lu, Z.; Liao, L.; Li, H.; Hao, L. A Graph Convolutional Incorporating GRU Network for Landslide Displacement Forecasting Based on Spatiotemporal Analysis of GNSS Observations. *Remote Sens.* **2022**, *14*, 1016. [CrossRef]

28. Zhang, X.; Zhu, C.; He, M.; Dong, M.; Zhang, G.; Zhang, F. Failure Mechanism and Long Short-Term Memory Neural Network Model for Landslide Risk Prediction. *Remote Sens.* **2022**, *14*, 166. [CrossRef]

29. Chung, J.; Gulcehre, C.; Cho, K.; Bengio, Y. Empirical Evaluation of Gated Recurrent Neural Networks on Sequence Modeling. *arXiv* **2014**, arXiv:1412.3555.

30. Wang, B.; Wang, J. Energy Futures and Spots Prices Forecasting by Hybrid SW-GRU with EMD and Error Evaluation. *Energy Econ.* **2020**, *90*, 104827. [CrossRef]

31. Gao, S.; Huang, Y.; Zhang, S.; Han, J.; Wang, G.; Zhang, M.; Lin, Q. Short-Term Runoff Prediction with GRU and LSTM Networks without Requiring Time Step Optimization during Sample Generation. *J. Hydrol.* **2020**, *589*, 125188. [CrossRef]

32. Shannon, C.E. A Mathematical Theory of Communication. *Mob. Comput. Commun. Rev.* **1997**, *5*, 53.

33. Kingma, D.P.; Ba, J. Adam: A Method for Stochastic Optimization. *arXiv* **2017**, arXiv:1412.6980.

34. Zhu, X.; Xu, Q.; Tang, M.; Li, H.; Liu, F. A Hybrid Machine Learning and Computing Model for Forecasting Displacement of Multifactor-Induced Landslides. *Neural Comput. Appl.* **2018**, *30*, 3825–3835. [CrossRef]

35. Lian, C.; Zeng, Z.; Yao, W.; Tang, H. Multiple Neural Networks Switched Prediction for Landslide Displacement. *Eng. Geol.* **2015**, *186*, 91–99. [CrossRef]

36. Li, X.; Zhang, N.X.; Liao, Q.L.; Wu, J. Analysis on Hydrodynamic Field Influenced by Combination of Rainfall and Reservoir Level Fluctuation. *Chin. J. Rock Mech. Eng.* **2004**, *23*, 3714–3720.

37. Yang, B.; Yin, K.; Lacasse, S.; Liu, Z. Time Series Analysis and Long Short-Term Memory Neural Network to Predict Landslide Displacement. *Landslides* **2019**, *16*, 677–694. [CrossRef]

38. Wu, Q.; Tang, H.M.; Wang, L.Q.; Lin, Z.H. Analytic Solutions for Phreatic Line in Reservoir Slope with Inclined Impervious Bed under Rainfall and Reservoir Water Level Fluctuation. *Rock Soil Mech.* **2009**, *30*, 3025–3031.

39. Guo, Z.; Chen, L.; Gui, L.; Du, J.; Yin, K.; Do, H.M. Landslide Displacement Prediction Based on Variational Mode Decomposition and WA-GWO-BP Model. *Landslides* **2020**, *17*, 567–583. [CrossRef]

40. Fang, Z.; Hang, D.; Xinyi, Z. Rainfall Regime in Three Gorges Area in China and the Control Factors: Rainfall regime in three gorges area in china. *Int. J. Climatol.* **2010**, *30*, 1396–1406. [CrossRef]

41. Yao, Y.; Luo, Y.; Zhang, J.; Zhao, C. Correlation Analysis between Haze and GNSS Tropospheric Delay Based on Coherent Wavelet. *Geomat. Inf. Sci. Wuhan Univ.* **2018**, *43*, 2131–2138.

42. Li, L.; Wu, Y.; Miao, F.; Liao, K.; Zhang, L. Displacement Prediction of Landslides Based on Variational Mode Decomposition and GWO-MIC-SVR Model. *Chin. J. Rock Mech. Eng.* **2018**, *37*, 1395–1406.

43. Feng, F.; Wu, X.; Niu, R.; Xu, S. A Landslide Deformation Analysis Method Using V/S and LSTM. *Landslide Deform. Anal. Method.* **2019**, *44*, 784–790.

44. Huang, F.; Yin, K.; Zhang, G.; Gui, L.; Yang, B.; Liu, L. Landslide Displacement Prediction Using Discrete Wavelet Transform and Extreme Learning Machine Based on Chaos Theory. *Env. Earth Sci.* **2016**, *75*, 1376. [CrossRef]
45. Huang, F.; Yin, K.; Yang, B.; Li, X.; Liu, L.; Fu, X.; Li, X. Step-like Displacement Prediction of Landslide Based on Time Series Decomposition and Multivariate Chaotic Model. *Earth Sci.* **2018**, *43*, 887–898.

Article

Quantifying Water Impoundment-Driven Air Temperature Changes in the Dammed Jinsha River, Southwest China

Xinzhe Li [1,2], Jia Zhou [1,2], Yangbin Huang [3], Ruyun Wang [2,4] and Tao Lu [1,*]

[1]   Chengdu Institute of Biology, Chinese Academy of Sciences, Chengdu 610041, China; lixz1@cib.ac.cn (X.L.); zhoujia@cib.ac.cn (J.Z.)

[2]   University of Chinese Academy of Sciences, Beijing 100049, China; wangruyun@iie.ac.cn

[3]   State Key Laboratory of Hydro-Science and Engineering, Department of Hydraulic Engineering, Tsinghua University, Beijing 100084, China; hyb22@mails.tsinghua.edu.cn

[4]   Institute of Information Engineering, Chinese Academy of Sciences, Beijing 100195, China

*   Correspondence: lutao@cib.ac.cn

Abstract: A number of previous studies have contributed to a better understanding of the thermal impacts of dam-related reservoirs on stream temperature, but very few studies have focused on air temperature, especially at the catchment scale. In addition, due to the lack of quantitative analysis, the identification of the effects of water impoundment on regional air temperature is still lacking. We investigated the impacts of reservoirs on the regional air temperature changes before and after two large dam constructions in the lower Jinsha River located in southwest China, by using a 40 year record of reanalysis data at 90 m resolutions. Furthermore, the long short-term memory (LSTM) model was also employed to construct an impoundment effect on the temperature (*IET*) index. Research results indicate that compared to the pre-impoundment period (1980–2012), the variations in the air temperature at the catchment scale were reduced during the post-impoundment period (2013–2019). The annual maximum air temperature decreased by 0.4 °C relative to the natural regimes. In contrast, the cumulative effects of dam-related reservoirs increased the annual mean and minimum air temperature by 0.1 °C and 1.0 °C, respectively. Warming effects prevailed during the dry season and in the regions with high elevations, while cooling effects dominated within a 4 km buffer of the reservoirs. Therefore, this study offers important insights about the impacts of anthropogenic impoundments on air temperature changes, which could be useful for policymakers to have a more informed and profound understanding of local climate changes in dammed areas.

Keywords: dams and reservoirs; climate change; ANUSPLIN; LSTM; trend analysis

Citation: Li, X.; Zhou, J.; Huang, Y.; Wang, R.; Lu, T. Quantifying Water Impoundment-Driven Air Temperature Changes in the Dammed Jinsha River, Southwest China. *Remote Sens.* **2023**, *15*, 4280. https://doi.org/10.3390/rs15174280

Academic Editor: Prasad S. Thenkabail

Received: 16 July 2023
Revised: 23 August 2023
Accepted: 29 August 2023
Published: 31 August 2023

1. Introduction

Anthropogenic impoundments (e.g., dams, reservoirs, and ponds) are often considered important contributors to the development of human societies by providing multiple services including hydropower generation, flood control, and water supply [1]. Since the 1950s, anthropogenic impoundments have been flourishing as a response to the ever-growing requirements for water and energy [2]. However, despite their societal significance, anthropogenic impoundments are also associated with several serious environmental problems, including river system fragmentation [3], water diversion [4], changes in flow conditions [5], decreased nutrient deliveries [6], interruptions of fish migrations [7], reservoir flushing [8], alterations in bio-geochemical cycling [9], and disturbing climate regimes [10]. Among them, the thermal variations associated with the dammed reservoirs are of considerable interest. This is because the dams and reservoirs substantially alter the hydrological and thermal regimes of the river, as well as the surface energy balance of the region [11].

The dammed hydropower, as one of the typical anthropogenic impoundments, generates power by using a dam that changes the natural flow of a river. It provides multiple benefits and values to economic prosperity and social well-being by acting as a primary and low-cost renewable source of energy worldwide [12]. Approximately two-thirds of the rivers in the world have dammed reservoirs [13]. It is physically reasonable to expect a gradual air and water temperature variation in the impounded river basin. This is directly attributed to the changing atmospheric conditions due to heat transfers, which, in turn, occur because of the surface and internal thermodynamic phenomena of reservoirs [10]. As a result, numerous studies have been carried out to analyze the thermal responses caused by the generation of dammed hydropower. Nevertheless, these previous studies usually focused on water temperature variations associated with dams and reservoirs, including the reduction in magnitude, frequency, duration, and variability of water temperature, and weakening interactions at the air/water interface due to the altered heat exchange rate [14,15]. Although reservoirs exhibit an enormous capacity to regulate air temperature regimes, the relationship between water impoundment and air temperature remains poorly understood [10,16,17].

Air temperature is a sensitive indicator of the surface energy balance, as well as a crucial parameter in biogeochemical and biophysical feedbacks to the climate system [18]. Several investigations have been carried out to realize the impact of dams and associated reservoirs on air temperature. For instance, Zeng et al. [19] found that the annual average air temperature increased in the Three Gorges Dam reservoir area. Another study performed by Fonseca and Santos [20] on the Sabor River demonstrated that the daily and seasonal temperature decreased in the post-impoundment period. Miller et al. [21] modeled the climate changes for eight weeks in the Three Gorges Dam area and suggested that the decreased local surface and air temperature likely occurred because the reservoir acted as a potential evaporating surface. Irambona et al. [22] found that reservoirs induced localized 2 m temperature warming in winter and cooling in the summer in the La Grande River watershed. Wang et al. [18] revealed that the reservoir had both warming and cooling effects on the surface temperature in different parts of the dry–hot valley. Zhao et al. [23] explored the potential connection between reservoir characteristic factors and meteorological variables at the global scale, and suggested that they possibly have contrasting effects on the air temperature. However, compared to the well-informed water temperature variations, a significant delay in the knowledge of air temperature changes is noticeable. Hence, there is still a major research gap in the regional-scale understanding of dam-induced air temperature alterations, and it is critically required to determine how artificial reservoirs influence the air temperature at the catchment scale [17].

The lower Jinsha River, as the upper reaches of the Yangtze River, is one of the largest hydropower production regions in China [24]. It has a hydropower generation capacity of 40 million kW, which is twice that of the Three Gorges Reservoir [25]. As a consequence, the lower Jinsha River has become one of the most highly modified rivers in the world [18]. The dams and related reservoirs in the lower Jinsha River may change the local air temperature patterns and further affect the thermal regime of the middle and lower Yangtze River. Moreover, these cascade reservoirs in the lower Jinsha River are more pronounced than those of an individual dam [26], and, thus, they will further change the annual or seasonal thermal regimes of the river basin. By mostly considering the Gezhouba Dam and the Three Gorges Dam in the middle river, previous investigations have examined the effects of dam-related reservoirs on the thermal regime in the Yangtze River, however, the effects of water impoundment on the local air temperature in the lower Jinsha River remain largely unclear [18]. Therefore, if water impoundments inadvertently modify climate patterns in impounded river basins, understanding the influence exerted by reservoirs in the lower Jinsha River will be key.

Over the past two decades, a long short-term memory (LSTM) model has been successfully developed and extensively employed for temperature forecasting [27]. As a deep learning model, LSTM is composed of multiple simple nonlinear models, and the original input data are learned layer-by-layer to establish complex equations between variables [28,29]. Thus, to quantify the impact of water impoundment on air temperature, LSTM can be utilized to predict future temperature under the no impoundment scenario through historical data, and the difference between actual and predicted data can be used to define the impacts.

Hence, in this study, we tried to find an answer to the following open question: What is the probable influence of large dams and related reservoirs on local air temperature changes? The novelty and primary contribution of this study were to highlight complexities in the response of the local air temperature to water impoundments. This has not been previously fully investigated in the literature, and only site-specific analyses have already paid attention to it. Specifically, the objectives of this study were as follows: (a) explaining spatiotemporal patterns of local air temperature in pre- and post-impoundment periods; (b) developing a predictive model based on the LSTM method and reconstructing air temperatures in the absence of impoundments; (c) assessing the influence of dams and related reservoirs on local air temperatures by comparing observed and predicted values in the post-impoundment period. The findings of this study can be beneficial to develop a more general understanding of the impacts of anthropogenic impoundments on air temperature variations, because global change changes will likely exacerbate them.

2. Materials and Methods

2.1. Study Area

Four large hydropower stations including Xiangjiaba, Xiluodu, Baihetan, and Wudongde have been built in the lower Jinsha River. They are all ranked among the top five largest hydropower stations in China, with a total installed capacity of 42.96 gigawatt [30]. Specifically, the Xiangjiaba, Xiluodu, Baihetan, and Wudongde reservoirs started the impoundment in December 2012, May 2013, April 2021, and January 2020, respectively [31]. Consequently, by considering the late impoundment time of the Baihetan and Wudongde hydropower stations, we only focused on the thermal disturbance of the Xiangjiaba and Xiluodu reservoirs. The water storage capacity is 5.16 billion m^3 for Xiangjiaba Reservoir and 12.67 billion m^3 for Xiluodu Reservoir.

The river basin (27.37°N–28.90°N, 102.83°E–104.42°E) where Xiangjiaba and Xiluodu hydropower stations are located was selected as a case study area, spanning two administrative provinces of Sichuan and Yunnan, and is approximately 11,167.70 km^2 in size (Figure 1). Elevations vary greatly ranging from around 240 m in the northeastern region to more than 4000 m in the southwestern region, showing complicated topographic structures. The study area has a subtropical climate with the dry season and the wet season [32]. The dry season runs from November to April, and the wet season stretches from May to October. Due to the great changes in terrain, elevations, and latitudes, the climatic conditions in the catchment are complex. Its mean annual precipitation is 732–1028 mm and its mean temperature is 13–18 °C, presenting a decreasing and increasing trend from northeast to southwest of the basin, respectively [33]. The study area is characterized by two distinct climate areas and vegetation, ranging from the monsoon region with subtropical evergreen broad-leaved vegetation in the northeastern zone, to the dry–hot valley with Savanna-like vegetation in the southwestern zone [34].

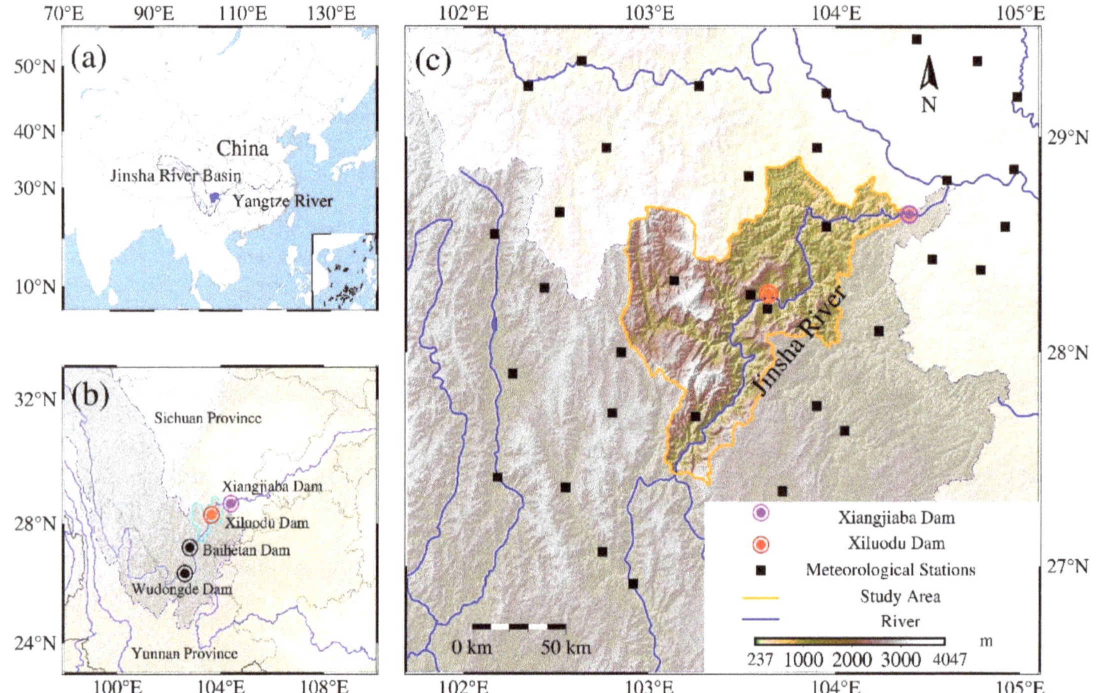

Figure 1. Location of the study area showing (**a**) the location of Jinsha River Basin in China, (**b**) the location of dams including Xiangjiaba and Xiluodu, and (**c**) the elevation of the study area and the distribution of the meteorological stations.

2.2. Data Sources

2.2.1. Meteorological Data

In our study region within the 100 km buffer zone [35], daily mean (Tmean), maximum (Tmax), and minimum (Tmin) air temperatures of 47 meteorological stations from January 1980 to December 2019 were collected from the China Meteorological Data Service Centre (http://data.cma.cn, accessed on 8 June 2022). After excluding four stations with discontinuous time series for one or more calendar months, we reserved 43 meteorological stations for analysis as indicated by the black rectangles in Figure 1.

To ensure the reliability and continuity of the air temperature data, the Pauta criterion method was employed to detect outliers, with 37 (0.24%), 335 (2.18%), and 10 (0.07%) outliers detected for Tmean, Tmax, and Tmin, respectively. The linear interpolation method was further utilized to replace them and the original missing values [36], and the results can be seen in Figure 2. The annual and seasonal air temperatures were acquired by averaging the corresponding month's temperatures. Hence, four air temperature datasets at monthly, annual, dry season, and wet season scales were generated. It is pertinent to note that the statistical analysis or mapping in this study was all implemented using Python 3.9 software.

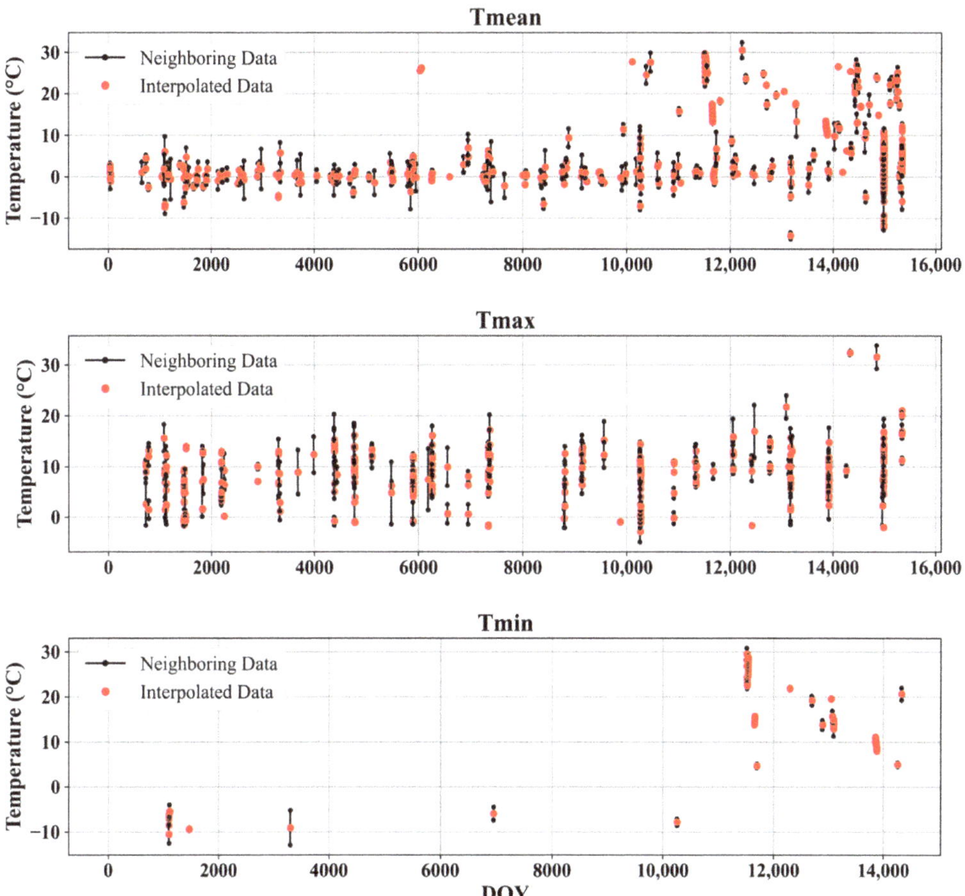

Figure 2. Results of linear interpolation for Tmean, Tmax, and Tmin. Only the interpolated data (red dots) and their two neighboring data (black dots) are displayed to ensure visualization clarity. DOY indicates the day of year.

2.2.2. Terrain Morphology Data

The digital elevation model (DEM) at the spatial resolution of 90 m was used to represent the terrain of the study area, which was obtained from the NASADEM digital elevation dataset via the Google Earth Engine (GEE) platform [37].

2.3. Methods

The methodological framework of the whole research is depicted in Figure 3a. The specific methods employed are described as follows.

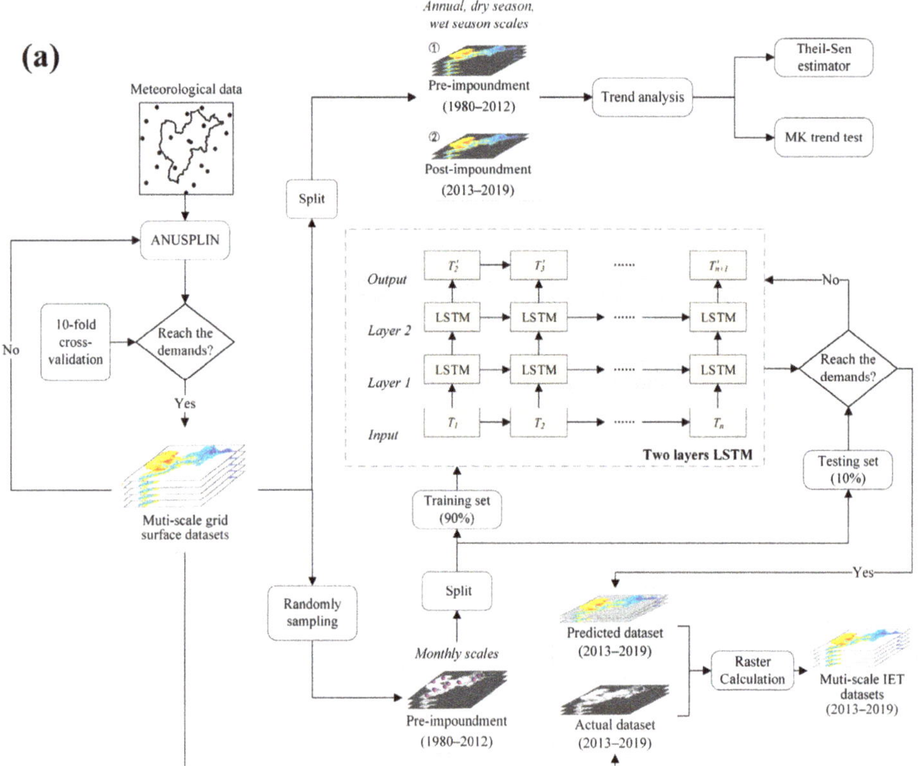

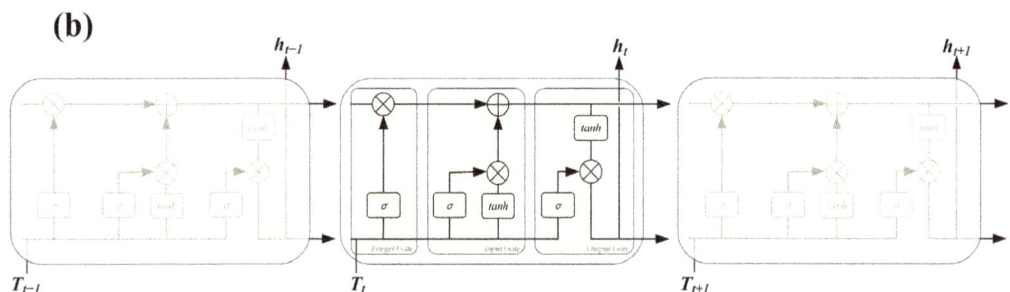

Figure 3. (**a**) Methodological framework applied in the present analysis, and (**b**) the detailed structure of the LSTM model. T_t and h_t are the input temperature data to the LSTM, the hidden state of the current time step, respectively. T_{t-1} and h_{t-1} correspond to the previous time step, while T_{t+1} and h_{t+1} are applied for the next time step, respectively. σ and tanh denote the sigmoid function and hyperbolic tangent function.

2.3.1. Interpolation Methods

To generate the air temperature grid surface datasets at annual, dry season, and wet season scales based on the meteorological data, the Australian National University Spline (ANUSPLIN) model was adopted. The ANUSPLIN model provides a means for the interpolation of noisy multi-variate data using thin-plate smoothing splines. It has been widely employed to develop geographically continuous climate grid surface data from weather station data and topographic variables that primarily include longitude, latitude, and elevation. The ANUSPLIN calculation formula was described by Hutchinson [38] as follows:

$$Z_i = f(x_i) + b^T y_i + \varepsilon_i (i = 1, \ldots, n) \tag{1}$$

where x_i represents a p-dimensional vector of spline independent variables, f denotes a smoothing function of the x_i, each y_i represents a p-dimensional vector of independent covariates, b stands for an unknown p-dimensional vector of coefficients of the y_i, and each ε_i represents an independent, zero mean error term.

2.3.2. Trends Analysis

The Theil–Sen estimator and Mann–Kendall (MK) trend test were integrated to identify the air temperature trend variations on the pixel scale during both the pre-impoundment period (1980–2012) and the post-impoundment period (2013–2019). The combined results were classified into the following four distinct categories: non-significant increase (Ni), significant increase (Si), non-significant decrease (Nd), and significant decrease (Sd). The Theil–Sen estimator is a robust nonparametric statistics method for estimating the slope of a trend [39]. Its calculation formula is as follows:

$$S = Median\left(\frac{T_j - T_i}{j - i}\right), i < j \tag{2}$$

where S signifies the Theil–Sen median slope, and T_i and T_j symbolize the air temperature at time i and j of each pixel, respectively. When $S > 0$, it indicates an increasing trend, and when $S < 0$, it indicates a decreasing trend.

The MK trend test is a nonparametric test method for quantifying the significance of a trend. The Z statistic is usually calculated initially. By checking within the standard normal distribution table, the p-value can be obtained according to the Z statistic. The p-value is utilized as a likelihood indicator to measure the significance of a trend. The smaller the p-value, the greater the likelihood for the temperature series to have a significant trend. We set the significance level α to 5%, and when $p < 0.05$, the trend was considered significant; otherwise, it was considered not significant. Specific details on the calculation procedure can be found in Liu and Menzel [40].

Additionally, to detect and identify the time when significant changes in air temperature occurred in the time series of 1980–2019, the MK mutation test was also adopted. UF_K was used to test the mutation of trends in the air temperature series, while UB_K was calculated with the reverse series of UF_K. The UF_K and UB_K curves were plotted under annual, dry season, and wet season scales. The mutation point of the temperature series can be identified as the point where the curves of UF_K and UB_K intersect within the confidence interval ($UF_{\partial=0.05/2} = 1.96$). Further detailed information regarding the MK mutation test can be found in Hong and Zhang [41].

2.3.3. Quantitative Analysis of Impoundment Effects

We developed an impoundment effect on the temperature (IET) index to further quantify the impact of water impoundments on regional air temperature. We defined the IET index as the difference value between actual air temperature and predicted air temperature. The actual air temperature was simulated using the ANUSPLIN model in the post-impoundment period (2013–2019). Predicted air temperature in the post-impact period was simulated by the LSTM model, by using the actual air temperature data during the pre-impoundment period (1980–2012). Thus, the effects of water impoundment on air temperature were simulated as follows:

$$IET = ATV - PTV \tag{3}$$

where ATV stands for the actual temperature value (°C), and PTV represents the predicted temperature value (°C). When $IET > 0$, the water impoundment has a warming effect on air temperatures, whereas when $IET < 0$, the impoundment has a cooling effect.

LSTM model can capture the nonlinear features of long time-series data and can avoid the gradient vanishing and exploding issues encountered in the conventional recurrent neural network (RNN) [42]. It consists of an input layer, an output layer, and a hidden layer. The input data are sequentially processed through these three layers to acquire pertinent information and eliminate extraneous information, thereby achieving an accurate prediction. The hidden layer is composed of four units: the forget gate, the input gate, and the output gate, as well as the memory state unit. The structure of the hidden layer is critical for filtering the data and can be composed of multiple layers (Figure 3b). For more detailed information regarding the LSTM model, please refer to Hochreiter and Schmidhuber [43].

To enhance the accuracy of our prediction, the LSTM model was designed with two layers, in which the output of the first LSTM layer acted as the input for the second LSTM layer [44]. The input dataset of the LSTM model aggregated 10,000 temperature series samples, which were randomly extracted from all pixels of the historical actual grid surface dataset (384 months, from 1980 to 2012). It was then partitioned into a training set (90%) and a testing set (10%) based on the sample locations, which were employed to train the model and assess the model performance, respectively. After training and testing, the predicted air temperature dataset (2013–2019) was generated via the constructed LSTM model by inputting the historical dataset (1980–2012). It should be noted that the predicted grid surface datasets of Tmean, Tmax, and Tmin were generated by the three LSTM models.

2.3.4. Model Assessment

We employed the 10-fold cross-validation method to evaluate the effectiveness of the ANUSPLIN model, which has been broadly employed to evaluate the model performance [45]. The original observation data at 43 meteorological stations were randomly divided into ten subsets. We further used nine of those ten subsets as the training datasets, which were inputted into ANUSPLIN models, and reserved one-tenth as the observed datasets to test the models. The simulated datasets were extracted through the training outputs at the corresponding locations of the observed datasets. These processes were iterated ten times. Hence, ten different combinations of stimulated and observed datasets were generated, and the final performance of the ANUSPLIN model was determined by the average discrepancy between each pair.

Furthermore, we calculated the mean absolute error (MAE), root mean square error ($RMSE$), and coefficient of determination (R^2) to evaluate the performance of the ANUSPLIN model and LSTM model [46]. The calculation formulas were as follows:

$$RMSE = \sqrt{\frac{1}{n}\sum\nolimits_{i=1}^{n}(T_s - T_o)} \tag{4}$$

$$MAE = \frac{1}{n}\sum\nolimits_{i=1}^{n}|T_s - T_o| \tag{5}$$

$$R^2 = \frac{\sum_{i=1}^{n}\left(T_s - \overline{T_o}\right)^2}{\sum_{i=1}^{n}\left(T_o - \overline{T_o}\right)^2} \tag{6}$$

where n represents the number of samples, T_s denotes the simulated data obtained from the model, and T_o stands for the observed data from the testing sets. The distribution range of $RMSE$ and MAE values is [0, +∞), while the range of R^2 values is [0, 1]. A higher level of precision is deduced in the evaluated model when the $RMSE$ and MAE values approach 0 and the R^2 value approaches 1.

3. Results

3.1. Evaluation of the ANUSPLIN Model Performance

Figure 4 shows the MAE, $RMSE$, and R^2 values of each air temperature indicator at annual and seasonal scales. For the interpolated air temperature grid datasets at the annual scale, the mean $RMSE$ was 0.45 °C (the values ranged from 0.31 °C to 0.60 °C), the mean MAE was 0.07 °C (the values were about 0.1 °C), and the R^2 was 0.94 (the values were within the range of 0.93–0.96). The mean RMSE for the interpolated air temperature grid datasets for the dry season was 0.40 °C (the values were within the range of 0.31–0.50 °C), and for the wet season was 0.45 °C (the values ranged from 0.37 °C to 0.52 °C). Compared to Tmean and Tmin, Tmax had a higher interpolation accuracy, with smaller $RMSE$ and MAE values. In addition, as the magnitude of air temperature varied for different seasons, the $RMSEs$ (or $MAEs$) of the interpolated grid datasets were smaller in the dry season than in the wet season, and the R^2 remained reasonably constant with all values greater than 0.93 and close to 1 across the study period. The results prove that the ANUSPLIN method is reliable for the air temperature estimation in the lower Jinsha River Basin.

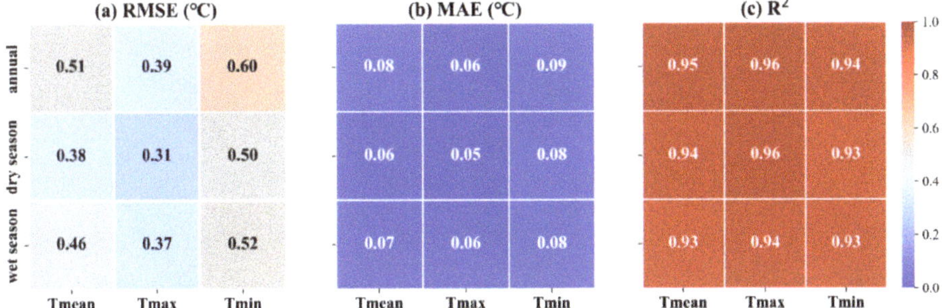

Figure 4. Ten-fold cross-validation results of the ANUSPLIN model. (**a**) Root mean square error ($RMSE$), (**b**) mean absolute error (MAE), and (**c**) coefficient of determination (R^2) between observed and simulated temperature grid datasets.

3.2. Air Temperature Changes before and after Impoundments

3.2.1. Spatiotemporal Patterns

Figure 5 exhibits the comparison of air temperature change trends for the pre-impoundment period (1980–2012) and the post-impoundment period (2013–2019). It can be observed that the air temperature shows an overall significant increase before the water impoundment, while after the impoundments, the Tmean and Tmax experience an average decrease, and the Tmin displays an average increase. All air temperature indicators generally transform from a significant trend to a non-significant trend in most areas of the catchment. Note that some regions still show a significant increase for Tmin after the impoundment.

As for annual Tmean, 88.43% of the study area experience a non-significant decrease, particularly in the central and eastern regions (Figure 5a). A similar spatial pattern also occurs in the dry season (52.59%, Figure 5b) and wet season (68.85%, Figure 5c). However, it is worth noting that some western regions exhibit a non-significant increase in the dry season (0.47 °C year^{-1}), and a non-significant decrease in the wet season (0.69 °C year^{-1}), which results in a non-significant decrease for annual scale. In addition, 97.61% of the catchment's Tmax shows a non-significant decrease after water impoundments (Figure 5d). This phenomenon is attributed to the presence of a non-significant decrease in both dry and wet seasons (Figure 5e,f). Finally, Tmin exhibits a more conspicuous spatial heterogeneity than Tmean and Tmax after water impoundments. The regions with a Tmin decrease (i.e., a non-significant decrease of 54.13%, Figure 5g) predominantly appear around reservoirs in both dry and wet seasons (17.79% and 16.92%, respectively, Figure 5h,i). However, the annual and wet seasonal Tmin maintains a significant increase similar to the trend before impoundments (14.56% and 12.37%, respectively, Figure 5g), which indicates small variations in Tmin in comparison to Tmean and Tmax.

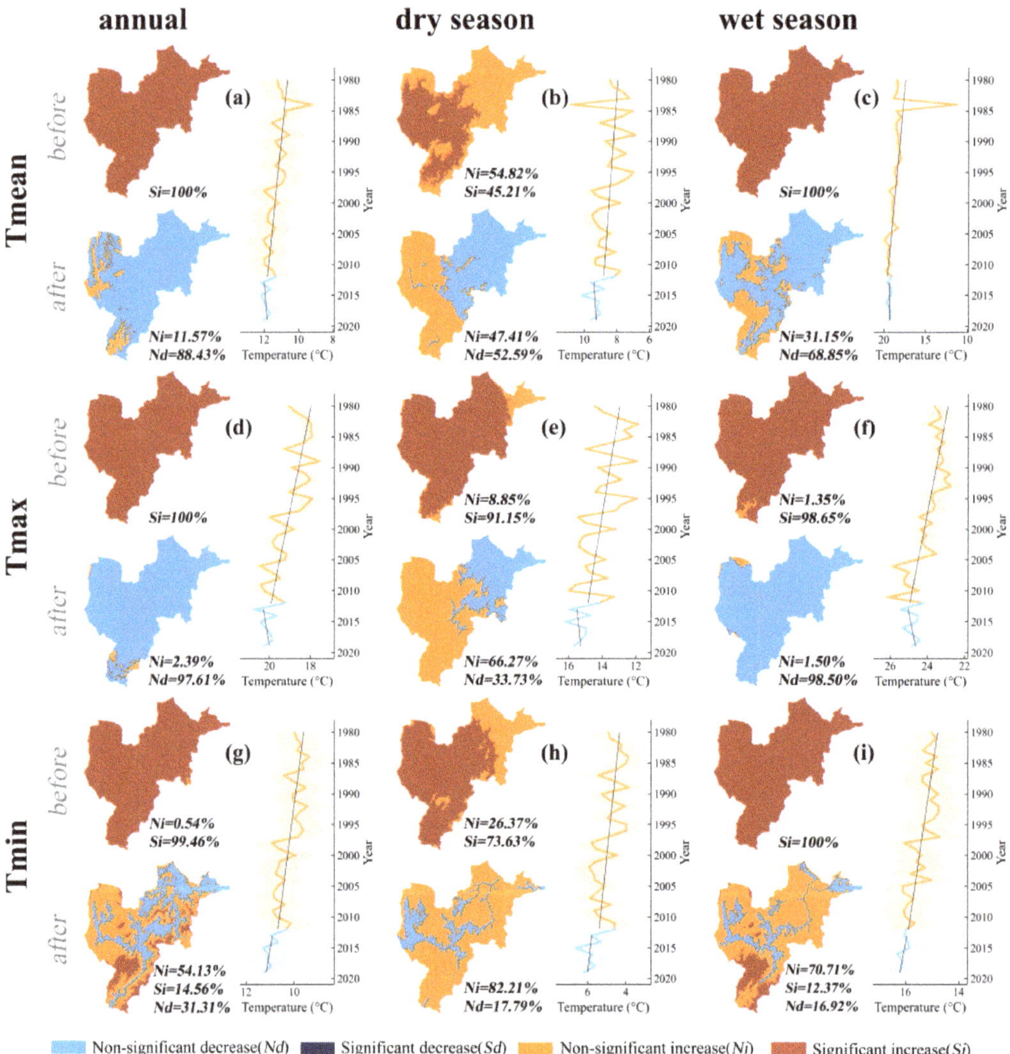

Figure 5. The spatial variations of Tmean (**a–c**), Tmax (**d–f**), and Tmin (**g–i**) for the pre-impoundment period (1980–2012) and the post-impoundment period (2013–2019) at the annual, dry season, and wet season scales. The orange and blue thick lines in the right axis show the average temperature variations by the ensemble mean of each year during the pre- and post-impoundment period, respectively, and the black dashed lines are their linear fits. The proportions of four classification categories are noted in subscripts.

3.2.2. Change-Points Detection

To detect and identify the time in which significant changes in air temperatures occurred, the temperature indicators were identified by using the MK mutation test at both annual and seasonal scales. It can be observed that the annual changes for Tmean, Tmax, and Tmin are quite similar to the seasonal changes (Figure 6). There is no significant change point for each indicator at the annual scale, while mutations occur at both the dry and wet season scales. In particular, the significant changes occurred in 2013 for Tmean, and in 2006 and 2008 for Tmax. Even though UF_K and UB_K curves of Tmin intersect, the UF_K statistics do not exceed the critical line (Figure 6g–i), thus, no mutations could be confirmed for the Tmin. Notably, the mutation time for the Tmean during the dry season in 2012 coincided with the impounding time of the Xiluodu Reservoir (Figure 6b).

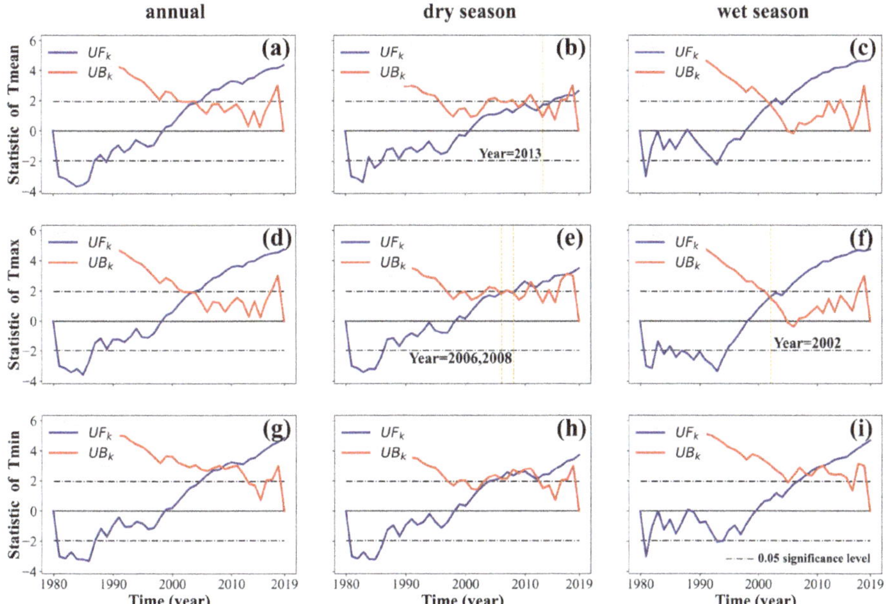

Figure 6. Mutation analysis of annual (**a,d,g**), dry season (**b,e,h**), and wet season (**c,f,i**) air temperature changes during the 1980–2019 period. The yellow dashed lines in each subfigure mark the intersection points of the UF_K and UB_K curves, indicating the observed mutation points, and the specific mutation years are labeled. Subfigures without mutation points are not labeled with relevant information. The black dashed lines indicate 0.05 significance level.

Overall, only one significant change point was observed at the dry season scale after the water impoundments. This phenomenon implied that after the water impoundment in the lower Jinsha River Basin, the air temperature variations were alleviated.

3.3. Effects of Water Impoundments on Air Temperature

3.3.1. Evaluation of the LSTM Model Performance

To evaluate the performance of LSTM models in predicting each air temperature grid surface dataset (Figure 7), we used the testing sets from monthly temperature grid surface datasets of the 1980–2012 period. The monthly LSTM forecast was well supported by actual data. It can be seen that both the MAE and $RMSE$ values of the three models hover around 0.1 °C and the R^2 values are close to 1, reflecting the robustness of LSTM models. We further utilized the least squares method to determine the best linear fit in each scatterplot. The LSTM models tend to slightly underestimate the air temperature (the fitting slope for the Tmean is 1.00062, for the Tmax is 1.00058, and for the Tmin is 1.00091). However, they are all close to 1, reflecting the remarkable similarities between actual data and LSTM forecast, which further confirms the high reliability and accuracy of the LSTM models in predicting Tmean, Tmax, and Tmin datasets.

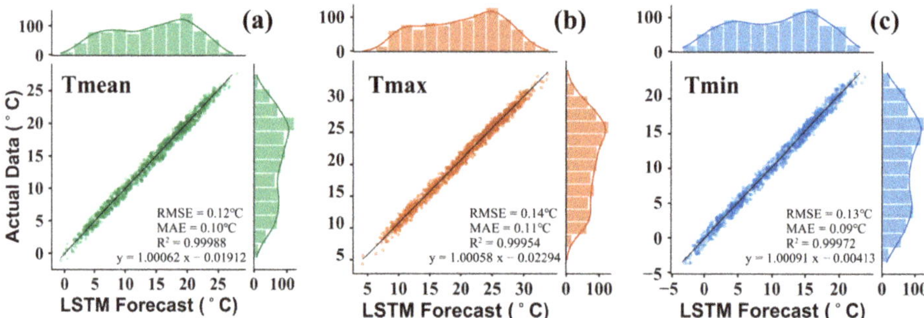

Figure 7. Performance evaluations of the LSTM model for Tmean (**a**), Tmax (**b**) and Tmin (**c**) by comparing the observed temperature grid surface datasets (actual data) and the predicted temperature grid surface datasets (LSTM forecast) using the testing sets from monthly temperature grid surface datasets (1980–2012). The position of each point is jittered by an amount of 0.8 for visual purposes. The grey line indicates the linear fit by using the least squares method, with y and x indicating actual data and LSTM forecast in the linear equations, respectively. The histograms depict the distribution of actual data and LSTM forecast, with univariate Kernel density estimation curves on the top and right side, respectively.

3.3.2. Patterns of the *IET* Index

Figure 8 displays the spatial distribution of the *IET* index at annual and seasonal scales. According to Figure 8, the *IET* index had both positive and negative values across the catchment and at different scales, suggesting that both warming and cooling effects existed. Generally, there was a similar pattern for Tmean and Tmax, while there was a different pattern for Tmean. In particular, for the annual Tmean, the average *IET* was 0.1 °C, with 51% of the catchment showing a warming effect (*IET* > 0) after water impoundment (chiefly in the western region), while the remaining 49% (principally in the eastern region) exhibited a cooling effect with *IET* < 0 (Figure 8a). The cooling effect in the eastern region was mainly attributed to the negative *IET* index in both the dry and wet seasons. Specifically, 44% (out of 80%) of the regions show a substantially positive reduction during the wet season, while 41% (out of 74%) of the regions exhibit a substantially positive increase in the western region during the dry season (Figure 8b,c).

The average *IET* value for the annual Tmax was −0.4 °C, indicating that water impoundment gave rise to an overall 0.4 °C reduction. This was mostly caused by significant temperature decreases in 81% of the basin, especially during the wet season with a magnitude of around 2 °C reduction primarily distributed in surrounding areas of the reservoirs (Figure 8d–f). It is worth mentioning that the annual Tmax did not pass the significance test around the reservoir areas, although it displayed a reduced pattern. This was largely because the area around the reservoirs exhibited insignificant warming during the dry season, which partially compensated for the cooling effects shown during the wet season on an annual scale.

As for the annual Tmin, the average *IETs* were 1.0 °C, 1.2 °C, and 0.8 °C at annual, dry season, and wet season scales, respectively. The reservoir impoundment caused 98% of the basin to show an air temperature increase (*IET* > 0), of which 85% of the catchment exhibited a significant increase. At the same time, we also found an interesting phenomenon, that is, the increase in the Tmin in the surrounding areas of the reservoirs was less than that in the high altitudes.

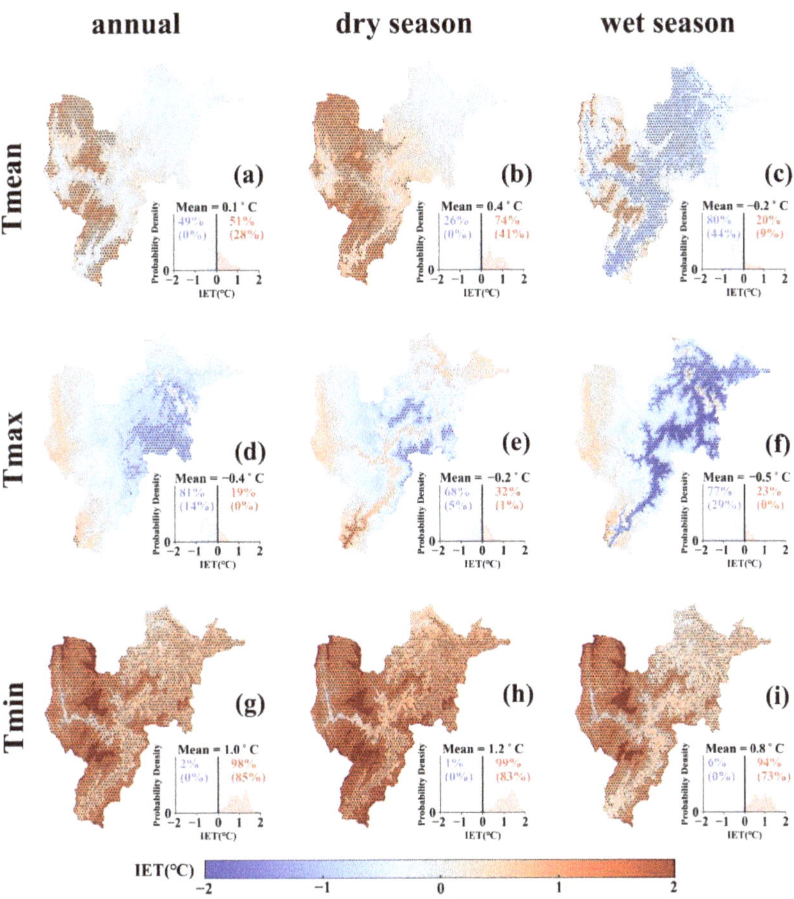

Figure 8. Spatial distribution of the *IET* index under annual (**a,d,g**), dry season (**b,e,h**), and wet season (**c,f,i**)scales during the 2013–2019 period. The dots on the colorful regions indicate the impacts of impoundments on temperatures are significant ($p < 0.05$), which were conducted by the paired sample *t*-test between predicted and actual grid surface datasets (2013–2019). The lower right probability density function shows the distribution of the *IET* index at different values. The red area denotes an air temperature increase (*IET* > 0), while the blue area denotes an air temperature decrease (*IET* < 0). The text in brackets shows the proportion of temperatures that are significant ($p < 0.05$).

4. Discussion

4.1. Impacts of Reservoir Impoundment on Air Temperature

We observed that reservoir impoundment led to simultaneous warming and cooling effects at different time scales (Figure 8). The reasons for this phenomenon can be explained by two biophysical mechanisms, namely, evaporative cooling and thermal inertia regulation. First, the water impoundment in the Xiangjiaba and Xiluodu hydropower plants has resulted in the conversion of approximately 243.90 km^2 of land to the water surface [47]. The increase in the water area, in turn, considerably increases local evaporation while absorbing heat through the water, which leads to a decrease in air temperature, especially in regions that surround the reservoirs [48]. Second, the water bodies in the reservoirs have substantial thermal inertia and can act as both a "thermal energy sink" and "thermal energy source" under different conditions, thus, moderating the degree of air temperature variability. For example, this characteristic of reservoir waters can cool the air temperature by absorbing heat from the air in summer and can warm the air temperature by releasing heat in winter [49]. In our study, we noticed an overall decreasing trend for Tmax, with 81%, 68%, and 77% of the whole catchment showing significantly decreasing temperatures at the annual, dry season, and wet season scales, respectively. However, Tmin showed an overall increasing trend, with 98%, 99%, and 94% of the area exhibiting greatly increasing temperatures at the annual, dry season, and wet season scales, respectively (Figure 8d–i). These results can be justified by the thermal inertia mechanism.

Similar to other studies, our study also demonstrates that the cooling impact of the reservoir occurs predominantly in the surrounding area of the reservoirs [10]. In particular, in the lower Jinsha River Basin, the effect of reservoir impoundment on Tmean at the annual scale was roughly confined to an area within about 4 km from the reservoirs (Figures 8 and 9d). Thus, the cooling effect of water impoundment within this distance range can be justified by evaporative cooling theory.

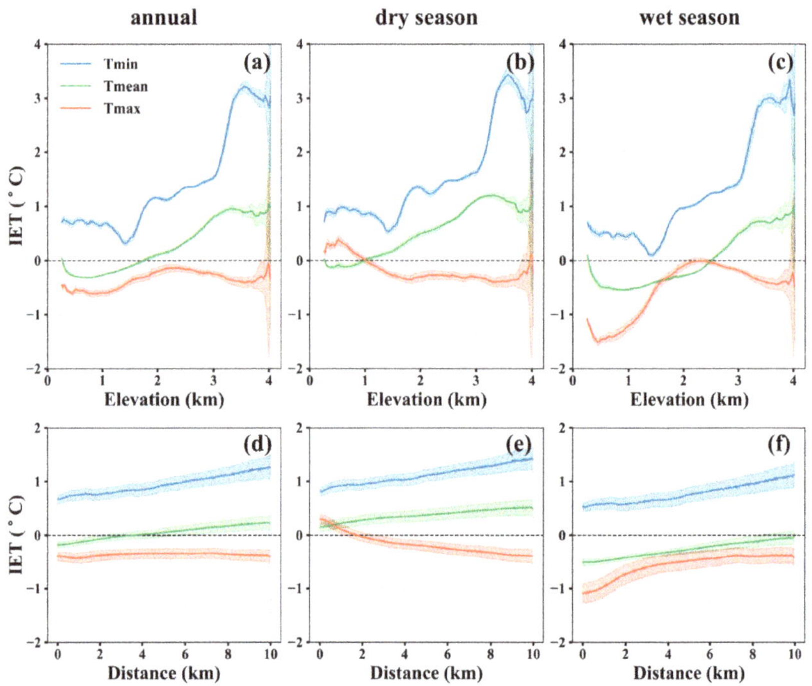

Figure 9. The relationship between muti-scale *IET* and elevation (**a–c**) and Euclidean distance (**d–f**) from the reservoirs. Thick lines refer to the average *IETs* in 200 elevation or distance bins that are created based on group statistics, whereas shaded areas display their 95% confidence interval magnified by a factor of 10 for visual purposes.

Interestingly, we realized that the modification exerted by water impoundments at longer distances (around 4 km away from the reservoirs) was opposite to that at shorter distances, with a dominant warming effect (Figure 9d). We speculated that the reason for this phenomenon was probably the thermal inertia regulation of the impoundments. We found that the thermal inertia of the large volume of water resulted in a warming effect at high elevations where the temperature was supposed to be low (Figure 9a). That is, the evaporative cooling effect was dominant at low elevations within about 4 km from the reservoirs, while the heat source effect of the thermal inertia was dominant at high elevations.

The relationship between the *IET* index and elevation further supported this view. For instance, we realized that the elevation threshold for the cooling or warming effect of the reservoirs in the lower Jinsha River region was approximately 1.8 km above sea level at the annual scale (Figure 9a). In other words, in regions with elevation smaller than 1.8 km, the effect of reservoirs was chiefly reflected as a cooling effect. However, in regions where the elevation was greater than 1.8 km, the warming effect was dominant, and the higher the elevation, the more obvious the warming effect. Furthermore, the threshold became about 1 km and 2.5 km in the dry season and the wet season scale (Figure 9b,c), respectively, suggesting that the area showing the warming effect in the wet season was less than that in the dry season.

Xiangjiaba and Xiluodu hydropower plants are characterized by high water level operation during the dry season, and low water level operation during the wet season [50]. This suggests the water area and volume of reservoirs are higher in the dry season than in the wet season. Therefore, owing to the thermal inertia mechanism of the reservoirs, milder temperature variations occur in the dry season than in the wet season. For example, in our study, the percentage of the area where the temperature was significantly affected was as high as 53% in the wet season, while this percentage was only 41% in the dry season (Figure 8b,c). This finding is similar to that of Wang et al. [17], who also indicated that air temperatures were more moderate during the dry season than in the wet season after reservoir impoundment.

4.2. Limitations and Future Research Directions

Assessing regional air temperature changes is very challenging because of regional heterogeneity and the non-linear characteristics of air temperature itself [51]. Generally, air temperatures are largely influenced by the broad climatic context, topographic factors, vegetation types, human activities, and complex interactions among them [19]. Despite the fact our study has been theoretically conducted to eliminate the influence of natural factors on air temperature and solely assess the effect of water impoundments, the impact of other anthropogenic factors has not been assessed. Except for the effects of the reservoirs studied in this study, many other factors may all contribute to the variation in air temperature [52]. For instance, the study by Ospina Noreña et al. [53] indicated that Tmean, Tmax, and Tmin around the hydroelectric power plants in Sinú-Caribbean Basin all showed an increasing trend in the context of global warming. Moreover, other factors such as the implementation of ecological restoration projects [16], and the reduction in carbon emissions due to the out-migration of the population after the construction of hydropower plants [54], may also have an impact on the air temperature change. Hence, in the future, it is necessary to consider all these various factors together when assessing the effect of reservoir impoundment on regional air temperature variations.

Another limitation of this research is that its study period after impoundment is only seven years. As the neighboring Baihetan hydropower station at the upstream of Jinsha river impounded water in January 2020, our study had to be limited to 2019 in order to exclude its impact on the air temperature in our study area. However, 7 years is a relatively short period of time for air temperature changes, which may result in certain long-term regular features not being effectively reflected.

5. Conclusions

This study provided an improved method for quantifying the effects of dam-related reservoir impoundment on regional temperature. We found that in the lower Jinsha River catchment, the reservoirs exerted an impact on regional air temperatures. Specifically, temperature variability in the river basin decreased after water impoundment. At the annual scale, Tmean increased by an average of 0.1 °C, Tmin increased by an average of around 1.0 °C, and Tmax decreased by an average of about 0.4 °C. The impact of reservoirs on air temperature was principally within about 4 km of the reservoirs and was dominated by the cooling effect. Temperature variations were less pronounced during the dry season than during the wet season. The warming effect was more pronounced and the air temperatures were more moderate during the dry season compared to that in the wet season after reservoir impoundment. Our research results contribute to a better understanding of the impact of dam-related reservoirs on regional climate, which may, in turn, affect the existing ecological ecosystems adversely. Therefore, more attention to dam-related reservoirs may be necessary to minimize the negative impacts of thermal regime changes on the terrestrial ecosystem and the method presented in this paper can achieve this goal.

Author Contributions: Software, formal analysis, investigation, resources, data curation, visualization, writing—original draft, X.L., J.Z., Y.H. and R.W.; conceptualization, methodology, validation, writing—review and editing, supervision, project administration, funding acquisition, T.L. All authors have read and agreed to the published version of the manuscript.

Funding: This research was supported by the Second Tibetan Plateau Scientific Expedition and Research Program (STEP) (grant no. 2019QZKK0402) and the National Natural Science Foundation of China (grant no. 42071238).

Data Availability Statement: Data will be made available on request.

Conflicts of Interest: The authors declare that they have no known competing financial interest o personal relationships that could have appeared to influence the work reported in this paper.

References

1. Boulange, J.; Hanasaki, N.; Yamazaki, D.; Pokhrel, Y. Role of dams in reducing global flood exposure under climate change. *Nat Commun.* **2021**, *12*, 417. [CrossRef] [PubMed]
2. Seyedhashemi, H.; Moatar, F.; Vidal, J.P.; Diamond, J.S.; Beaufort, A.; Chandesris, A.; Valette, L. Thermal signatures identify the influence of dams and ponds on stream temperature at the regional scale. *Sci. Total Environ.* **2021**, *766*, 142667. [CrossRef] [PubMed]
3. Kuriqi, A.; Pinheiro, A.N.; Sordo-Ward, A.; Bejarano, M.D.; Garrote, L. Ecological impacts of run-of-river hydropower plants–Current status and future prospects on the brink of energy transition. *Renew. Sustain. Energy Rev.* **2021**, *142*, 110833. [CrossRef]
4. Tian, S.; Xu, M.; Jiang, E.; Wang, G.; Hu, H.; Liu, X. Temporal variations of runoff and sediment load in the upper Yellow River China. *J. Hydrol.* **2019**, *568*, 46–56. [CrossRef]
5. Tao, Y.; Wang, Y.; Rhoads, B.; Wang, D.; Ni, L.; Wu, J. Quantifying the impacts of the Three Gorges Reservoir on water temperature in the middle reach of the Yangtze River. *J. Hydrol.* **2020**, *582*, 124476. [CrossRef]
6. Schmadel, N.M.; Harvey, J.W.; Schwarz, G.E.; Alexander, R.B.; Gomez-Velez, J.D.; Scott, D.; Ator, S.W. Small Ponds in Headwater Catchments Are a Dominant Influence on Regional Nutrient and Sediment Budgets. *Geophys. Res. Lett.* **2019**, *46*, 9669–9677 [CrossRef]
7. Wang, F.; Maberly, S.C.; Wang, B.; Liang, X. Effects of dams on riverine biogeochemical cycling and ecology. *Inland Waters* **2018**, *8*, 130–140. [CrossRef]
8. Asthana, B.N.; Khare, D. Reservoir sedimentation. In *Recent Advances in Dam Engineering*; Springer: Cham, Switzerland, 2022; pp. 265–288. [CrossRef]
9. Maavara, T.; Chen, Q.; Van Meter, K.; Brown, L.E.; Zhang, J.; Ni, J.; Zarfl, C. River dam impacts on biogeochemical cycling. *Nat. Rev. Earth Environ.* **2020**, *1*, 103–116. [CrossRef]
10. Degu, A.M.; Hossain, F.; Niyogi, D.; Pielke, R.; Shepherd, J.M.; Voisin, N.; Chronis, T. The influence of large dams on surrounding climate and precipitation patterns. *Geophys. Res. Lett.* **2021**, *38*, L04405. [CrossRef]
11. Albalasmeh, A.; Mohawesh, O.; Zeadeh, D.; Unami, K. Robust optimization of shading types to control the performance of water reservoirs. *J. Clean. Prod.* **2023**, *415*, 137730. [CrossRef]
12. Xu, Z.X.; Mo, L.; Zhou, J.Z.; Zhang, X. Optimal dispatching rules of hydropower reservoir in flood season considering flood resources utilization: A case study of Three Gorges Reservoir in China. *J. Clean. Prod.* **2023**, *388*, 135975. [CrossRef]
13. Grill, G.; Lehner, B.; Thieme, M.; Geenen, B.; Tickner, D.; Antonelli, F.; Babu, S.; Borrelli, P.; Cheng, L.; Crochetiere, H.; et al. Mapping the world's free-flowing rivers. *Nature* **2019**, *569*, 215–221. [CrossRef] [PubMed]
14. Olden, J.D.; Naiman, R.J. Incorporating thermal regimes into environmental flows assessments: Modifying dam operations to restore freshwater ecosystem integrity. *Freshw. Biol.* **2010**, *55*, 86–107. [CrossRef]
15. Rheinheimer, D.E.; Null Sarah, E.; Lund Jay, R. Optimizing Selective Withdrawal from Reservoirs to Manage Downstream Temperatures with Climate Warming. *J. Water Resour. Plan. Manag.* **2015**, *141*, 04014063. [CrossRef]
16. Song, Z.; Liang, S.; Feng, L.; He, T.; Song, X.P.; Zhang, L. Temperature changes in Three Gorges Reservoir Area and linkage with Three Gorges Project. *J. Geophys. Res. Atmos.* **2017**, *122*, 4866–4879. [CrossRef]
17. Wang, D.; Wang, F.; Huang, Y.; Duan, X.; Liu, J.; Hu, B.; Sun, Z.; Chen, J. Examining the Effects of Hydropower Station Construction on the Surface Temperature of the Jinsha River Dry-Hot Valley at Different Seasons. *Remote Sens.* **2018**, *10*, 600. [CrossRef]
18. Hörhold, M.; Münch, T.; Weißbach, S.; Kipfstuhl, S.; Freitag, J.; Sasgen, I.; Lohmann, G.; Vinther, B.; Laepple, T. Modern temperatures in central–north Greenland warmest in past millennium. *Nature* **2023**, *613*, 503–507. [CrossRef]

19. Zeng, Y.; Zhou, Z.; Yan, Z.; Teng, M.; Huang, C. Climate Change and Its Attribution in Three Gorges Reservoir Area, China. *Sustainability* **2019**, *11*, 7206. [CrossRef]
20. Fonseca, A.; Santos, J.A. The Impact of a Hydroelectric Power Plant on a Regional Climate in Portugal. *Atmosphere* **2021**, *12*, 1400. [CrossRef]
21. Miller, N.L.; Jin, J.; Tsang, C.F. Local climate sensitivity of the Three Gorges Dam. *Geophys. Res. Lett.* **2005**, *32*, L16704. [CrossRef]
22. Irambona, C.; Music, B.; Nadeau, D.F.; Mahdi, T.F.; Strachan, I.B. Impacts of boreal hydroelectric reservoirs on seasonal climate and precipitation recycling as simulated by the CRCM5: A case study of the La Grande River watershed, Canada. *Theor. Appl. Climatol.* **2018**, *131*, 1529–1544. [CrossRef]
23. Zhao, Y.; Liu, S.; Shi, H. Impacts of dams and reservoirs on local climate change: A global perspective. *Environ. Res. Lett.* **2021**, *16*, 104043. [CrossRef]
24. Huang, X.R.; Gao, L.Y.; Yang, P.P.; Xi, Y.Y. Cumulative impact of dam constructions on streamflow and sediment regime in lower reaches of the Jinsha River, China. *J. Mt. Sci.* **2018**, *15*, 2752–2765. [CrossRef]
25. Li, D.; Lu, X.X.; Yang, X.; Chen, L.; Lin, L. Sediment load responses to climate variation and cascade reservoirs in the Yangtze River: A case study of the Jinsha River. *Geomorphology* **2018**, *322*, 41–52. [CrossRef]
26. Dos Santos, N.C.L.; García-Berthou, E.; Dias, J.D.; Lopes, T.M.; Affonso, I.d.P.; Severi, W.; Gomes, L.C.; Agostinho, A.A. Cumulative ecological effects of a Neotropical reservoir cascade across multiple assemblages. *Hydrobiologia* **2018**, *819*, 77–91. [CrossRef]
27. Toharudin, T.; Pontoh, R.S.; Caraka, R.E.; Zahroh, S.; Lee, Y.; Chen, R.C. Employing long short-term memory and Facebook prophet model in air temperature forecasting. *Commun. Stat. Simul. Comput.* **2020**, *52*, 279–290. [CrossRef]
28. Espeholt, L.; Agrawal, S.; Sønderby, C.; Kumar, M.; Heek, J.; Bromberg, C.; Gazen, C.; Carver, R.; Andrychowicz, M.; Hickey, J.; et al. Deep learning for twelve hour precipitation forecasts. *Nat. Commun.* **2022**, *13*, 5145. [CrossRef]
29. Lecun, Y.; Bengio, Y.; Hinton, G. Deep learning. *Nature* **2015**, *521*, 436–444. [CrossRef]
30. Hu, G.; Tian, S.; Chen, N.; Liu, M.; Somos-Valenzuela, M. An effectiveness evaluation method for debris flow control engineering for cascading hydropower stations along the Jinsha River, China. *Eng. Geol.* **2020**, *266*, 105472. [CrossRef]
31. Zhang, M.; Ge, S.; Yang, Q.; Ma, X. Impoundment-Associated Hydro-Mechanical Changes and Regional Seismicity Near the Xiluodu Reservoir, Southwestern China. *J. Geophys. Res. Solid Earth* **2021**, *126*, e2020JB021590. [CrossRef]
32. Chen, Q.; Chen, H.; Wang, J.; Zhao, Y.; Chen, J.; Xu, C. Impacts of Climate Change and Land-Use Change on Hydrological Extremes in the Jinsha River Basin. *Water* **2019**, *11*, 1398. [CrossRef]
33. Xiong, D.H.; Zhou, H.Y.; Yang, Z.; Zhang, X.B. Slope lithologic property, soil moisture condition and revegetation in dry-hot valley of Jinsha River. *Chin. Geogr. Sci.* **2005**, *15*, 186–192. [CrossRef]
34. Gong, Z.L.; Tang, Y. Impacts of reforestation on woody species composition, species diversity and community structure in dry-hot valley of the Jinsha River, southwestern China. *J. Mt. Sci.* **2016**, *13*, 2182–2191. [CrossRef]
35. Ma, J.; Yan, X.; Hu, S.; Guo, Y. Can monthly precipitation interpolation error be reduced by adding periphery climate stations? A case study in China's land border areas. *J. Water Clim. Chang.* **2016**, *8*, 102–113. [CrossRef]
36. Xie, H.; Zhao, A.; Huang, S.; Han, J.; Liu, S.; Xu, X.; Luo, X.; Pan, H.; Du, Q.; Tong, X. Unsupervised hyperspectral remote sensing image clustering based on adaptive density. *IEEE Geosci. Remote. Sens. Lett.* **2018**, *15*, 632–636. [CrossRef]
37. NASA JPL. NASADEM Merged DEM Global 1 arc Second V001 [Data Set]. NASA EOSDIS Land Processes DAAC. 2020. Available online: https://doi.org/10.5067/MEaSUREs/NASADEM/NASADEM_HGT.001 (accessed on 7 July 2022).
38. Hutchinson, M.F. Interpolating mean rainfall using thin plate smoothing splines. *Int. J. Geogr. Inf. Sci.* **1995**, *9*, 385–403. [CrossRef]
39. Bian, Y.; Yue, J.; Gao, W.; Li, Z.; Lu, D.; Xiang, Y.; Chen, Y. Analysis of the Spatiotemporal Changes of Ice Sheet Mass and Driving Factors in Greenland. *Remote Sens.* **2019**, *11*, 862. [CrossRef]
40. Liu, Z.; Menzel, L. Identifying long-term variations in vegetation and climatic variables and their scale-dependent relationships: A case study in Southwest Germany. *Glob. Planet. Chang.* **2016**, *147*, 54–66. [CrossRef]
41. Hong, B.; Zhang, J. Long-Term Trends of Sea Surface Wind in the Northern South China Sea under the Background of Climate Change. *J. Mar. Sci.* **2021**, *9*, 752. [CrossRef]
42. Gers, F.A.; Schmidhuber, J.; Cummins, F. Learning to Forget: Continual Prediction with LSTM. *Neural Comput.* **2000**, *12*, 2451–2471. [CrossRef]
43. Hochreiter, S.; Schmidhuber, J. Long Short-Term Memory. *Neural Comput.* **1997**, *9*, 1735–1780. [CrossRef] [PubMed]
44. Wang, Q.; Liu, S.; Chanussot, J.; Li, X. Scene Classification With Recurrent Attention of VHR Remote Sensing Images. *IEEE Trans. Geosci. Remote. Sens.* **2019**, *57*, 1155–1167. [CrossRef]
45. Peng, K.; Radivojac, P.; Vucetic, S.; Dunker, A.K.; Obradovic, Z. Length-dependent prediction of protein intrinsic disorder. *BMC Bioinform.* **2006**, *7*, 208. [CrossRef]
46. Chen, H.; Ren, J.; Sun, W.; Hou, J.; Miao, Z. Mosquito swarm counting via attention-based multi-scale convolutional neural network. *Sci. Rep.* **2023**, *13*, 4215. [CrossRef] [PubMed]
47. Pekel, J.F.; Cottam, A.; Gorelick, N.; Belward, A.S. High-resolution mapping of global surface water and its long-term changes. *Nature* **2016**, *540*, 418–422. [CrossRef] [PubMed]
48. Wu, L.; Zhang, Q.; Jiang, Z. Three Gorges Dam affects regional precipitation. *Geophys. Res. Lett.* **2006**, *33*, L13806. [CrossRef]
49. Fink, G.; Schmid, M.; Wahl, B.; Wolf, T.; Wüest, A. Heat flux modifications related to climate-induced warming of large European lakes. *Water Resour. Res.* **2014**, *50*, 2072–2085. [CrossRef]

50. Wang, X.; Wang, F.; Feng, T.; Zhang, S.; Guo, Z.; Lu, P.; Liu, L.; Yang, F.; Liu, J.; Rose, N.L. Occurrence, sources and seasonal variation of $PM_{2.5}$ carbonaceous aerosols in a water level fluctuation zone in the Three Gorges Reservoir, China. *Atmos. Pollut. Res.* **2020**, *11*, 1249–1257. [CrossRef]

51. Arent, D.J.; Tol, R.S.; Faust, E.; Hella, J.P.; Kumar, S.; Strzepek, K.M.; Tóth, F.L.; Yan, D.; Abdulla, A.; Kheshgi, H.; et al. Key economic sectors and services. In *Climate Change 2014 Impacts, Adaptation and Vulnerability: Part A: Global and Sectoral Aspects*; Cambridge University Press: Cambridge, UK, 2015; pp. 659–708. [CrossRef]

52. Solaun, K.; Cerdá, E. Climate change impacts on renewable energy generation. A review of quantitative projections. *Renew. Sustain. Energy Rev.* **2019**, *116*, 109415. [CrossRef]

53. Ospina Noreña, J.; Gay García, C.; Conde, A.; Magaña, V.; Sánchez Torres Esqueda, G.J.A. Vulnerability of water resources in the face of potential climate change: Generation of hydroelectric power in Colombia. *Atmósfera* **2009**, *22*, 229–252.

54. Roy, J.; Pal, S. Lifestyles and climate change: Link awaiting activation. *Curr. Opin. Environ. Sustain.* **2009**, *1*, 192–200. [CrossRef]

remote sensing

Article

Spatio-Temporal Changes in Ecosystem Service Value and Its Coordinated Development with Economy: A Case Study in Hainan Province, China

Jie Fu [1,2], Qing Zhang [1,*], Ping Wang [2], Li Zhang [1], Yanqin Tian [1] and Xingrong Li [3]

[1] Key Laboratory of Earth Observation of Hainan Province, Hainan Research Institute, Aerospace Information Research Institute, Chinese Academy of Sciences, Sanya 572029, China; jeff101@sdust.edu.cn (J.F.); zhangli@aircas.ac.cn (L.Z.); tianyq@jou.edu.cn (Y.T.)

[2] School of Geomatics and Spatial Information, Shandong University of Science and Technology, Qingdao 266590, China; skd990058@sdust.edu.cn

[3] Jiangsu Key Laboratory for Information Agriculture, National Engineering and Technology Center for Information Agriculture (NETCIA), Nanjing Agricultural University, Nanjing 210095, China; 2021201094@stu.njau.edu.cn

* Correspondence: zhangqing202017@aircas.ac.cn; Tel.: +86-135-5257-6637

Citation: Fu, J.; Zhang, Q.; Wang, P.; Zhang, L.; Tian, Y.; Li, X. Spatio-Temporal Changes in Ecosystem Service Value and Its Coordinated Development with Economy: A Case Study in Hainan Province, China. *Remote Sens.* **2022**, *14*, 970. https://doi.org/10.3390/rs14040970

Academic Editors: Jun Li, Xinyi Shen, Qiusheng Wu and Chengye Zhang

Received: 23 December 2021
Accepted: 15 February 2022
Published: 16 February 2022

Publisher's Note: MDPI stays neutral with regard to jurisdictional claims in published maps and institutional affiliations.

Abstract: Ecosystem service value is crucial to people's intuitive understanding of ecological protection and the decision making with regard to ecological protection and economic green development. This study improved the benefit transfer method to evaluate ESV in Hainan Province, proposed the coupling analysis method of economic and environmental coordination, and explored the relationship between ESV and economic development based on the medium-resolution remote sensing land use data and socio-economic data from 2000 to 2020. The results show that Hainan Province's ESV decreased by 33.305 billion CNY from 2000 to 2020. The highest ESV per unit area was found in the water system and forest ecosystem, mainly distributed in the central mountainous area. The overall condition of EEC decreased from a basic coordination state to a moderate disorder state. Areas with high economic development had better EEC, such as Haikou and Sanya. Meanwhile, we analyzed the driving force of ESV and EEC by Geodetector. The results show that land use intensity was the most important driving factor affecting ESV, with a contribution rate of 0.712. Total real estate investment was the most important driving factor affecting EEC, with a contribution rate of 0.679. These results provide important guidance for the coordinated development of regional economy and ecosystem protection.

Keywords: ecosystem service value; degree of economic and environmental coordination; land use change; sustainable development; analysis of driving force

1. Introduction

Ecosystem services refer to the benefits that humans derive directly or indirectly from ecosystems [1], which include a range of goods and services that are extremely important to human well-being [2,3]. The concept of ecosystem services was defined in the Millennium Ecosystem Assessment published by the United Nations in 2005 [1], which includes supply services, support services, regulatory services and cultural services [4]. Subsequently, ecosystem services have received increasing attention from scientists, policymakers and the public [5–8]. It is often difficult to obtain a clear picture of the importance and richness/scarcity of ecosystem services from the point of view of physical quantity alone [9]. The monetary valuation of ecosystem services has greatly facilitated the assessment of global ecosystem services [3]. A growing number of studies have attempted to use ecosystem service value (ESV) monetary evaluation methods to evaluate ESV at different scales, including the global [3], national [10] and regional [8]. Much research has been carried out on ESV evaluation methods at home and abroad, but a unified evaluation system has not been developed [2,6,11]. ESV can be broadly divided into two categories: price methods based on unit service functions, and value-equivalent factor methods based on unit areas [6]. The former method has many input parameters, has a complicated calculation process and has difficulty achieving a unified standard, so it has not generated a breakthrough in the specific theory and method [12,13]. Compared with price methods, the equivalent factor method is easier and more intuitive to use, is more comprehensive, has fewer data needs and has high comparability. It is mainly influenced by Costanza, who divided the world's ecosystems into 16 types and 17 subtypes based on their service functions [3,5]. It is particularly suitable for assessing the ESV at the regional and global scales [2,6,8–10]. In addition, a series of models for calculating ESV have been proposed based on existing research, such as the Integrated Valuation of Ecosystem Services and Tradeoffs (INVEST) model [14], Multi-Scale Integrated Models of Ecosystem Services (MIMES) model [15], etc. These models explore the evaluation of ecosystem services from different perspectives, focusing on ecological data, and can be applied to the global, watershed or landscape scale [16]. Using modeling methods to evaluate ecosystem services has scientific and objective advantages, but the result of the assessment is the quality of ecosystem services. In order to better serve ecological compensation and ecosystem management, the quality of ecosystem services needs to be converted into value. However, the quality of ecosystem services may have various values, and using these methods it is difficult to fully evaluate all ecosystem service functions [17]. Additionally, these methods rely too much on the quality and quantity of satellite imagery.

Many researchers have evaluated ESV on a national or regional scale based on the benefit transfer method [18,19]. However, their results have dramatically differed from the results of applying this approach to China. For example, deviations in some cases may result in an underestimate of farmland ESV and an overestimate of wetland ESV [20]. The ESV that other researchers have assessed reflects the economies of developed countries (such as the US and European countries) rather than developing countries (such as China) [10,21]. Xie et al. [4,6,7] improved this approach and calculated the basic equivalent of ESV in China based on Costanza's evaluation model, which has been adopted by most researchers. It has been widely used to evaluate ESV at different spatial scales, such as country [6], province [22], urban agglomeration [13] and watershed [23]. However, this may not be enough for all regional cases in China. As a static coefficient table, the evaluation based upon it cannot reflect the physical geographic and socio-economic characteristics of the region [24]. Generally speaking, people's willingness to pay and ability to pay increas with the improvement of the socio-economic level. Moreover, ecosystem service functions vary with different physical geographical conditions [25]. This is especially true in China. China has a vast territory with diverse terrain and natural features. The above facts indicate that the value coefficient should be dynamically adjusted to reflect the physical geographic and socio-economic characteristics of the region [26,27].

In the face of increasingly severe global issues, such as a rapid population explosion, food shortages, resource depletion and ecological degradation, increasing attention is being paid to clarifying the relationship between ecological protection and economic development [28]. Coordinating the relationship between economic growth and environmental degradation and resource depletion is a basic requirement of sustainable development and the only way to achieve it [29]. Therefore, it is necessary to analyze the coordination between ecosystem services and regional economic development [30]. Ecosystem services can reflect the relationship between economic development and the ecological environment (for example, economic development can improve the ecological environment by optimizing industrial structure and improving population quality, or through population growth and the expansion of built-up land lead to the deterioration of the ecological environment [30,31]) as bridges and bonds that couple natural and social processes, which provides new theoretical support for studying the coupling of human and natural systems [32]. Analyzing the relationship between ecosystem services and economic development can better elucidate the impact of socioeconomic activities on ecosystem services, which plays an important role in adjusting the structure of land use, improving the efficiency of land distribution, and protecting the ecosystem. It is a realistic step towards realizing the coordination and optimization of regional economic development and ecological protection.

The co-growth of ESV and the economy is the goal and direction of current social and economic development. Previous studies were mostly limited to single-factor one-way studies, and lack the analysis of the degree of coordination between ESV and regional economic development, especially the spatial distribution characteristics of its coupling and coordination characteristics [33,34]. The research of domestic and foreign scholars on the coordination degree between the environment and economy mainly focuses on two aspects. The first aspect is the theory of the coordinated development of the environment and economy. Scholars believe that there is an organic connection between the environment and economy, and that ecological construction and economic development should adapt to and promote each other [35,36]. The second aspect is evaluating the coordinated development of the environment and economy. Scholars at home and abroad have carried out a large number of studies by comprehensively applying a variety of research methods, including the comprehensive index evaluation method [37], energy analysis method [38], and the law of space–time change method [39]. Foreign scholars have introduced land use change into the study of the coordination degree between economic development and ecological environment, focusing on the interaction between the economy and environment [40,41]. Chinese scholars focus on measuring, analyzing and evaluating this coordination from the multiple perspectives of economy, environment, society, culture, information and urbanization [42].

Hainan Province is an important strategic fulcrum of the "21st Century Maritime Silk Road". Since 2018, the Central Committee of the Communist Party of China and the State Council have proposed to establish a Chinese pilot-free trade zone and build a national ecological civilization pilot zone in Hainan. Therefore, in the context of the national strategy, how to achieve the optimal balance of ecosystem services on the province based on the premise of not destroying the ecological environment has important theoretical significance and practical value for understanding the development and utilization of land resources and ecological environmental protection. In this paper, normalized vegetation index (NDVI) and net primary productivity (NPP) data were used for the first time to improve the biomass correction factor to represent the spatial heterogeneity of physical geography. In addition, we increased the willingness to pay, the ability to pay, and the scarcity of resources to construct a socio-economic adjustment factor to reflect economic development. We constructed a regional ESV evaluation model to obtain more accurate evaluation results based on the actual situation of the study area. Meanwhile, this paper quantitatively reflects the coordinated development degree of environment and economy from the perspective of coupling analysis, and we used the new coupling method of the coordination degree of economy and environment to study their coordinated development. This paper chose Hainan Province as a research case to analyze the distribution pattern of Hainan Province's ecosystem, the characteristics of ESV temporal and spatial changes and the temporal and spatial relationship of the coordinated development of the economic environment based on these studies. This can provide a scientific reference for the rational allocation of land resources and the coordinated development of the economic environment. To this end, this study set five research goals, as follows:

(1) To explore the temporal and spatial variation characteristics of ecosystems in Hainan Province from 2000 to 2020;

(2) To explore the spatial–temporal variation characteristics of the ESV in Hainan Province from 2000 to 2020;

(3) To analyze the temporal and spatial characteristics of coordinated economic and environmental development in Hainan Province from 2000 to 2020;

(4) To analyze the driving factors of the ESV and economic environment coordination degree (EEC) in Hainan Province;

(5)   To put forward policy suggestions for coordinated economic and environmental growth.

2. Materials and Methods

2.1. Study Area

Hainan Province (18.80°–20.10°N, 108.37°–111.03°E) is located at the northern end of the continental shelf of the South China Sea, which has 18 cities and counties and covers an area of approximately 34,000 km². The province's landforms are diverse. Its topography is high in the middle and low on all sides. The terrain is a peak, with Wuzhishan (1867 m) and Yingge Ling (1811 m) as the core of the uplift. This gradually descends from the middle to the surrounding mountains, hills, platforms and plains (Figure 1). Hainan Province has a typical tropical monsoon maritime climate. It has a hot, rainy climate with long summers and no winter. The annual precipitation of 1000–2500 mm is unevenly distributed between the rainy and dry seasons, and across the province. The average annual temperature is 23.8–26.2 °C. Hainan Province is dry in the west and wet in the east [43,44]. Hainan has witnessed rapid social and economic development, rapid urban expansion and drastic land changes since the construction of Hainan International tourism island in 2009, which have profoundly affected the structure and function of the entire island ecosystem. This profoundly affects the structure and function of the entire province ecosystem. At present, the province's land use is mainly forest and farmland, which account for nearly 90% of the total area. The province is rich in forest resources and diverse in ecosystems and has the most intact tropical natural forest in the country, which has laid a solid foundation for the development strategy of Hainan's ecological province and the construction of the national ecological civilization experimental zone.

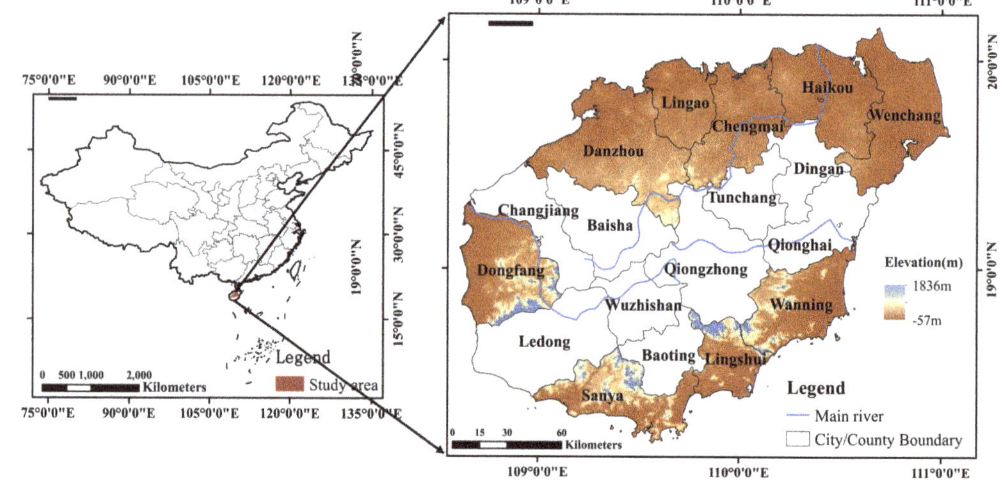

Figure 1. Overview of the study area.

2.2. Data Sources and Preprocessing

The land use classification data of Hainan Province in 2000, 2005, 2010, 2015 and 2020 came from the Resource Science and Satellite Center (http://www.resdc.cn/, accessed on 31 August 2021), with a spatial resolution of 30 m and a classification accuracy of more than 85% [45]. The Resource Science and Satellite Center's classification system includes 6 primary classes and 22 subclasses. On this basis, this study established an ecosystem classification system table (Table 1). We divided ecosystems into 8 primary classes: farmland, forest, grassland, shrubs, wetlands, bare land, river, and construction land.

Table 1. Ecosystem classification system.

Ecosystems Classes	LUCC Classes	
Primary Classes	**Subclasses**	**Codes**
Forest	Forestland	21
	Sparse woods	22
	Other forestland	24
Shrubs	Shrubs	23
Farmland	Paddy field	11
	Dry farmland	12
Grassland	High-covered grassland	31
	Medium-covered grassland	32
	Low-covered grassland	33
River	Rivers and canals	41
	Lakes	42
	Reservoir pond	43
Wetlands	Tidal flat	45
	Beach	46
	Marshland	64
Construction land	Urban residential area	51
	Rural residential area	52
	Other construction land	53
Bare land	Desert sand	61
	Saline-alkali land	63
	Bare land	65
	Bare rock	66

The socioeconomic data such as food crop yield, area and agricultural product prices, agricultural product price index, population, and per capita GDP needed to evaluate the ESV were derived from the "China Statistical Yearbook", "Hainan Statistical Yearbook", and "National Agricultural Products" compilations of cost–benefit information and national economy and development reports of various years. The administrative division data and digital elevation model (DEM) data came from the geographic and national conditions monitoring cloud platform (http://www.dsac.cn, accessed on 28 August 2021).

The NPP data were mosaicked by the MODIS satellite MOD17A3HGF.006 in the Google Earth Engine (GEE) platform based on the shapefile data of Hainan Province, with a spatial resolution of 500 m. The NDVI data were calculated by using the annual maximum value synthesis method for Landsat TM/OLI image data after cloud removal on the GEE cloud platform, with a spatial resolution of 30 m.

2.3. Methods

Figure 2 is the flow chart of the calculation and driving force analysis of ESV and EEC in Hainan Province. It mainly includes five steps: (1) Using land use data to classify ecosystem data; the spatial and temporal variation characteristics of ecosystem data were analyzed and the ecosystem service value equivalent per unit area of Hainan Province was constructed based on the ecosystem service value equivalent per unit area of China. (2) Calculating a standard equivalent value and socio-economic adjustment coefficient using social and economic statistics data; the annual NPP and NDVI data were calculated by the GEE platform and administrative division data to improve the biomass factor adjustment factor. (3) Calculating the ESV results of Hainan Province and analyzing the spatiotemporal variation characteristics of ESV. (4) Calculating the economic environment coordination index through the coupling model of ESV per unit area and GDP per capita, and analyzing the spatiotemporal variation characteristics of EEC. (5) Driving factors of ESV and EEC are analyzed by using the GeoDetector model (9 driving factors, such as DEM and cultivated land area).

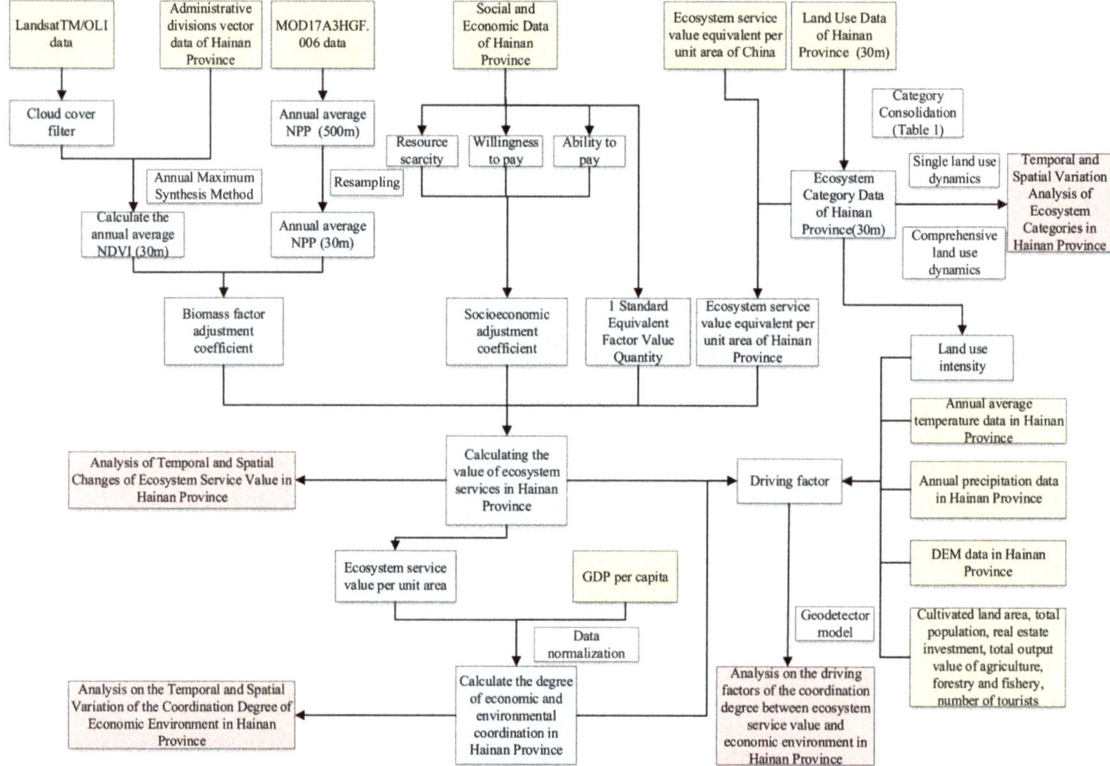

Figure 2. Technology roadmap for computational and driving force analysis of ESV and EEC.

2.3.1. Ecosystem Category Analysis Method

We used single land use dynamics [46] and comprehensive land use dynamics [47] to analyze the number and degree of ecosystem types and regional differences in ecosystem changes. These results reflected the severity of changes in ecosystem types in the assessment area.

The state of the ecosystem was expressed by the land use intensity (LUI), which reflects the effects of human activities on the natural world [48]. The calculation method is as follows:

$$LUI = \frac{S_{CLE}}{S} \times 100\% \tag{1}$$

$$S_{CLE} = \sum_{i=1}^{n}(SL_i \times EC_i) \tag{2}$$

where LUI is the land use intensity; S_{CLE} is the equivalent area of construction land; S is the total area of the region; SL_i is the area of land use/cover type i; EC_i is the equivalent conversion coefficient of construction land for land use/cover type i; and n is the land use/cover type in the area.

According to the two-level stepwise algorithm: (1) The first level. We set the eigenvalue of the unchanged surface natural property and unused natural cover as 0, and the eigenvalue of the artificial surface layer and blocked water, nutrient, air and heat exchange above and below the surface as 1. In other words, we used natural cover change and the exchange of water, nutrients, air and heat above and below the surface as indicators, and divided them into 5 equal parts from 0 to 1. (2) The second level. We marked the unchanged surface natural cover being used, the changed natural cover—planting perennial plants, and the changed natural cover—planting 1-year-old crops, and divided them into 3 equal grades from 0 to 0.2. We obtained the conversion coefficient of the construction land equivalent for different land use/cover types according to the two-level stepwise algorithm [48], as shown in Table 2.

Table 2. Equivalent conversion coefficients of different land use types.

Land Use Type	Equivalent Conversion
Farmland	0.200
Forest	0.067
Shrubs	0.060
Grassland	0.100
Wetlands	0.200
Construction land	1.000
Bare land	0.001
River	0.002

2.3.2. Dynamic ESV Evaluation Method

Xie et al. revised the ecosystem service classification and equivalent table according to the specific situation in China [49]. On the basis of Xie et al.'s research, we added the socioeconomic adjustment coefficient and spatial heterogeneity adjustment coefficient [50]. Then, the equivalence factors were matched according to the status of Hainan's ecosystem. Finally, the Hainan Ecosystem Service Value Equivalent Table of Unit Area was formulated (Table 3).

Specifically, the standard equivalent coefficient is defined as 1/7 of the economic value of farmland's annual grain. To eliminate the impact caused by price fluctuations over the years, we increased the Consumer Price Index (CPI). We used formula (3) and related statistics to convert all prices based on 2015 prices. According to relevant statistical data, we collected the national average price and national output per unit area of each grain crop in 2000, 2005, 2010, 2015 and 2020. We found that the standard equivalent factor value of Hainan Province in each year was 1223.98 CNY/hm^2, 1289.59 CNY/hm^2, 1533.58 CNY/hm^2, 1627.53 CNY/hm^2, and 1472.49 CNY/hm^2, respectively. Finally, the average value of the data in the five periods was 1429.44 CNY/hm^2, as the standard equivalent factor value.

With the development of the economy, people will pay more attention to protecting ecological systems and the ecological environment, which means that the costs people must pay to protect the ecosystem are becoming increasingly affordable. We adjusted for this by adding the socioeconomic development coefficient [51]. The socioeconomic development coefficient includes the ability to pay represented by GDP, the willingness to pay represented by Engel's coefficient, and the resource scarcity coefficient represented by population density.

As the ecosystem changes in different years and different regions, the ESV also changes accordingly. Biomass has a great impact on ecosystems and it is positively correlated with ESV. In this paper, annual NPP and annual NDVI were used as biomass adjustment coefficients reflecting natural biomass. We used these data to modify ESV as a coefficient of spatial heterogeneity.

Some studies assert that construction land provides different ESV (such as soil conservation and resource consumption). Therefore, this paper refers to the relevant literature [47,52,53], and we think that construction land contributes to ESV. Tables 3 and 4 show the unit area equivalent factor and unit area equivalent value of Hainan Province, respectively. The specific formula for calculating ESV is as follows:

$$y_n = \begin{cases} 100, n = 2000 \\ x_n, n = 2001 \\ x_n \times y_{n-1}/100, n = 2002, 2003, \ldots \end{cases} \tag{3}$$

$$ESV = \sum VC \times P_j \times A_j \times PI \tag{4}$$

$$VC = \frac{1}{7} \times \sum_{i=1}^{n} m_i p_i q_i / M \times \frac{R}{R_0} \tag{5}$$

$$P_j = \left(\frac{B_j}{\overline{B}} + \frac{N_j}{\overline{N}} \right) / 2 \tag{6}$$

$$PI = W_t \times A_t \times S \tag{7}$$

$$W_t = \left(\frac{2}{1 + e^{-\frac{1}{En_r \times (1-P_u) + En_u \times P_u} + 2.5}} \right)_A \Big/ \left(\frac{2}{1 + e^{-\frac{1}{En_r \times (1-P_u) + En_u \times P_u} + 2.5}} \right)_N \tag{8}$$

$$A_t = \frac{pGDP_A}{pGDP_N} \times \frac{U_A}{U_N} \tag{9}$$

$$S = \ln P_A / \ln P_N \tag{10}$$

where x_n refers to the CPI in the nth year relative to the previous year = 100; y_n refers to CPI in the nth year relative to the base year 2000 = 100; n = 2000, … ,2020; ESV is the total value of ecosystem service functions (CNY); VC is the value coefficient of ecosystem service function (CNY/hm^2); P_j is the adjustment coefficient of spatial heterogeneity; A_j is the area of type j ecosystem (hm^2); and PI is the social and economic adjustment coefficient. i is the type of grain in Hainan Province, of which the main grain types are rice, soybeans, etc.; n is the number of food categories; m_i is the area of type i grain (hm^2); p_i is the national yield per unit area of type i grain (kg/hm^2); q_i is the national average price of category i grain (CNY/kg); M is the total area of n kinds of grain (hm^2); R is the grain yield per unit area of farmland in Hainan Province (t/hm^2); and R_0 is the national grain output per unit area of farmland (t/hm^2). B_j is the average annual NPP (t/hm^2) of ecosystem type j in Hainan Province; $\overline{B}$ is the average annual NPP of this type of ecosystem nationwide (t/hm^2); N_j is the average annual NDVI value of the category j ecosystem in Hainan Province; and $\overline{N}$ is the average annual NDVI value of this type of ecosystem across the country. W_t is the willingness to pay; En_u is the urban and rural Engel's coefficient in year t representing the urban Engel's coefficient and the rural Engel's coefficient, respectively; and P_u is the proportion (%) of the urban population in year t. A_t is the ability to pay; $pGDP_A$ is the GDP per capita of the study area (CNY); $pGDP_N$ is the national GDP per capita (CNY); U_A is the urbanization rate of the study area (%); and U_N is the national urbanization rate (%). S is the resource scarcity index; P_A is the population density of the study area (person/km^2); and P_N is the national population density (person/km^2).

Table 3. Ecosystem service value equivalent per unit area of Hainan Province.

Categories	Sub-Categories	Farmland	Forest	Shrubs	Grassland	Wetlands	Bare Land	River	Construction Land	Total
Supplying services	Food production	1.105	0.29	0.19	0.38	0.51	0	0.8	0.01	3.285
	Raw material production	0.245	0.66	0.43	0.56	0.5	0	0.23	0	2.625
	Water supply	−1.305	0.34	0.22	0.31	2.59	0	8.29	−7.51	2.935
Regulating services	Gas regulation	0.89	2.17	1.41	1.97	1.9	0.02	0.77	−2.42	6.71
	Climate regulation	0.465	6.5	4.23	5.21	3.6	0	2.29	0	22.295
	Environment purification	0.135	1.93	1.28	1.72	3.6	0.1	5.55	−2.46	11.855
	Hydrological regulation	1.495	4.74	3.35	3.82	24.23	0.03	102.24	0	139.905
Supporting services	Soil conservation	0.52	2.65	1.72	2.4	2.31	0.02	0.93	0.02	10.57
	Maintenance of nutrient circulation	0.155	0.2	0.13	0.18	0.18	0	0.07	0	0.915
	Biodiversity	0.17	2.41	1.57	2.18	7.87	0.02	2.55	0.34	17.11
Cultural services	Recreation and culture	0.075	1.06	0.69	0.96	4.73	0.01	1.89	0.01	9.425
	Total	3.95	22.95	15.22	19.69	52.02	0.2	125.61	−12.01	227.63

Table 4. Ecosystem service value per unit area in Hainan Province (CNY/hm^2).

Categories	Sub-Categories	Farmland	Forest	Shrubs	Grassland	Wetlands	Bare Land	River	Construction Land	Total
Supplying services	Food production	1579.53	414.54	271.59	543.19	729.01	0.00	1143.55	14.29	7232.97
	Raw material production	350.21	943.43	614.66	800.49	714.72	0.00	328.77	0.00	4774.33
	Water supply	−1865.42	486.01	314.48	443.13	3702.25	0.00	11,850.06	−10,735.09	1829.68
Regulating services	Gas regulation	1272.20	3101.88	2015.51	2816.00	2715.94	28.59	1100.67	−3459.24	13,022.20
	Climate regulation	664.69	9291.36	6046.53	7447.38	5145.98	0.00	3273.42	0.00	37,494.21
	Environment purification	192.97	2758.82	1829.68	2458.64	5145.98	142.94	7933.39	−3516.42	18,611.31
	Hydrological regulation	2137.01	6775.55	4788.62	5460.46	34,635.33	42.88	146,145.95	0.00	206,411.14
Supporting services	Soil conservation	743.31	3788.02	2458.64	3430.66	3302.01	28.59	1329.38	28.59	18,196.77
	Maintenance of nutrient circulation	221.56	285.89	185.83	257.30	257.30	0.00	100.06	0.00	1772.51
	Biodiversity	243.00	3444.95	2244.22	3116.18	11,249.69	28.59	3645.07	486.01	26,544.70
Cultural services	Recreation and culture	107.21	1515.21	986.31	1372.26	6761.25	14.29	2701.64	14.29	14,394.46
	Total	5646.29	32,805.65	21,756.08	28,145.67	74,359.47	285.89	179,551.96	−17,167.57	325,383.43

2.3.3. Calculation Method of EEC

The coordination index can reflect the strength of the interaction between different systems and the degree of coordination. In this paper, the ESV per unit area was used to characterize the ecological environment in different regions [42], and the per capita GDP was used to characterize the level of economic development in different regions [54]. We constructed a coordination index between the ESV and GDP per capita and established different coordination levels [29] (Table 5).

Table 5. Level of coordination degree between economic development and environmental conditions.

Level	High Maladjustment	Mid Maladjustment	Basic Coordination	Mid Coordination	High Coordination
D	(0,0.2]	(0.2,0.4]	(0.4,0.6]	(0.6,0.8]	(0.8,1]

(a) Data Standardization

To eliminate the influence of different dimensions on the calculation results of the coordination index, this paper used the range method to standardize the ESV per unit area and GDP per capita data in 2000, 2005, 2010, 2015, and 2020 [36], as follows:

$$QESV = \frac{QESV_0 - QESV_{\min}}{QESV_{\max} - QESV_{\min}} \tag{11}$$

$$QGDP = \frac{QGDP_0 - QGDP_{\min}}{QGDP_{\max} - QGDP_{\min}} \tag{12}$$

where $QESV$ and $QGDP$ are the standardized values of ESV per unit area and GDP per capita; $QESV$ and $QGDP_0$ are the original value of ESV per unit area and GDP per capita; $QESV_{\max}$ and $QGDP_{\max}$ are their maximum values; and $QESV_{\min}$ and $QGDP_{\min}$ are their minimum values.

(b) Coordination index

The degree of coordination index can not only reflect the strength of the interaction between economic development and regional environment, but also reflect the degree of coordination between them. This paper constructs the coordination index between ESV and GDP per capita.

$$T = \alpha \times QESV + \beta \times QGDP \tag{13}$$

$$C = \sqrt{\frac{QGDP \times QESV}{(QGDP + QESV/2)^2}} \tag{14}$$

$$D = \sqrt{T \times C} \tag{15}$$

where D is the EEC; the larger the value of D, the better the EEC in the region. C is the degree of coupling; T is the composite index; $QESV$ and $QGDP$ are the ESV per unit area and GDP per capita, respectively; and α and β are undetermined weights for the ESV per unit area and GDP per capita, respectively. Economic development is based on environmental quality, and environmental quality is affected by economic development. Therefore, we believe that economic development is as important as the ecological environment ($\alpha = \beta = 0.5$) [29].

2.3.4. Driving Force Analysis Method

To explore the driving factors of ESV (Y1) and EEC (Y2), we chose six socioeconomic and human activity factors: cultivated land area (X1); total population (X2); total real estate investment (X3); total output value of agriculture, forestry and fishery (X4); number of tourists (X5); LUI (X6); and three natural factors: altitude (X7); precipitation (X8); and temperature (X9) based on relevant research and the actual situation in Hainan Province. We used these factors for driving force analysis.

We used Geodetector to analyze the driving forces of ESV and EEC and used the bivariate correlation method to analyze the correlation between each driving force and ESV and EEC. Geodetector is a set of statistical methods to detect spatial differentiation and reveal the driving force behind it [55]. Factor detection in the Geodetector can detect the spatial divergence of variable Y, which is measured by the q value. The calculation method is as follows:

$$q = 1 - \sum_{h=1}^{L} N_h \sigma_h^2 / N \sigma^2 \qquad (16)$$

where $h = 1, \dots ; L$ is the stratification of variable Y or factor X; N_h and N are the number of units in layer h and the whole area, respectively; and σ_h^2 and σ^2 are the variance of the Y value of layer h and the whole area, respectively. The value range of q is [0, 1]. The larger the value of q, the more obvious the spatial divergence of Y.

3. Results

3.1. Ecosystem Category Changes

The spatial distribution of different ecosystems in Hainan Province over the past 20 years is shown in Figure 3a–e. The ecosystem of Hainan Province was dominated by forest, which accounted for more than 55% of the entire province and was mainly distributed in the province's central and southern regions. Forest was followed by farmland, which accounted for approximately 26% of the entire province and was mainly distributed in the north and around the mountains. The total area of the two ecosystems was close to 82% of the total area of Hainan Province. Other types of ecosystems accounted for a relatively small area. As seen from the spatial distribution, the transformation of Hainan Province's ecosystem was mainly concentrated in cities, counties, urban areas and coastal areas. Among them, the Haikou, Sanya, Danzhou and Dongfang city urban areas, Yangpu Economic Development Zone, and coastal areas northeast of Wenchang caused major changes in ecosystem types due to urban construction. The main feature of these changes was the conversion of farmland and forest to construction land. The construction of large-scale reservoirs has led to the conversion of farmland, forest, and grassland into wetlands. Second, as a result of Hainan's afforestation activities on Treasure Island, grassland, farmland and other suitable forestland in Wenchang, Dongfang and other cities and counties were converted into forests in large areas. This somewhat guaranteed the balance of forest resources across the province.

The area classifications and proportions of ecosystems in Hainan Province from 2000 to 2020 are shown in Table 6. The single land use dynamics and change rates of each ecosystem during the study period are shown in Figure 4. From 2000 to 2020, only river, wetlands, and construction land were in a state of increase, while the rest of the ecosystem types were in a state of decline. Among them, the area of construction land had the largest increase and increase rate. Its rate of increase reached 87.28%, with an additional area of 65,307 hm^2. Construction land showed a continuous increasing trend and explosive growth in 2010–2015. The newly increased areas of river and wetlands were 8287 hm^2 and 5364 hm^2, with increases of 53.15% and 4.71%, respectively. Forest was the most important type of ecosystem in Hainan Province; it showed a trend of first increasing and then decreasing, reducing its area by 25,396 hm^2, and its area ratio dropped by 1.32% from 2000 to 2020. Farmland, grassland, shrubs and bare land showed a fluctuating downward trend. Their area reductions were 30,556 hm^2, 7940 hm^2, 9936 hm^2, and 5134 hm^2, respectively, and their area proportions decreased to 3.85%, 2.58%, 6.43%, 3.94%, and 40.15%, respectively. In terms of area, farmland was the most reduced, followed by forest, shrubs, grasslands and bare land. Bare ground had the largest decline because the area was actively small. In terms of the overall degree of change, construction land had the largest change, followed by bare land. Forest had the smallest change, which meant that forest was the most stable in the ecosystem of Hainan Province.

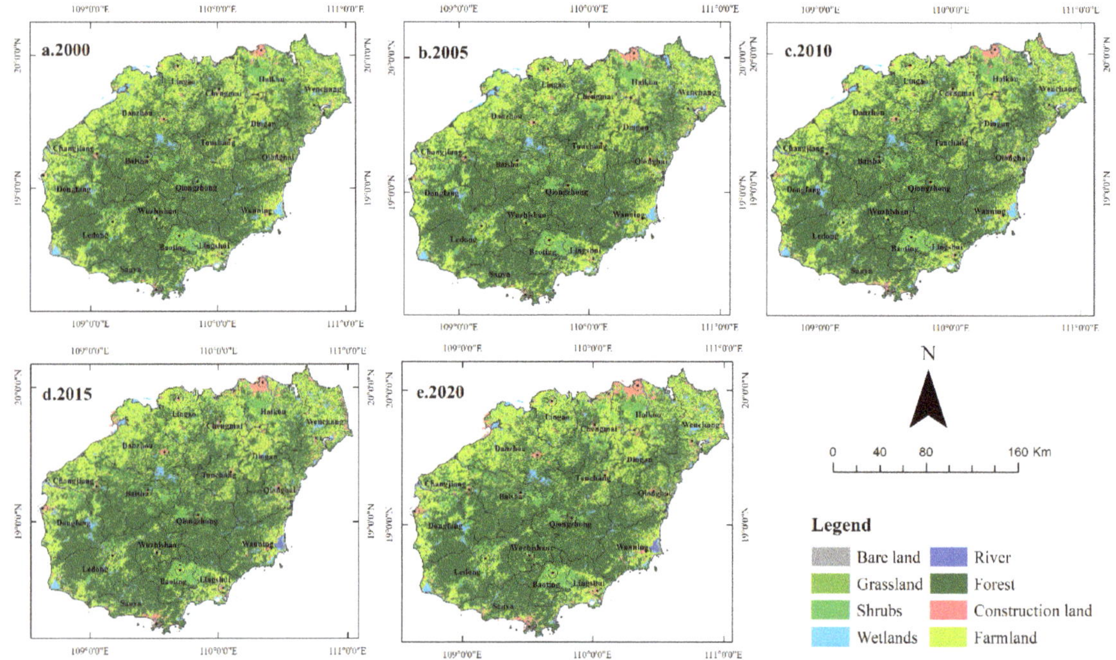

Figure 3. Spatial pattern of ecosystems in Hainan Province from 2000 to 2020. Note: (**a**) spatial pattern of ecosystems in Hainan Province in 2000; (**b**) spatial pattern of ecosystems in Hainan Province in 2005; (**c**) spatial pattern of ecosystems in Hainan Province in 2010; (**d**) spatial pattern of ecosystems in Hainan Province in 2015; and (**e**) spatial pattern of ecosystems in Hainan Province in 2020.

Table 6. Areas and percentages of different ecosystem classes in Hainan from 2000 to 2020.

Ecosystem Class	2000		2005		2010		2015		2020	
	Area (hm²)	Percentage (%)	Area (hm²)	Percentage (%)	Area (hm²)	Percentage (%)	Area (hm²)	Percentage (%)	Area (hm²)	Percentage (%)
Forest	1,929,453	56.38	1,933,173	56.41	1,926,740	56.32	1,910,746	55.86	1,904,057	55.63
Farmland	900,232	26.30	894,627	26.11	888,736	25.98	871,892	25.48	869,676	25.41
Construction land	74,821	2.19	81,821	2.39	90,870	2.66	127,502	3.73	140,128	4.09
Shrubs	252,229	7.37	249,423	7.28	247,214	7.23	244,369	7.14	242,293	7.08
Grassland	123,404	3.61	117,008	3.41	113,545	3.32	115,130	3.37	115,464	3.37
Wetlands	113,959	3.33	124,929	3.65	127,032	3.71	120,021	3.51	119,323	3.49
River	15,593	0.46	16,383	0.48	18,521	0.54	24,105	0.70	23,880	0.70
Bare land	12,788	0.37	9624	0.28	8578	0.25	7569	0.22	7654	0.22

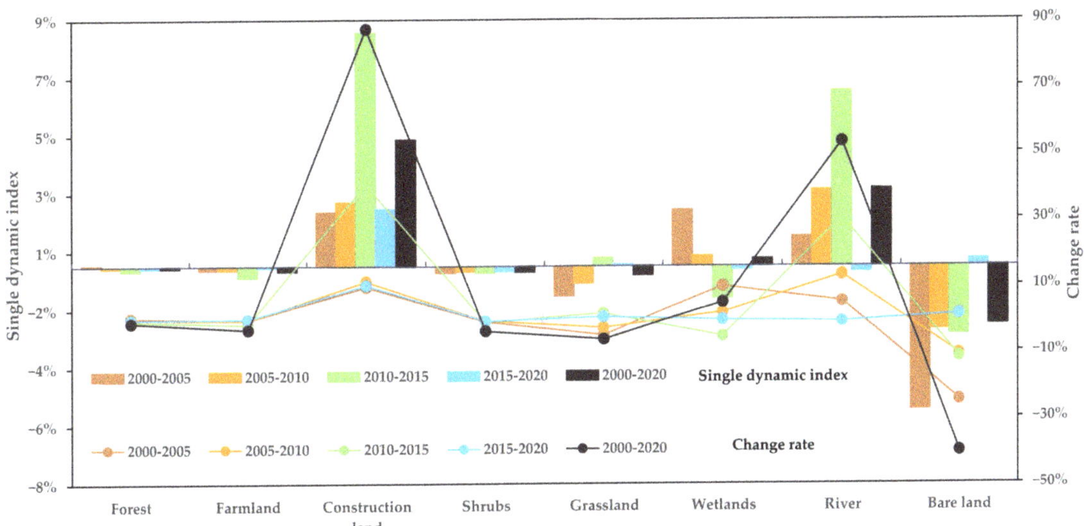

Figure 4. Single dynamic index and change rate of land use types in Hainan from 2000 to 2020.

3.2. Spatial and Temporal Analysis of ESV

As seen in Tables 7 and 8, the ESV in Hainan Province showed a trend of first decreasing, then increasing, and then decreasing, and it showed a downward trend as a whole. The ESV dropped from 206.842 billion CNY in 2000 to 175.615 billion CNY in 2005, then increased to 177.286 billion CNY in 2010, increased to 184.399 billion CNY in 2015, and finally decreased to 173.537 billion CNY in 2020. In the past 20 years, the ESV decreased by 33.305 billion CNY, a decrease of 16.1%. The ESV was the highest in 2000, which was mainly due to the biomass adjustment coefficient, the socioeconomic coefficient was the highest (Figure 5), and this year had the least construction land area. In terms of ecosystem types, forest had the highest ESV contribution rate, as high as 77.7%. This was followed by farmland, shrubs, wetlands, grassland, and river; their proportions were 5.9%, 5.8%, 5.3%, 3.8%, and 2.4%, respectively. The ESV contribution rate of bare land was the lowest, at only 0.02%, which showed that the ESV of bare land had the least impact on the overall ESV of Hainan Province. In the past 20 years, only the ESV of wetlands and river showed an upward trend, which was mainly due to the increase in the area of wetlands and river for the construction of reservoirs and fishponds. The ESV of the remaining ecosystems all showed a downward trend. The reason was not only the decline in the area of the remaining ecosystems but also the decline in the biomass adjustment coefficient and the socioeconomic coefficient (Figure 5). Construction land always had a negative effect on the ESV. Although the biomass and socioeconomic regulation factors declined from 2000 to 2020, the ESV of construction land dropped by 68.72%, which showed that the construction land in Hainan Province had increased significantly.

Figure 6 shows the ESV per unit area of different ecosystems. As shown in Figure 6, the river had the highest ESV per unit area, followed by wetlands, forest, grassland, shrubs, farmland and bare land. The ESV per unit area of construction land was always negative. This result shows that to increase the ESV in Hainan Province, we should increase the areas of river, wetlands, and forest to ensure the steady progress of the policy of returning farmland to forest and wetland protection.

Table 7. ESV and proportion by ecosystem class in Hainan Province (unit: 10^6 CNY).

Year	Factor	Forest	Shrubs	Wetlands	Grassland	Farmland	River	Bare Land	Construction Land	Total
2000	ESV	1625.58	124.74	94.22	79.59	122.87	38.31	0.43	−17.32	2068.42
	Percentage	78.59%	6.03%	4.56%	3.85%	5.94%	1.85%	0.02%	−0.84%	100.00%
2005	ESV	1381.84	101.81	87.55	65.41	101.32	34.16	0.37	−16.30	1756.16
	Percentage	78.69%	5.80%	4.99%	3.72%	5.77%	1.94%	0.02%	−0.93%	100.00%
2010	ESV	1378.82	104.17	93.88	65.67	105.68	42.34	0.40	−18.09	1772.87
	Percentage	77.77%	5.88%	5.30%	3.70%	5.96%	2.39%	0.02%	−1.02%	100.00%
2015	ESV	1433.40	107.46	101.35	70.95	109.88	48.64	0.38	−28.06	1843.99
	Percentage	77.73%	5.83%	5.50%	3.85%	5.96%	2.64%	0.02%	−1.52%	100.00%
2020	ESV	1348.13	100.60	98.28	67.29	103.81	46.12	0.36	−29.22	1735.38
	Percentage	77.69%	5.80%	5.66%	3.88%	5.98%	2.66%	0.02%	−1.68%	100.00%

Table 8. ESV change rate by ecosystem class from 2000 to 2020 in Hainan Province.

Year	Factor	Forest	Shrubs	Wetlands	Grassland	Farmland	River	Bare Land	Construction Land	Total
2000–2005		−14.99%	−18.38%	−7.08%	−17.82%	−17.54%	−10.85%	−14.93%	−5.90%	−15.10%
2005–2010	Change rate	−0.22%	2.31%	7.23%	0.40%	4.30%	23.97%	9.05%	11.01%	0.95%
2010–2015		3.96%	3.16%	7.95%	8.05%	3.97%	14.87%	−5.64%	55.10%	4.01%
2015–2020		−5.95%	−6.38%	−3.03%	−5.16%	−5.52%	−5.18%	−3.49%	4.14%	−5.89%
2000–2020		−17.07%	−19.35%	4.31%	−15.45%	−15.51%	20.38%	−15.52%	68.72%	−16.10%

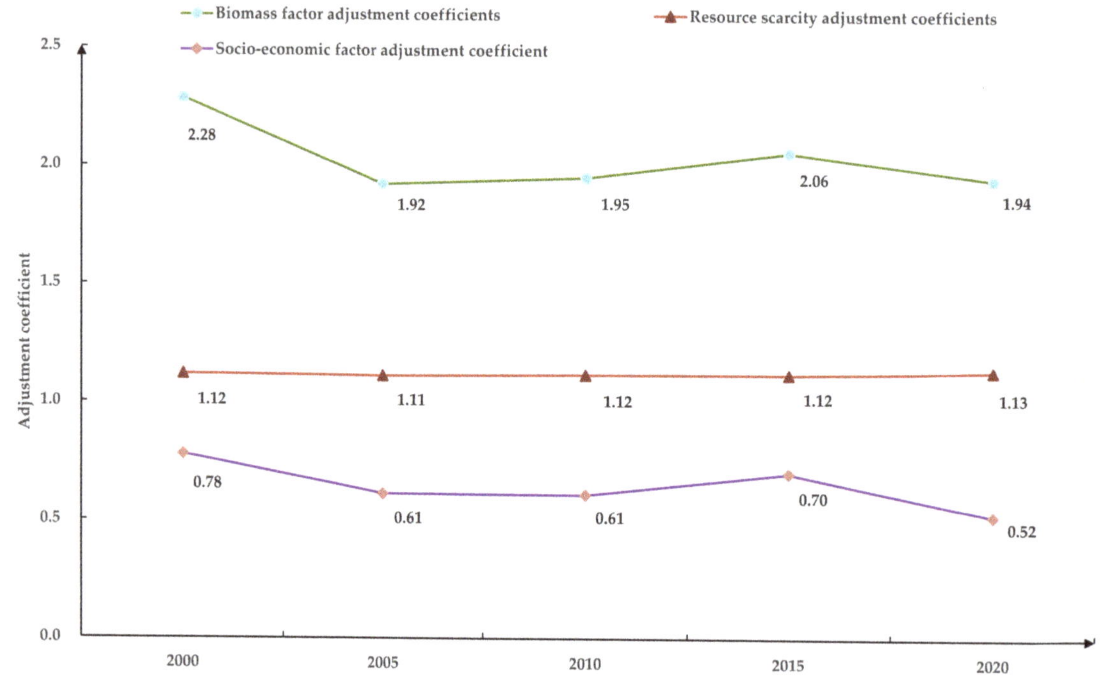

Figure 5. Adjustment coefficient of Hainan province from 2000 to 2020.

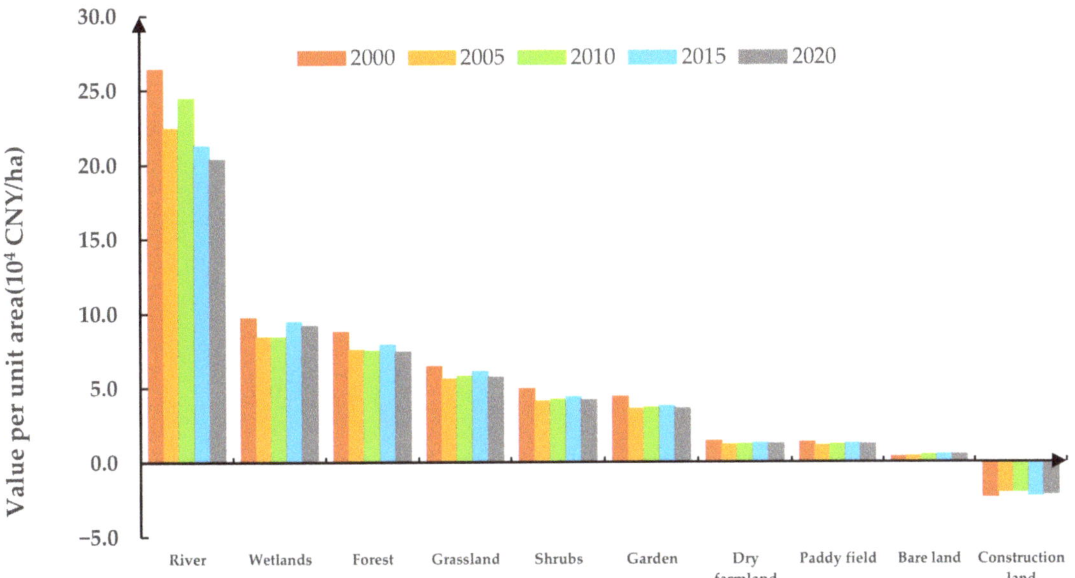

Figure 6. ESV value per unit area (CNY/ha) of ecosystem classes from 2000 to 2020.

Tables 9 and 10 showed the value and change rate of different ecological services from 2000 to 2020. Climate regulation and hydrological regulation were the main functions affecting ecosystem services, and their values accounted for more than 50% of the total value of Hainan Province. The value of maintenance of nutrient circulation and water supply accounted for the smallest proportion. Among these, water supply resources always had a negative impact on the ESV. This was mainly due to the excessive consumption of water resources on construction land and farmland, which showed that water conservation and protection in Hainan Province was urgent. From 2000 to 2020, the value of all individual ecosystem services decreased due to reductions in the biomass adjustment coefficient and socioeconomic adjustment coefficient, and the impact of changes in different ecosystem types. The value of the water supply continuously decreased, mainly because of the rapid expansion of construction land. Although the increase in river and wetlands had a positive effect on the water supply, the increase in the area of the two was much smaller than the increase in construction land, which led to a continuous decrease in the ESV. The reduction in raw material production, hydrological regulation, climate regulation, environmental purification, gas regulation, soil conservation, and recreation and culture was related to the continuous reduction in forest, grassland and shrubs. In terms of the rate of change, the rate in hydrological regulation was the lowest (−12.11%), followed by biodiversity (−15.03%). The change rates of hydrological regulation and biodiversity were lower than those of other ecosystem services mainly due to the increase in the area of wetlands and river, which indicates that the change in ecosystem types had an important impact on the function of ecological services.

Table 9. Proportion of ESV and ESV by different categories of ecosystem services in Hainan Province (unit: 10^6 CNY).

Year	Categories		Supplying Services			Regulating Services				Supporting Services			Cultural Services
	Sub-Categories		Raw Material Production	Food Production	Water Supply	Climate Regulation	Hydrological Regulation	Gas Regulation	Environment Purification	Soil Conservation	Biodiversity	Maintenance of Nutrient Circulation	Recreation and Culture
2000	ESV		60.32	53.81	−12.61	539.82	508.87	198.93	165.73	232.38	217.76	20.54	82.86
	Percentage		2.92%	2.60%	−0.61%	26.10%	24.60%	9.62%	8.01%	11.23%	10.53%	0.99%	4.01%
2005	ESV		51.10	44.87	−10.33	453.85	436.17	168.19	141.00	197.36	186.22	17.35	70.37
	Percentage		2.91%	2.55%	−0.59%	25.84%	24.84%	9.58%	8.03%	11.24%	10.60%	0.99%	4.01%
2010	ESV		51.41	45.91	−11.47	456.56	447.17	168.65	141.34	198.12	187.33	17.49	70.33
	Percentage		2.90%	2.59%	−0.65%	25.75%	25.22%	9.51%	7.97%	11.18%	10.57%	0.99%	3.97%
2015	ESV		53.70	47.89	−17.87	473.72	471.64	173.96	145.81	206.99	196.52	18.25	73.39
	Percentage		2.91%	2.60%	−0.97%	25.69%	25.58%	9.43%	7.91%	11.22%	10.66%	0.99%	3.98%
2020	ESV		50.57	45.36	−18.98	446.69	445.58	163.19	136.65	194.82	185.33	17.21	68.96
	Percentage		2.91%	2.61%	−1.09%	25.74%	25.68%	9.40%	7.87%	11.23%	10.68%	0.99%	3.97%

Table 10. Rate of change in ESV by different categories of ecosystem services in Hainan Province.

Year	Categories		Supplying Services			Regulating Services				Supporting Services			Cultural Services
	Sub-Categories		Raw Material Production	Food Production	Water Supply	Climate Regulation	Hydrological Regulation	Gas Regulation	Environment Purification	Soil Conservation	Biodiversity	Maintenance of Nutrient Circulation	Recreation and Culture
2000–2005	Change rate		−15.27%	−16.62%	−18.1%	−15.93%	−14.29%	−15.45%	−14.92%	−15.07%	−14.49%	−15.57%	−15.08%
2005–2010			0.61%	2.33%	11.04%	0.60%	2.52%	0.28%	0.24%	0.38%	0.60%	0.85%	−0.05%
2010–2015			4.44%	4.31%	55.83%	3.76%	5.47%	3.15%	3.16%	4.48%	4.90%	4.35%	4.35%
2015–2020			−5.82%	−5.29%	6.19%	−5.71%	−5.53%	−6.19%	−6.28%	−5.88%	−5.69%	−5.72%	−6.04%
2000–2020			−16.16%	−15.70%	50.53%	−17.25%	−12.44%	−17.97%	−17.55%	−16.17%	−14.89%	−16.24%	−16.78%

As shown in Figure 7, regulation services were the most important function of ecosystem services, with values of up to 68%. Second were support services, the value of which accounted for approximately 23%, and the value of supply services and cultural services accounted for approximately 5% and 4%, respectively. The change trend of the four services during the study period was the same as the total ESV, which showed a trend of first decreasing, then increasing, and then decreasing, and showed a downward trend as a whole.

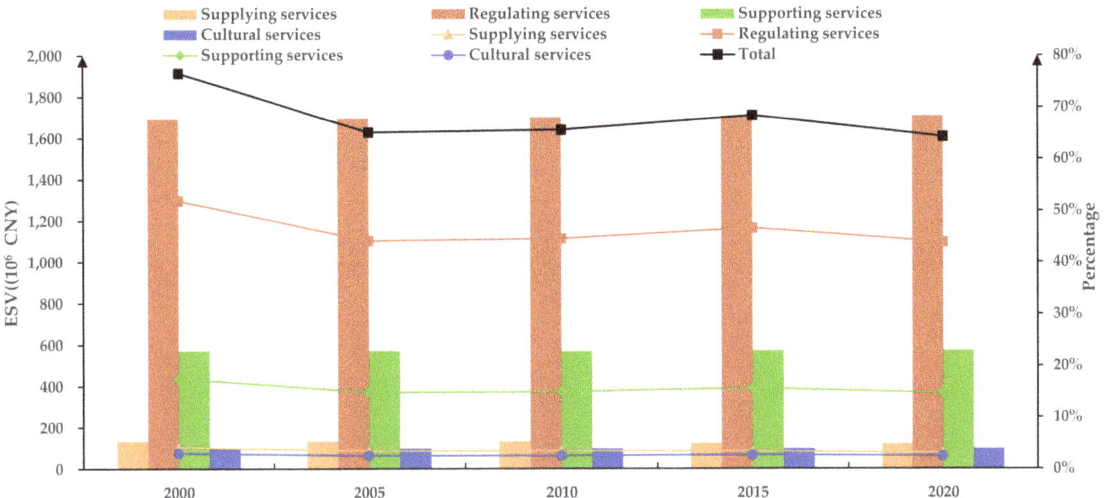

Figure 7. Four categories of ESV and the total ESV of Hainan.

In summary, the changes in ecosystems and the decline in the biomass adjustment coefficients and socioeconomic adjustment coefficients played important roles in the decline in ESV in Hainan Province. As shown in Figure 5, the biomass adjustment coefficient of Hainan Province dropped from 2.28 to 1.94, a decrease of 14.91% from 2000 to 2020. The advantages of Hainan Province's NPP and NDVI over China's average NPP and NDVI continue to decline. The socioeconomic adjustment coefficient dropped from 0.78 to 0.52, a decrease of 33.3%, which showed that the economic development speed of Hainan Province was slower than that of China.

Due to the unit area, the ESV value in some pixels (such as river ecosystem) was much higher than in other pixels. If we used the natural breakpoint method, it may result in poor picture display, while the artificial breakpoint method can control the picture display status and present a better display effect. Therefore, the artificial breakpoint method was used to divide the ESV into six categories, as shown in Figure 8. The ESV of Hainan Province conformed to the spatial distribution characteristics of being high in the middle and low in the surrounding area, which was highly consistent with the spatial distribution of elevation. The reason was the lower LUI in mountainous areas with higher altitudes, which were mainly distributed in forest and grassland ecosystems, and these areas had higher ESV. The surrounding areas with lower elevations had higher LUI, and they were mainly distributed in ecosystems such as construction land and farmland; these areas had lower ESV. The ecosystem with the highest ESV level in Hainan Province was mainly river, which was concentrated in major river areas, with some scattered in the middle. The lowest ESV ecosystem was mainly construction land, which was mainly concentrated in northern Hainan and coastal plains. The high-value areas of ESV in Hainan Province continued to slump and the low-value areas continued to expand from 2000 to 2020.

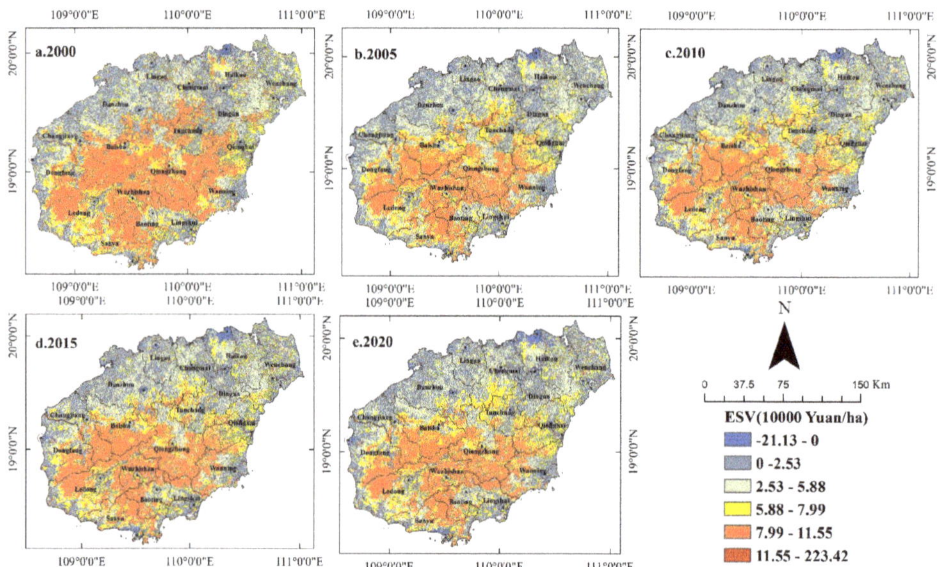

Figure 8. Distribution map of ESV in Hainan from 2000 to 2020. (**a**) Distribution map of ESV in Hainan in 2000; (**b**) distribution map of ESV in Hainan in 2005; (**c**) distribution map of ESV in Hainan in 2010; (**d**) distribution map of ESV in Hainan in 2015; and (**e**) distribution map of ESV in Hainan in 2020.

Figure 9 shows the ESV and ESV per unit area in different periods of cities and counties in Hainan Province. Figure 9b shows that the largest ESV in most cities and counties appeared in 2000, and the lowest ESV occurred in 2020. Among these cities and counties, Qiongzhong's ESV was always the largest, and Lingao's ESV was always the lowest. Figure 9b shows a clockwise direction from Lingao to Qiongzhong, and that the overall ESV of each city and county was on the rise. Figure 8a shows that the highest ESV per unit area of most cities and counties occurred in 2000, and the lowest ESV occurred in 2020, which was the same trend as the total ESV. Among these cities and counties, Wuzhishan always had the highest ESV per unit area, and Lingao always had the lowest ESV. Figure 9a shows a clockwise direction from Lingao to Wuzhishan, and the overall ESV per unit area of each city and county showed an upward trend. The ESV per unit area could be used to indicate the regional ecological quality, which showed that the ecological environment of Wuzhishan was the best and that of Lingao was the worst.

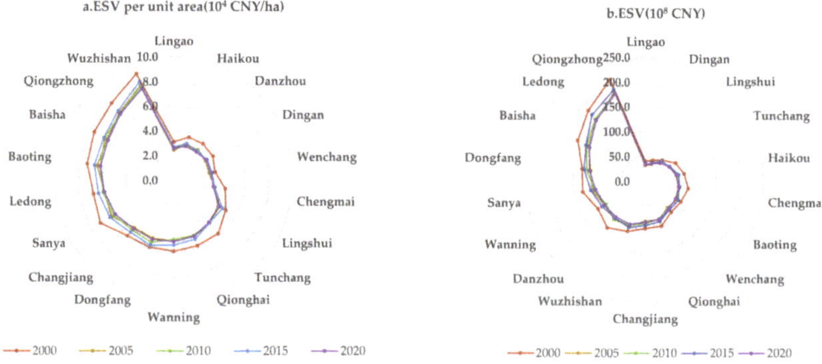

Figure 9. ESV and ESV per unit area change in 18 cities from 2000 to 2020. (**a**) ESV per unit area change in 18 cities from 2000 to 2020; (**b**) ESV area change in 18 cities from 2000 to 2020.

3.3. Changes in the EEC

The GDP of Hainan Province has increased significantly in the past 20 years. It increased from 52.682 billion CNY in 2000 to 553.239 billion CNY in 2020, which was an increase of 10.5 times and represents China's economic development speed since the beginning of the 21st century. The spatial distribution of GDP per capita is shown in Figure 10. The per capita GDP of Hainan Province changed dramatically in different years and increased rapidly from 2000 to 2020. Spatial distribution was contrary to the spatial distribution of ESV, which showed a distribution pattern of being low in the middle and high in the surrounding areas. Areas with high GDP per capita were mainly distributed in the coastal plains, with Haikou and Sanya having the highest GDP, and the central mountainous areas had low GDP per capita. In 2020, the GDP per capita of Sanya city was 88,900 CNY and that of Ledong County was only 31,000 CNY, which showed that the uneven economic development of Hainan Province was relatively obvious.

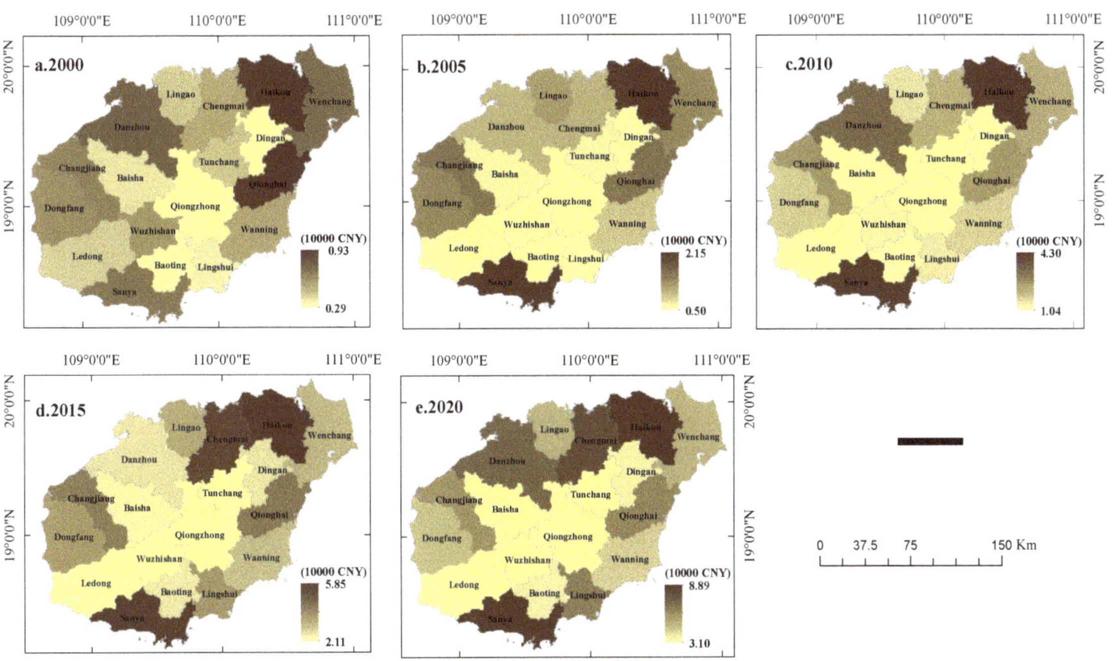

Figure 10. Spatial distribution of GDP per capita in Hainan Province. (**a**) Spatial distribution of GDP per capita in Hainan Province in 2000; (**b**) spatial distribution of GDP per capita in Hainan Province in 2005; (**c**) spatial distribution of GDP per capita in Hainan Province in 2010; (**d**) spatial distribution of GDP per capita in Hainan Province in 2015; and (**e**) spatial distribution of GDP per capita in Hainan Province in 2020.

The rates of change in GDP per capita in Hainan Province and each city and county are shown in Figure 11. The GDP per capita of Hainan Province rose rapidly. From 2000 to 2020, the province's average GDP per capita increased by 861%. Haikou, Sanya, Tunchang, and Wanning's growth rates exceeded 1000%. Wenchang and Qiongzhong had the lowest growth rates, at approximately 500%. The GDP per capita of cities and counties grew the fastest in 2005–2010 and 2010–2015.

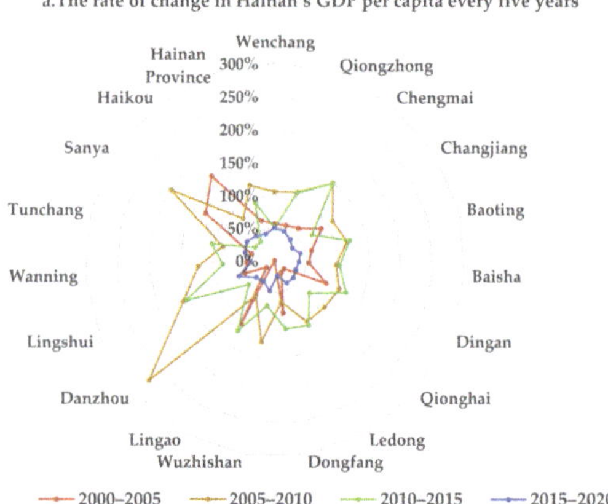

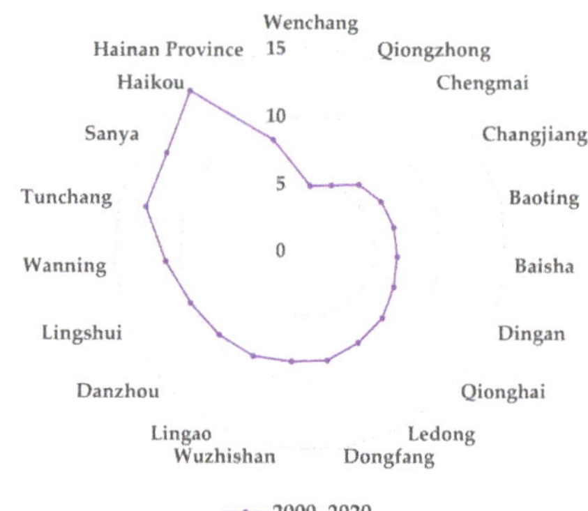

Figure 11. GDP per capita change rate of each city and county in Hainan. (**a**) The rate of change in Hainan's GDP per capita every five years, and (**b**) the rate of change in Hainan's GDP per capita from 2000 to 2020.

The EEC calculation result is shown in Table 10. The overall EEC of Hainan Province declined from 0.42 in 2000 to 0.36 in 2020. It first decreased, then increased, and then decreased, which was consistent with the change trend of the ESV. This was mainly because the GDP per capita was in a state of rapid development, but the decrease in the ESV caused the EEC to show a negative growth trend, and the increase in the ESV caused the EEC to show a positive growth trend. Hainan Province was in a state of basic coordination in only 2000, and it was in a state of mid maladjustment in 2005, 2010, 2015, and 2020.

According to the grading method in Table 5, we divided the EEC into five grades. Therefore, the spatial distribution map of the Hainan EEC was obtained (Figure 12), and the area of each grade was measured (Table 11). As seen in the figure, the EEC in the central part of Hainan Province was low in all years. This result was mainly because the GDP per capita in these areas was low but the ESV value was high, so the economic environment development was not coordinated. The EEC was relatively high in the round-island areas, which was mainly because the GDP per capita and ESV in these areas were both high, and the economic and environmental development was more coordinated. In 2000, 2010, 2015, and 2020, Hainan Province had the most land at the basic coordination level. In 2005, Hainan Province had the most land at the mid-maladjustment level. This result was mainly due to the large decline in ESV in 2005 and the rapid growth of GDP per capita. From 2000 to 2020, the change trend of land area in a maladjusted state was the opposite to that of ESV, and the change trend of land area in a coordinated state was the same as that of the ESV. There was very little land with mid and high levels of coordination, scattered around rivers with high ESV values. Since 2015, the area of moderately coordinated land has increased in Sanya, which shows that the EEC of Sanya has improved.

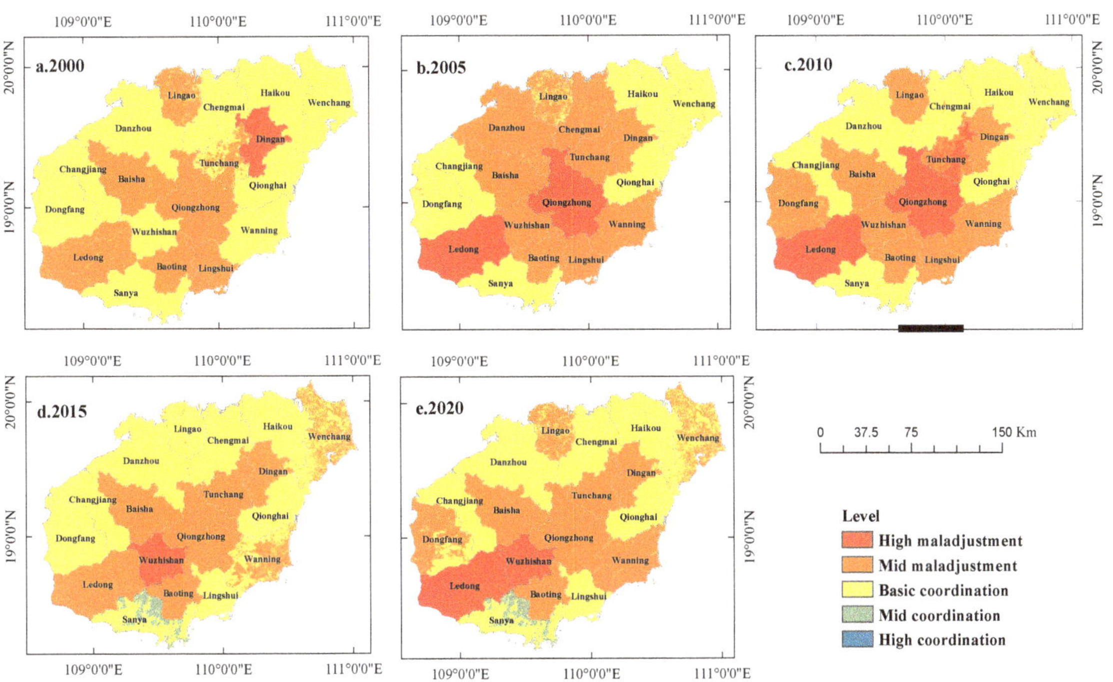

Figure 12. Spatial distribution map of the EEC in Hainan. (**a**) Spatial distribution map of the EEC in Hainan in 2000; (**b**) spatial distribution map of the EEC in Hainan in 2005; (**c**) spatial distribution map of the EEC in Hainan in 2010; (**d**) spatial distribution map of the EEC in Hainan in 2015; and (**e**) spatial distribution map of the EEC in Hainan in 2020.

Table 11. Area statistics of EEC level (km^2).

Year	Coordination Level	High Maladjustment	Mid Maladjustment	Basic Coordination	Mid Coordination	High Coordination
2000	Area	1175.43	11,619.16	20,967.4	89.23	1.09
	Percentage	3.472%	34.323%	61.938%	0.264%	0.003%
2005	Area	5414.21	16,236.91	12,139.39	60.28	1.52
	Percentage	15.994%	47.964%	35.860%	0.178%	0.004%
2010	Area	5866.15	13,038.4	14,863.18	83.14	1.45
	Percentage	17.329%	38.516%	43.906%	0.246%	0.004%
2015	Area	1188.39	13,357.01	18,591.9	712.46	2.41
	Percentage	3.511%	39.457%	54.921%	2.105%	0.007%
2020	Area	3895.89	14,133.5	15,181.4	640.2	1.32
	Percentage	11.508%	41.750%	44.846%	1.891%	0.004%

Table 12 shows counties and cities in ascending order of coordination. Sanya's EEC was the highest in 2015 and 2020, Haikou's EEC was the highest in 2005 and 2020, and Qionghai's EEC was the highest in 2000. They were all in a basic coordinated state. Sanya, Haikou, Wenchang, and Changjiang were all in a basically coordinated state over the past 20 years, which showed that the GDP growth of these regions also improved the ESV. Ledong, Qiongzhong, Baisha, Tunchang, Ding'an, and Baoting were all in a state of imbalance in the past 20 years due to the low GDP per capita and high ESV in these areas, so it was necessary to vigorously increase the speed of economic development.

Table 12. Environmental and economic coordination index in Hainan.

Coordination Index	2000	2005	2010	2015	2020
Hainan Province	0.42	0.32	0.35	0.39	0.36
Ledong	0.38	0.12	0.13	0.24	0.11
Wuzhishan	0.47	0.24	0.29	0.01	0.14
Qiongzhong	0.28	0.12	0.17	0.24	0.23
Baisha	0.36	0.28	0.25	0.29	0.24
Tunchang	0.39	0.27	0.20	0.23	0.24
Dingan	0.11	0.24	0.25	0.34	0.30
Baoting	0.30	0.22	0.26	0.35	0.32
Wanning	0.44	0.35	0.37	0.40	0.37
Dongfang	0.46	0.43	0.38	0.45	0.39
Lingao	0.39	0.39	0.34	0.42	0.39
Wenchang	0.49	0.42	0.42	0.40	0.40
Changjiang	0.48	0.45	0.45	0.48	0.45
Qionghai	0.57	0.45	0.45	0.48	0.46
Lingshui	0.33	0.23	0.33	0.45	0.47
Danzhou	0.50	0.36	0.49	0.50	0.48
Chengmai	0.43	0.37	0.42	0.53	0.51
Haikou	0.54	0.56	0.54	0.54	0.53
Sanya	0.50	0.50	0.57	0.58	0.58

3.4. Analysis of the Driving Force of the Coordination between ESV and EEC

The correlation coefficients between ESV and EEC and each driving force factor are shown in Table 13. The p value of each driving force factor was 0.000, which indicated that the results passed the hypothesis test with a significance level of 0.05. The ESV was negatively correlated with X1, X2, X3, X4, X5, X6 and X8. Among them, the ESV had the greatest correlation with LUI, with a value of -0.821. The ESV was positively correlated with X7 and X9. Among them, the ESV was highly correlated with X7, with was 0.61. This result showed that the ESV had a negative correlation with the impact of human activities. The ESV was lower in areas with more severe human activities, while it was higher in biomass-rich areas such as the mountainous areas of Hainan Province. The EEC was positively correlated with X1, X2, X3, X4, X5, X6 and X9. Among them, the EEC had a high correlation with socioeconomic factors and had a negative correlation with X7 and X8. The driving forces of EEC and ESV had opposite effects, which showed that the EEC was higher in the economically developed areas of Hainan Province.

Table 13. Correlation analysis of 9 driving forces of EEC and ESV.

Index		X1	X2	X3	X4	X5	X6	X7	X8	X9
Y1	Correlation	-0.300	-0.253	-0.203	-0.278	-0.095	-0.821	0.610	-0.010	0.032
	p value	0.000	0.000	0.000	0.000	0.000	0.000	0.000	0.000	0.000
Y2	Correlation	0.499	0.666	0.726	0.579	0.581	0.150	-0.452	-0.079	0.030
	p value	0.000	0.000	0.000	0.000	0.000	0.000	0.000	0.000	0.000

The ESV and EEC driving force results obtained by using the Geodetector are shown in Table 14. The p value of each driving force factor was 0.000, which showed that the results passed the hypothesis test with a significance level of 0.05. Among the driving force factors of ESV, the contribution rate of X6 was the highest, with a value of 0.712. Next, X7 had a contribution rate of 0.411. X4, X2 and X1 had contribution rates of 0.231, 0.159 and 0.137, respectively. The contribution rates of the remaining driving force factors were all below 0.1, and the contribution rates of X8 and X9 were the lowest. The ESV was mainly affected by human activity factors and natural factors, and socioeconomic factors had little influence. Among the driving force factors of the EEC, X3 had the highest contribution rate, with a value of 0.679. It was followed by the contribution rates of X2, X1, X4 and X5, which had values of 0.635, 0.573, 0.561, and 0.555, respectively. X7 and X8 had contribution rates of 0.328 and 0.194, respectively. The contribution rate of X6 was the lowest. The EEC was mainly affected by socioeconomic factors, and the contribution rate of each socioeconomic driving force was above 0.55. The EEC was less affected by natural factors.

Table 14. Driving force spatial differentiation.

Index		X1	X2	X3	X4	X5	X6	X7	X8	X9
Y1	q statistic	0.137	0.159	0.087	0.231	0.093	0.712	0.411	0.057	0.056
	p value	0.000	0.000	0.000	0.000	0.000	0.000	0.000	0.000	0.000
Y2	q statistic	0.573	0.635	0.679	0.561	0.555	0.057	0.328	0.194	0.074
	p value	0.000	0.000	0.000	0.000	0.000	0.000	0.000	0.000	0.000

4. Discussion

4.1. Improvement of Dynamic ESV Assessment Method

ESV assessments have been widely used. However, most studies have not considered the spatial heterogeneity of ESV [30,45,56,57]. In addition, the monetary estimation of ESV is affected by different factors, including resource scarcity, ability to pay, willingness to pay, market price and inflation [58], which will greatly affect the accuracy of the results. This paper comprehensively considered the social and economic development status and regional differences. We used biomass adjustment coefficients, socioeconomic adjustment coefficients and resource scarcity adjustment coefficients to modify the value of the equivalent factor and obtained accurate ESV assessment results. This method allowed us to better explore the degree of EEC in Hainan Province and understand the driving forces that affect ESV and EEC. The monetization of ecosystem services can enable people to better understand the important contribution of ecosystem services to society [5]. The ESV is beneficial to society as a whole, but it is not owned by individuals [59].

We used the biomass adjustment coefficient for the ESV spatial heterogeneity adjustment, and NPP was used to express the area's net primary production capacity. Hainan Province is at the forefront of the country in terms of ecological quality and it is rich in biological resources [60]. In addition, the equivalent factor of Xie et al. was based on the national level, and biological resources in northwestern China are very scarce [6]. Thus, the ratio of Hainan's NPP to the national NPP is very large. As an index to evaluate vegetation status, the NDVI can reflect the regional biomass status. To more accurately reflect the biomass ratio between Hainan Province and the country, we used the ratio of the average value of Hainan Province's NDVI and NPP to the national average value of the NDVI and NPP as the biomass adjustment coefficient to obtain a more accurate spatial heterogeneity distribution situation.

4.2. Ecosystem Changes

As this paper differentiated shrub ecosystem from forest ecosystem, the proportion of forest area is different from the proportion of forest in other studies, which accounts for more than 62% [61]. This study found that the ecosystem of Hainan Province was dominated by forest and farmland. From 2000 to 2020, the structure of the ecosystems was relatively stable, and its overall change was small, with a comprehensive rate of change of approximately 3.22%. Among the ecosystems, the area of construction land and river increased significantly, from 74,821 hm^2 to 140,128 hm^2 and 15,593 hm^2 to 23,880 hm^2, respectively. The 20-year rates of change were 87.28% and 53.15%, respectively. Due to the promotion of policies such as Hainan International Tourism Island, urban construction and land development further expanded from 2010 to 2015. The growth rate of construction land was the fastest in the past five years. The extensive development model was the main reason for the significant differences in the urbanization level among various regions in Hainan Province. Therefore, we recommend that the government further optimize the allocation of land resources and appropriately limit the expansion of urban construction land. We should rationally lay out the space, structure and scale of industry and urban development in accordance with the requirements of high-quality development and scientifically plan and improve the matching degree with various resources. In addition, we need to further optimize the urban spatial structure.

The higher productivity of aquaculture fishponds [62] and the construction of large reservoirs led to an increase in the wetland area, with a 20-year change rate of 4.71%. The remaining ecosystems all showed a decreasing trend. In the past 20 years, the ecosystem of Hainan Province was mainly transformed from farmland, forest, and wetlands, with contribution rates of 29.82%, 27.07%, and 8.18%, respectively. The transformed ecosystems were mainly construction land and wetlands, with contribution rates of 33.49% and 12.18%, respectively. Among them, the transfer-out and transfer-in ratios of wetlands were relatively large. The reason was that natural wetlands were constantly decreasing, while artificial wetlands such as fishponds were constantly increasing [63,64]. Ecosystem transformation was mainly manifested in the conversion of farmland, forest, and shrubs into construction land. In the past 20 years, their contribution rate was 64.51%, which showed that the rapid expansion of construction land had become the main factor in the loss of farmland and forest [56,65,66]. Construction land was mainly transferred out of farmland, which was similar to the results of previous studies [31,65,67]. This resulted in the reduction of a large amount of high-quality farmland and increased the fragmentation of farmland, leading to prominent food security problems [68]. There was also the conversion of wetlands to farmland and the conversion of natural wetlands to constructed wetlands. This was because although the overall area of wetlands has shown an increasing trend, the artificial wetlands in Hainan Province were expanding rapidly, and reclamation and breeding and port development [63] had severely damaged the natural wetland resources in Hainan Province. Therefore, natural wetlands urgently need to be protected.

4.3. ESV Status and Changes

Studies have shown that the ESV of Hainan Province dropped from 206.842 billion CNY in 2000 to 173.537 billion CNY in 2020 (Table 7), and the overall ESV biology has shown a downward trend, which is very consistent with the results of Sui et al. and Sun et al. [43,69]. Ecosystem changes and a decline in biomass adjustment coefficients and socioeconomic adjustment coefficients played important roles in the decline in ESV in Hainan Province.

As the dominant category of Hainan Province, forest ecosystem had the highest contribution rate to ESV, accounting for 78%. Forest's ESV was higher due to not only its larger area but also its higher ESV per unit area. It was the most important ecosystem category in Hainan Province, which is very consistent with the results of Deng et al. [70]. Rivers and wetlands had the highest ESV per unit area, which is very consistent with the results of Xie et al., Hasan et al., etc. [6,20,42,56,67,71]. However, their contribution rates to ESV were low, because the river and wetland areas of Hainan Province are small. Therefore, policies such as increasing river and wetland areas, returning farmland to wetland, and returning farmland to forest are important means to improve ESV [72]. The ESV of construction land had a negative value, and the rapid expansion of construction land in the past 20 years has provided increasingly negative values. Wuzhishan, Qiongzhong, Baisha and other places had relatively large areas of forest and wetlands, and the area of construction land was relatively small, so the unit ESV in these areas was relatively high (Figure 8). Many studies regard construction land as worthless in terms of research [49,73]. Some studies have also shown that construction land can provide some ecosystem services, such as culture and entertainment [53]. However, the ecological harm it brings will exceed the ecological benefits [52], which will cause the ecological environment to be destroyed, so we agree that construction land provides a negative value.

The first-level category of ecosystem services (Figure 6) showed that regulation services were the most important function, with the ESV accounting for up to 68%, followed by support services, with the ESV accounting for approximately 23% and cultural services accounting for the least, at approximately 4%, which is highly similar to many studies [58]. The secondary classification showed that climate regulation, hydrological regulation, soil conservation and biodiversity were the main ecological service functions of Hainan Province. Among them, water supply provided a negative value. The reason was that the high level of water consumption of construction land and farmland exceeded the water supply service of the ecosystem. Although Hainan Province is located on the equator and has plenty of rainfall, there is a significant engineering water shortage due to the large slope of the province area and the difficulty of storing freshwater resources [74]. Therefore, the problem of water shortages in Hainan urgently needs to be addressed.

4.4. EEC Status and Changes

Studies have shown that Hainan's GDP grew rapidly, increasing by 550.557 billion CNY. Its ESV dropped by 33.305 billion CNY, and the growth rate of ESV was lower than that of GDP from 2000 to 2020 [29]. The EEC declined from 0.42 to 0.36 and degraded from the basic coordinated state to the mid-maladjusted state. This result shows that the economic development of Hainan Province and the protection of its ecosystem were not well coordinated, and it was necessary to rationally allocate land resources and the development of the ecological economy. The change trends of EEC and ESV were consistent. The reason was that GDP per capita was in a state of rapid development, but the decline in ESV caused the EEC to show a negative growth trend. The economic development of Hainan Province was unevenly distributed on the province. The GDP per capita of the eastern, western and central parts of Hainan Province showed a stepwise downward trend (Figure 9). The distribution of the ESV was also uneven (Figure 7). The ESV was higher in the central region and lower in the east and west. These caused the uneven distribution of the EEC in Hainan Province. As Hainan Province led the country in ecological quality and its environment was relatively good, the ecological economy was a prominent shortcoming in the construction of Hainan's ecological civilization [75]. Therefore, the conclusion of this study is different from the literature, which believes that the low degree of economic environment coordination is due to the better regional economic development and the poorer environment [29,30]. As a key ecological protection area, Hainan was subject to multiple protections at the national and provincial levels [76]. We thought that the reason for the low degree of EEC in the central region of Hainan Province was that its environmental condition was good but the economic level was too low, and the economic level could not keep up with the environmental conditions. The reason for the higher degree of EEC in the eastern and western regions of Hainan was that these regions had higher economic levels, which could be coordinated with good environmental conditions. Therefore, the improvement of EEC in Hainan Province requires not only the rational allocation of land resources and the protection of forest, wetlands, and other ecosystems but also the vigorous development of ecological economy, considering the organic integration of economic construction and the various ecological elements of mountains, water, farmland, forest, lakes and grasses. Hainan should develop a characteristic ecological economy while protecting the ecology so that the economy and the environment can develop in harmony.

4.5. Drivers of ESV and EEC

Studies have shown that LUI was the most important factor affecting ESV. The LUI in Geodetector contributed the most to the ESV, with a value of 0.712 (Table 14). LUI was related to human activities and was the highest for construction land and farmland, which is highly similar to Xu et al. [48]. There was a significant negative correlation between LUI and ESV, which indicated that the expansion of construction land was an important factor in the reduction of ESV. These results are highly consistent with previous studies [43,47,56–58,67]. The ESV had a significant positive correlation with the DEM, which was affected by the distribution characteristics of Hainan's landforms. The high-altitude mountainous areas in the central part were mainly distributed in forest, wetlands, and other ecosystems with low LUI, so the ESV in these areas was high. The low-altitude coastal plains were mainly distributed in ecosystems with high LUI, such as farmland and construction land, so the ESV of these areas was low. The selected social and economic factors, such as cultivated land area and total real estate investment, had a low contribution rate to ESV. They were mostly negatively correlated with ESV. This result shows that social and economic development, such as urban expansion, population growth, and real estate development, will cause a certain degree of deterioration in ESV, which is very consistent with the results of Zhang et al., Lei et al., etc. [43,45,77]. Therefore, we need to control the intensity of land use and rationally allocate land resources to strengthen sustainable land use to improve Hainan's ESV.

The driving factor that had the greatest impact on EEC was the total real estate investment, with a value of 0.679 (Table 14). In addition, socioeconomic factors such as cultivated land area and number of tourists all had a greater impact on EEC, with values above 0.55. All socioeconomic factors and EEC were positively correlated, while natural factors such as the DEM were negatively correlated with low contribution rates. This result shows that Hainan Province's EEC was better in areas where people's living standards were higher due to the better overall environmental quality. The areas with the best EEC in Hainan were the places with the best development of ecotourism. Therefore, to improve the EEC of Hainan Province, it is necessary to vigorously develop ecotourism and other characteristic economic industries to maintain the rapid growth of people's income while not damaging the environment.

4.6. Policy Suggestion

To improve the EEC of Hainan Province, the ESV status has also been steadily improved. Therefore, the relationship between development and protection should be handled well in the process of regional land development and utilization. We recommend that the government draft and implement strict and effective policies to control the erosion of land from construction land to natural ecosystems. In addition, the government should rationally plan the spatial distribution of urban space, agricultural space and ecological space [78]. For example, the government should: (1) strengthen the planning of green space systems to increase the value of urban ecosystem services; (2) build multiple green space ecosystems, including parks, nurseries, and shelterbelts, and build artificial lakes, wetlands, forest parks, suburban water conservation forests, forests around the city, and large areas of lawn to reduce the impact of urban expansion on ecosystem services; (3) strictly implement the overall spatial planning of "multiple regulations in one" and the red line management of marine and land ecological protection in Hainan Province and implement the land management policy based on maps; and (4) pay attention to the protection and restoration of high-ESV areas such as wetlands, river and forest; maintain and strengthen the continuity and integrity of the entire province ecosystem; and guide land use to develop in the direction of ESV preservation or appreciation.

In addition, Hainan Province must adhere to the principle of "ecology first, green development" to achieve the coordinated development of its economic environment [71]. Hainan Province must speed up the adjustment of its economic structure, promote industrial transformation and upgrading, guide the rationalization of the spatial layout of industrial functions, and promote green and high-quality development. First, local governments must adhere to the equal emphasis on environmental protection and economic growth and regard strengthening ecological protection as an important means to adjust economic structure, transform economic growth patterns, and seek development in environmental protection [79]. Second, they must change their agricultural production method from a single water-intensive agricultural planting method to a diversified modern agricultural model that consumes less water, such as ecological agriculture and sightseeing agriculture, to dramatically curb the large-scale consumption of agricultural water. The government should vigorously promote a reduction in the use of fertilizers and pesticides, continuously strengthen the utilization of agricultural waste resources and the improvement and utilization of arable land, and actively build an agricultural brand of ecological recycling. Third, the ecological economy must be vigorously developed. We must firmly grasp the huge advantages of Hainan Province's high ecological quality and the construction of a free trade province. It is necessary to lead comprehensive efforts in resources, environment, ecology, growth quality, production methods, lifestyles, and social development to focus on the transformation of social development, production methods and lifestyles, in order to achieve coordinated economic and environmental development. The government should give full play to the unique advantages of vacation tourism; make full use of ethnic villages, ancient towns, historic sites, natural landscapes and other resources; and adjust production, living and ecological space in a reasonable way according to its own natural landscape, regional characteristics, cultural traditions, ethnic characteristics and industrial planning. We will promote the organic integration of urban and rural construction with investment attraction, industrial development, tourism and vacation, health and medical care, cultural cultivation and ecological protection; turn Hainan's tourism industry into the most competitive industry with the greatest potential; and build a green ecological urban and rural area that is livable and suitable for working, traveling and nourishing. Fourth, the government must promote the construction of an ecological civilization, strengthen education on ecological civilization, raise awareness of ecological protection and cultivate a low-carbon and green lifestyle for the whole people.

5. Conclusions

This paper used NDVI and NPP data to improve the biomass adjustment factor, and used the ability to pay, willingness to pay, and resource scarcity to construct the socioeconomic adjustment factor based on the benefit transfer method and LUCC data of Hainan Province. We measured the temporal and spatial characteristics of Hainan's dynamic ESV from 2000 to 2020. This article proposed a new coupling model of economic and environmental conditions. We used unit ESV as an environmental factor and GDP per capita as an economic factor. Then, we used the coupling analysis method to measure the EEC of Hainan Province. Finally, a detailed analysis of the driving force factors of the ESV and EEC was carried out. A powerful exploration of the assessment of the coordination of ESV and economic and environmental development in regional areas has been realized.

The results show that Hainan Province has the largest forest area and the least bare land area. The area of forests, farmland and grassland continued to decrease from 2000 to 2020. These ecosystems had positive values of ESV. Forests provided the highest ESV (above 77%), and rivers had the highest unit ESV (above 200,000 CNY/ha). Construction land was increasing rapidly, but it had a negative ESV value. Moreover, the advantages of Hainan Province's NPP and NDVI relative to the national average of NPP and NDVI continued to decline, and the economic development speed of Hainan Province slowed down relative to the economic development speed of the whole country. All these factors led to a decline in the ESV of the ecosystem, reduced to 33.305 billion CNY. The overall EEC of Hainan Province dropped by 0.06, and the state of EEC dropped from basic coordination to a moderately unbalanced state due to the good environmental quality and the slowdown in economic growth. In terms of driving force analysis, the LUI was the most important factor affecting ESV, with a value of 0.712. Socioeconomic factors and human activities were negatively correlated with ESV. The total real estate investment was the most important factor affecting the EEC, with a value of 0.679. Socioeconomic factors and human activities had a significant positive correlation with the EEC. The spatial distributions of GDP per capita and EEC were opposite to the spatial distribution of ESV. This result shows that economic growth will reduce ESV, but the reason for the EEC in Hainan Province was that the ecological economy was low and the ecological quality was high. Therefore, we must pay more attention to the coordination of economic development and ecological protection to seek green development in future land use planning and social and economic development.

There are still some shortcomings in this study, such as the spatial resolution of LUCC data and NPP data, the amount of GDP data not being very high, and the accuracy of ESV not being evaluated, etc. Therefore, in future research, we consider using high-resolution remote sensing data such as Sentinel satellites to interpret land use types; using the CASA model or downscaling models to obtain higher-resolution NPP data; adding soil conservation inverted by the RUSLE model data and precipitation data to improve spatial heterogeneity adjustment factors; further obtaining township-level GDP data; exploring the appropriate method to evaluate the accuracy of ESV results; using the Invest model to calculate ESV, and using this ESV to calculate the township-scale economic and environmental coordination; and exploring more effective methods to put forward more scientific and reasonable suggestions for the coordinated development of the economic environment.

Author Contributions: Conceptualization, J.F.; experiment, J.F., Q.Z., P.W., L.Z., Y.T. and X.L.; data collection, J.F. and Q.Z.; image processing, J.F.; data analysis, J.F., Q.Z., P.W., L.Z., Y.T. and X.L.; writing—original draft preparation, J.F., P.W. and Y.T.; writing—review and editing, Q.Z.; supervision, Q.Z. All authors have read and agreed to the published version of the manuscript.

Funding: This research received financial support from the Hainan Provincial Department of Science and Technology under Grant No. ZDKJ2019006 and the National Research Program of China under Grant No. 2020YFD1100204.

Data Availability Statement: Land use data with 1km spatial resolution, MODIS NPP data and Landsat data can be downloaded free of charge. Land use data with 30m spatial resolution can be purchased from funders by request.

Conflicts of Interest: The authors declare no conflict of interest.

References

1. Assessment, M.E. *Ecosystems and Human Well-Being*; Island Press: Washington, DC, USA, 2005; Volume 5, p. 563.
2. Kubiszewski, I.; Costanza, R.; Anderson, S.; Sutton, P. The future value of ecosystem services: Global scenarios and national implications. In *Environmental Assessments*; Edward Elgar Publishing: Northampton, MA, USA, 2020.
3. Costanza, R.; d'Arge, R.; De Groot, R.; Farber, S.; Grasso, M.; Hannon, B.; Limburg, K.; Naeem, S.; O'neill, R.V.; Paruelo, J. The value of the world's ecosystem services and natural capital. *Nature* **1997**, *387*, 253–260. [CrossRef]
4. Xie, G.; Lu, C.-X.; Cheng, S. Progress in evaluating the global ecosystem services. *Resour. Sci.* **2001**, *23*, 5–9.
5. Costanza, R.; De Groot, R.; Sutton, P.; Van der Ploeg, S.; Anderson, S.J.; Kubiszewski, I.; Farber, S.; Turner, R.K. Changes in the global value of ecosystem services. *Glob. Environ. Change* **2014**, *26*, 152–158. [CrossRef]
6. Xie, G.; Zhang, C.; Zhang, L.; Chen, W.; Li, S. Improvement of the evaluation method for ecosystem service value based on per unit area. *J. Nat. Resour.* **2015**, *30*, 1243.
7. Xie, G.; Zhen, L.; Lu, C.-X.; Xiao, Y.; Chen, C. Expert knowledge based valuation method of ecosystem services in China. *J. Nat. Resour.* **2008**, *23*, 911–919.
8. Pickard, B.R.; Van Berkel, D.; Petrasova, A.; Meentemeyer, R.K. Forecasts of urbanization scenarios reveal trade-offs between landscape change and ecosystem services. *Landsc. Ecol.* **2017**, *32*, 617–634. [CrossRef]
9. Zhou, J.; Wu, J.; Gong, Y. Valuing wetland ecosystem services based on benefit transfer: A meta-analysis of China wetland studies. *J. Clean. Prod.* **2020**, *276*, 122988. [CrossRef]
10. Song, W.; Deng, X. Land-use/land-cover change and ecosystem service provision in China. *Sci. Total Environ.* **2017**, *576*, 705–719. [CrossRef] [PubMed]
11. Zhongyuan, Y.; Hua, B. The key problems and future direction of ecosystem services research. *Energy Procedia* **2011**, *5*, 64–68. [CrossRef]
12. Jian, S. Research advances and trends in ecosystem services and evaluation in China. *Procedia Environ. Sci.* **2011**, *10*, 1791–1796. [CrossRef]
13. Zhang, Y.; Liu, Y.; Zhang, Y.; Liu, Y.; Zhang, G.; Chen, Y. On the spatial relationship between ecosystem services and urbanization: A case study in Wuhan, China. *Sci. Total Environ.* **2018**, *637*, 780–790. [CrossRef] [PubMed]
14. Ouyang, Z.; Zheng, H.; Xiao, Y.; Polasky, S.; Liu, J.; Xu, W.; Wang, Q.; Zhang, L.; Xiao, Y.; Rao, E. Improvements in ecosystem services from investments in natural capital. *Science* **2016**, *352*, 1455–1459. [CrossRef] [PubMed]
15. Capriolo, A.; Boschetto, R.; Mascolo, R.; Balbi, S.; Villa, F. Biophysical and economic assessment of four ecosystem services for natural capital accounting in Italy. *Ecosyst. Serv.* **2020**, *46*, 101207. [CrossRef]

16. Reining, C.E.; Lemieux, C.J.; Doherty, S.T. Linking restorative human health outcomes to protected area ecosystem diversity and integrity. *J. Environ. Plan. Manag.* **2021**, *64*, 2300–2325. [CrossRef]
17. Lei, J.; Wang, S.; Wang, J.; Wu, S.; You, X.; Wu, J.; Cui, P.; Ding, H. Effects of Land-Use Change on Ecosystem Services Value of Xunwu County. *Acta Ecol.* **2019**, *39*, 3089–3099.
18. Yan, F.; Zhang, S.; Su, F. Variations in ecosystem services in response to paddy expansion in the Sanjiang Plain, Northeast China. *Int. J. Agric. Sustain.* **2019**, *17*, 158–171.
19. Fei, L.; Shuwen, Z.; Jiuchun, Y.; Kun, B.; Qing, W.; Junmei, T.; Liping, C. The effects of population density changes on ecosystem services value: A case study in Western Jilin, China. *Ecol. Indic.* **2016**, *61*, 328–337. [CrossRef]
20. Hasan, S.; Shi, W.; Zhu, X. Impact of land use land cover changes on ecosystem service value—A case study of Guangdong, Hong Kong, and Macao in South China. *PLoS ONE* **2020**, *15*, e0231259. [CrossRef]
21. Wang, Z.; Wang, Z.; Zhang, B.; Lu, C.; Ren, C. Impact of land use/land cover changes on ecosystem services in the Nenjiang River Basin, Northeast China. *Ecol. Processes* **2015**, *4*, 11. [CrossRef]
22. Shifaw, E.; Sha, J.; Li, X.; Bao, Z.; Zhou, Z. An insight into land-cover changes and their impacts on ecosystem services before and after the implementation of a comprehensive experimental zone plan in Pingtan island, China. *Land Use Policy* **2019**, *82*, 631–642.
23. Lei, J.; Chen, Z.; Wu, T.; Li, X.; Yang, Q.; Chen, X. Spatial autocorrelation pattern analysis of land use and the value of ecosystem services in northeast Hainan island. *Acta Ecol. Sin.* **2019**, *39*, 2366–2377.
24. Yao, X.; Zhou, H.; Zhang, A.; Li, A. Regional energy efficiency, carbon emission performance and technology gaps in China: A meta-frontier non-radial directional distance function analysis. *Energy Policy* **2015**, *84*, 142–154. [CrossRef]
25. Wang, H.; Zhou, S.; Li, X.; Liu, H.; Chi, D.; Xu, K. The influence of climate change and human activities on ecosystem service value. *Ecol. Eng.* **2016**, *87*, 224–239. [CrossRef]
26. Chan, K.M.A.; Shaw, M.R.; Cameron, D.R.; Underwood, E.C.; Daily, G.C. Conservation planning for ecosystem services. *PLoS Biol.* **2006**, *4*, e379. [CrossRef]
27. Rodríguez, J.P.; Beard, T.D., Jr.; Bennett, E.M.; Cumming, G.S.; Cork, S.J.; Agard, J.; Dobson, A.P.; Peterson, G.D. Trade-offs across space, time, and ecosystem services. *Ecol. Soc.* **2006**, *11*, 1. [CrossRef]
28. Rozelle, S.; Huang, J.; Zhang, L. Poverty, population and environmental degradation in China. *Food Policy* **1997**, *22*, 229–251. [CrossRef]
29. Zhu, Y.; Yao, S. The coordinated development of environment and economy based on the change of ecosystem service value in Shaanxi province. *Acta Ecol. Sin.* **2021**, *41*, 3331–3342.
30. Chen, W.; Zeng, J.; Zhong, M.; Pan, S. Coupling Analysis of Ecosystem Services Value and Economic Development in the Yangtze River Economic Belt: A Case Study in Hunan Province, China. *Remote Sens.* **2021**, *13*, 1552. [CrossRef]
31. Xu, Q.; Yang, R.; Zhuang, D.; Lu, Z. Spatial gradient differences of ecosystem services supply and demand in the Pearl River Delta region. *J. Clean. Prod.* **2021**, *279*, 123849. [CrossRef]
32. Braat, L.C.; De Groot, R. The ecosystem services agenda: Bridging the worlds of natural science and economics, conservation and development, and public and private policy. *Ecosyst. Serv.* **2012**, *1*, 4–15. [CrossRef]
33. Li, G.; Fang, C. Global mapping and estimation of ecosystem services values and gross domestic product: A spatially explicit integration of national 'green GDP' accounting. *Ecol. Indic.* **2014**, *46*, 293–314. [CrossRef]
34. Zhang, K.-M.; Wen, Z.-G. Review and challenges of policies of environmental protection and sustainable development in China. *J. Environ. Manag.* **2008**, *88*, 1249–1261. [CrossRef] [PubMed]
35. Wei, W.; Shi, P.; Wei, X.; Zhou, J.; Xie, B. Evaluation of the coordinated development of economy and eco-environmental systems and spatial evolution in China. *Acta Ecol. Sin.* **2018**, *38*, 2636–2648.
36. Ma, L.; Jin, F.; Song, Z.; Liu, Y. Spatial coupling analysis of regional economic development and environmental pollution in China. *J. Geogr. Sci.* **2013**, *23*, 525–537. [CrossRef]
37. Gu, K.; Liu, J.; Chen, X.; Peng, X. Dynamic Analysis of Ecological Carrying Capacity of a Mining City. *J. Nat. Resour.* **2008**, *23*, 841–848.
38. Su, H.; Zhang, Z.; Zhang, X.; Wang, J. Energy analysis on sustainable development of ecological economic system in Shanxi Province. *Southwest China J. Agric. Sci.* **2019**, *32*, 1187–1193.
39. Wang, Z.; Fang, C.; Wang, J. Evaluation on the coordination of ecological and economic systems and associated spatial evolution patterns in the rapid urbanized Yangtze Delta Region since 1991. *Acta Geogr. Sin.* **2011**, *66*, 1657–1668.
40. Oliveira, C.; Antunes, C.H. A multi-objective multi-sectoral economy–energy–environment model: Application to Portugal. *Energy* **2011**, *36*, 2856–2866. [CrossRef]
41. Stern, D.I.; Common, M.S.; Barbier, E.B. Economic growth and environmental degradation: The environmental Kuznets curve and sustainable development. *World Dev.* **1996**, *24*, 1151–1160. [CrossRef]
42. Zhen, L.; Jinghu, P.; Yanxing, H. The spatio-temporal variation of ecological property value and eco-economic harmony in Gansu Province. *J. Nat. Resour.* **2017**, *32*, 64–75.
43. Sun, R.; Wu, Z.; Chen, B.; Yang, C.; Qi, D.; Lan, G.; Fraedrich, K. Effects of land-use change on eco-environmental quality in Hainan Island, China. *Ecol. Indic.* **2020**, *109*, 105777. [CrossRef]
44. Zhou, L.; Gao, S.; Yang, Y.; Zhao, Y.; Han, Z.; Li, G.; Jia, P.; Yin, Y. Typhoon events recorded in coastal lagoon deposits, southeastern Hainan Island. *Acta Oceanol. Sin.* **2017**, *36*, 37–45. [CrossRef]

45. Zhang, Z.; Xia, F.; Yang, D.; Huo, J.; Wang, G.; Chen, H. Spatiotemporal characteristics in ecosystem service value and its interaction with human activities in Xinjiang, China. *Ecol. Indic.* **2020**, *110*, 105826. [CrossRef]
46. Redo, D.J.; Aide, T.M.; Clark, M.L.; Andrade-Núñez, M.J. Impacts of internal and external policies on land change in Uruguay, 2001–2009. *Environ. Conserv.* **2012**, *39*, 122–131. [CrossRef]
47. Dai, X.; Johnson, B.A.; Luo, P.; Yang, K.; Dong, L.; Wang, Q.; Liu, C.; Li, N.; Lu, H.; Ma, L. Estimation of Urban Ecosystem Services Value: A Case Study of Chengdu, Southwestern China. *Remote Sens.* **2021**, *13*, 207. [CrossRef]
48. Xu, Y.; Xu, X.; Tang, Q. Human activity intensity of land surface: Concept, methods and application in China. *J. Geogr. Sci.* **2016**, *26*, 1349–1361. [CrossRef]
49. Xie, G.; Zhang, C.; Zhen, L.; Zhang, L. Dynamic changes in the value of China's ecosystem services. *Ecosyst. Serv.* **2017**, *26*, 146–154. [CrossRef]
50. Hu, M.; Li, Z.; Wang, Y.; Jiao, M.; Li, M.; Xia, B. Spatio-temporal changes in ecosystem service value in response to land-use/cover changes in the Pearl River Delta. *Resour. Conserv. Recycl.* **2019**, *149*, 106–114. [CrossRef]
51. Xing, L.; Xue, M.; Wang, X. Spatial correction of ecosystem service value and the evaluation of eco-efficiency: A case for China's provincial level. *Ecol. Indic.* **2018**, *95*, 841–850. [CrossRef]
52. Ye, Y.; Bryan, B.A.; Connor, J.D.; Chen, L.; Qin, Z.; He, M. Changes in land-use and ecosystem services in the Guangzhou-Foshan Metropolitan Area, China from 1990 to 2010: Implications for sustainability under rapid urbanization. *Ecol. Indic.* **2018**, *93*, 930–941. [CrossRef]
53. Yi, H.; Güneralp, B.; Filippi, A.M.; Kreuter, U.P.; Güneralp, İ. Impacts of land change on ecosystem services in the San Antonio River Basin, Texas, from 1984 to 2010. *Ecol. Econ.* **2017**, *135*, 125–135. [CrossRef]
54. Zhao, H.; Chen, Y.; Yang, J.; Pei, T. Ecosystem service value of cultivated land and its spatial relationship with regional economic development in Gansu Province based on improved equivalent. *Arid. Land Geogr.* **2018**, *4*, 851–858.
55. Wang, J.F.; Li, X.H.; Christakos, G.; Liao, Y.L.; Zhang, T.; Gu, X.; Zheng, X.Y. Geographical detectors-based health risk assessment and its application in the neural tube defects study of the Heshun Region, China. *Int. J. Geogr. Inf. Sci.* **2010**, *24*, 107–127. [CrossRef]
56. Wang, X.; Yan, F.; Zeng, Y.; Chen, M.; Su, F.; Cui, Y. Changes in Ecosystems and Ecosystem Services in the Guangdong-Hong Kong-Macao Greater Bay Area since the Reform and Opening Up in China. *Remote Sens.* **2021**, *13*, 1611. [CrossRef]
57. Zhao, Q.; Wen, Z.; Chen, S.; Ding, S.; Zhang, M. Quantifying land use/land cover and landscape pattern changes and impacts on ecosystem services. *Int. J. Environ. Res. Public Health* **2020**, *17*, 126. [CrossRef] [PubMed]
58. Su, K.; Wei, D.-Z.; Lin, W.-X. Evaluation of ecosystem services value and its implications for policy making in China–A case study of Fujian province. *Ecol. Indic.* **2020**, *108*, 105752. [CrossRef]
59. Wilson, M.A.; Howarth, R.B. Discourse-based valuation of ecosystem services: Establishing fair outcomes through group deliberation. *Ecol. Econ.* **2002**, *41*, 431–443. [CrossRef]
60. Zhao, L.; Li, L.; Wu, Y. Research on the coupling coordination of a sea–land system based on an integrated approach and new evaluation index system: A case study in Hainan Province, China. *Sustainability* **2017**, *9*, 859. [CrossRef]
61. Ren, H.; Li, L.; Liu, Q.; Wang, X.; Li, Y.; Hui, D.; Jian, S.; Wang, J.; Yang, H.; Lu, H. Spatial and temporal patterns of carbon storage in forest ecosystems on Hainan Island, southern China. *PLoS ONE* **2014**, *9*, e108163.
62. Yi, Z.-F.; Cannon, C.H.; Chen, J.; Ye, C.-X.; Swetnam, R.D. Developing indicators of economic value and biodiversity loss for rubber plantations in Xishuangbanna, southwest China: A case study from Menglun township. *Ecol. Indic.* **2014**, *36*, 788–797. [CrossRef]
63. Lei, J.; Chen, Z.; Chen, Y. Landscape pattern changesand driving factors analysis of wetland in Hainan island during 1990–2018. *Ecol. Environ. Sci.* **2020**, *29*, 63–74.
64. Ren, C.; Wang, Z.; Zhang, Y.; Zhang, B.; Chen, L.; Xi, Y.; Xiao, X.; Doughty, R.B.; Liu, M.; Jia, M. Rapid expansion of coastal aquaculture ponds in China from Landsat observations during 1984–2016. *Int. J. Appl. Earth Obs. Geoinf.* **2019**, *82*, 101902. [CrossRef]
65. Huang, Z.; Du, X.; Castillo, C.S.Z. How does urbanization affect farmland protection? Evidence from China. *Resour. Conserv. Recycl.* **2019**, *145*, 139–147. [CrossRef]
66. Lin, Y.; Qiu, R.; Yao, J.; Hu, X.; Lin, J. The effects of urbanization on China's forest loss from 2000 to 2012: Evidence from a panel analysis. *J. Clean. Prod.* **2019**, *214*, 270–278. [CrossRef]
67. Wang, X.; Yan, F.; Su, F. Impacts of urbanization on the ecosystem services in the Guangdong-Hong Kong-Macao greater bay area, China. *Remote Sens.* **2020**, *12*, 3269. [CrossRef]
68. Ting-yu, Z. An Empirical Analysis on Influencing Factors of Grain Output of Hainan Province. *J. Anhui Agric. Sci.* **2013**, *2013*, 17.
69. Sui, L.; Zhao, Z.; Jin, Y.; Guan, X.; Xiao, M. Dynamic Evaluation of Natural Ecosystem Service in Hainan Island. *Resour. Sci.* **2012**, *34*, 572–580.
70. Deng, X.; Huang, Z. Research on Forest Resources Health Evaluation Based on Analytic Hierarchy Process in Hainan. *Ecol. Econ.* **2017**, *6*, 201–204.
71. Hou, S.; Xu, B.; Lin, L.; Zhang, L.; Zhao, X.; Liu, J.; Pan, K. Research on Construction of Standard System for Development of Green Economy in Hainan Province. In Proceedings of the IOP Conference Series: Earth and Environmental Science, 2020 6th International Conference on Energy Materials and Environment Engineering, Tianjin, China, 24–26 April 2020.

72. Yan, F.; Zhang, S. Ecosystem service decline in response to wetland loss in the Sanjiang Plain, Northeast China. *Ecol. Eng.* **2019**, *130*, 117–121. [CrossRef]
73. Xiong, Y.; Zhang, F.; Gong, C.; Luo, P. Spatial-temporal evolvement of ecosystem service value in Hunan Province based on LUCC. *Resour. Environ. Yangtze Basin* **2018**, *27*, 1397–1408.
74. Qing, H.; Zhiming, F.; Yanzhao, Y.; Zhen, Y.; Ping, C.; Ling, D. Study of the population carrying capacity of water and land in Hainan Province. *J. Resour. Ecol.* **2019**, *10*, 353–361. [CrossRef]
75. Fu, G. Comparative Advantages of National Ecological Civilization Pilot Zone (Hainan) and Proposals. *Ecol. Econ.* **2020**, *36*, 216–220.
76. Sun, X.; Zhou, H. Draw a new picture of ecological civilization and promote the steady and long-term development of the free trade port. *Environ. Econ.* **2021**, *6*, 22–25.
77. Lei, J.; Chen, Z.; Chen, X.; Li, Y.; Wu, T. Spatio-temporal changes of land use and ecosystem services value in HainanIsland from 1980 to 2018. *Acta Ecol. Sin.* **2020**, *40*, 4760–4773.
78. Kuo, L.; Chang, B.-G. The affecting factors of circular economy information and its impact on corporate economic sustainability-Evidence from China. *Sustain. Prod. Consum.* **2021**, *27*, 986–997. [CrossRef]
79. Lian, X. Review on advanced practice of provincial spatial planning: Case of a western, less developed province. *Int. Rev. Spat. Plan. Sustain. Dev.* **2018**, *6*, 185–202. [CrossRef]

remote sensing

A Framework for the Construction of a Heritage Corridor System: A Case Study of the Shu Road in China

Fengting Yue, Xiaoqin Li *, Qian Huang and Dan Li

Tourism and Urban-Rural Planning College, Chengdu University of Technology, Chengdu 610059, China;
yuefengting@cdut.edu.cn (F.Y.); huangqian1@cdut.edu.cn (Q.H.); lidan2@cdut.edu.cn (D.L.)
* Correspondence: lixiaoqin@cdut.edu.cn; Tel.: +86-15390096056

Abstract: Heritage corridors are methods to effectively protect and utilize linear cultural heritage based on the concept of regional conservation. The construction of a heritage corridor system is extremely important to preserve the natural environment of the heritage corridor area as well as the history and culture alongside. The majority of the research on the construction of heritage corridors heretofore focused on the generation of corridors, whereas studies on the classification of corridors are relatively limited, without a complete system for the construction of heritage corridors. Therefore, this paper aimed to (1) establish a comprehensive system for the construction of heritage corridors, (2) provide new ideas for the construction of heritage corridors, and (3) guide the scientific development of heritage corridors combining conservation and tourism. In the first place, the minimum cumulative resistance (MCR) model was applied to analyze the spatial structure of the study area and explore site selection of the heritage corridors; secondly, spatial syntax was used to measure the heritage corridors and determine the level of the heritage corridors; last but not least, the kernel density analysis was used to classify the types of heritage corridors. The present study shows that the heritage corridor system is built in a scientific approach, covering all aspects including construction, protection, and development.

Keywords: heritage corridors; system construction; minimum cumulative resistance (MCR) model; spatial syntax; the Shu Road

Citation: Yue, F.; Li, X.; Huang, Q.; Li, D. A Framework for the Construction of a Heritage Corridor System: A Case Study of the Shu Road in China. *Remote Sens.* **2023**, *15*, 4650. https://doi.org/10.3390/rs15194650

Academic Editor: Prasad S. Thenkabail

Received: 21 July 2023
Revised: 11 September 2023
Accepted: 19 September 2023
Published: 22 September 2023

1. Introduction

Since the mid-19th century, the conservation of historical and cultural heritage has gradually become a focused problem around the world. The concern of heritage conservation has shifted from individual heritage to historical sites, with expanded range and incisive content [1]. In this context, the concept of heritage corridors was proposed. Heritage corridors originated in the United States in the 1980s, developed from the concept of the greenway [2]. Heritage corridors are linear cultural landscapes with special collections of cultural resources, and the conservation objects can be natural river valleys, canals, roads, and railway lines, or linear corridors of historical significance that connect individual heritage sites [3–5]. Until 2023, the United States had seven officially named heritage corridors and one named heritage canal within the 55 designated National Heritage Areas. Other typical heritage corridors include the Rideau Canal in Canada, the route of Santiago de Compostela in Spain, the Midi Canal in France, and the Kumano Kodo Trail in Japan, to name a few. Heritage corridors are of comprehensive conservation measure that promote the simultaneous and balanced development of nature, the economy, history, and culture. Therefore, the construction of heritage corridors is not only a meaningful way to realize the harmonious development of heritage conservation, tourism economic development, and ecological sustainability [6,7], but also a vigorous improvement for the cultural status and reputation of tourist places [8].

The construction of heritage corridors primarily includes generation, grading, and classification. (1) Various methods of corridor generation have been applied at home and abroad such as the qualitative analysis method [9–11], analytic hierarchy process [12,13], and minimum cumulative resistance (MCR) model [14–16] As the most prominent method to construct heritage corridors, the MCR model calculates the cost of a species' movement from the source to the destination, and primarily focuses on horizontal ecological processes. It was first applied to establish ecological security patterns for biodiversity conservation [17,18]. As the minimum cumulative resistance (MCR) model is being employed by a growing number of researchers, the method has been gradually applied for urban land use analysis [19,20], ecological pattern analysis [21], and corridor suitability analysis [22–24]. The model can calculate the least costly path based on the spatial unit resistance index, which takes full consideration of the geographical and behavioral characteristics with advanced operability and feasibility [25]. (2) In terms of the grading of heritage corridors, literature analysis [26] and connectivity indices [27–29] are usually used to assist with the analysis. The literature analysis method is mainly based on the researcher's analysis of a large amount of literature to obtain the grading results, of which the quality may be compromised by the skill of the associated researchers; the connectivity index is a measure of the corridor connectivity but lacks the visualization of the corridor characteristics. The concept of spatial syntax, officially introduced by Hillier et al. in 1984, is employed for the research of space as an independent element to outline the relationship among architectural, social, and cognitive domains and space [30]. The primary application field of spatial syntax includes spatial morphology [31,32], urban planning and design [33,34], and architectural design [35,36], while further development of the spatial syntax theory and method enables the spatial analysis of street and road networks [37,38]. In the road analysis, corridors can be characterized from multiple perspectives such as connectivity, choice, and integration. The results can be visualized in ArcGIS, which overcomes the drawbacks of the aforementioned and is more accurate and objective. (3) Most studies are deficient in the consideration of heritage corridor classification. Classification is a core component of heritage corridor studies regarding refined conservation, development themes, and the functional positioning of heritage corridors. Kernel density analysis is extensively applied in spatial clustering analysis, which can intuitively obtain the aggregation and dispersion characteristics of point data to describe the distribution characteristics [39]. By using kernel density analysis, the degree of clustering and dispersion for different types of heritage sites can be visualized, which provides scientific guidance for the classification of heritage corridors.

This paper introduces spatial syntax and kernel density analysis based on the least cumulative resistance model to conduct a hierarchical classification study of heritage corridors with the aim of providing new ideas for the construction of heritage corridors and scientific guidance for the protection and tourism development of heritage corridors. The objectives of this study are as follows: (1) determining resistance factors based on natural conditions and social environment, applying the minimum cumulative resistance (MCR) model for suitability analysis, and generating heritage corridors; (2) using the spatial syntax to measure heritage corridors and determine heritage corridors levels; (3) classifying the heritage corridor types with the assistance of kernel density analysis.

2. Materials and Methods

2.1. Study Area

The Shu Road is a major transportation route that stretched for more than 2000 years from the Warring States Period to the era of the Republic of China, with long historical standing and usage, a complicated terrain environment, and great historical and social influence. The Shu Road is a symbol of the ancient road system in terms of transportation, military activities, cultural communications, economics, and politics. In 2009, Sichuan Province launched the nomination of the Shu Road to the World Heritage List. In 2013, the Shu Road (Guangyuan section of the Jinniu Road) was honored on the Preparatory List of World Cultural Heritage in China. In November 2015, it was inscribed on the UNESCO World Heritage Center's Tentative List of World Heritage Sites and has been included in the scope of nominated sites for declaration of World Natural and Cultural Heritage. Exploring the generation and classification of the Shu Road heritage corridors is a key step for the nomination of the Shu Road to the World Heritage List and the economic and social development of the region. The study area included Chaotian, Lizho, Zhaohua, Jiang, Zitong, and Langzhong, which are located at the edge of the Sichuan Basin. The terrain transitions from mountains to gentle hills from the north to the south, with numerous remains along the road (as shown in Figure 1). Four types of heritage, namely road remains, plants, buildings, and ancient ruins, have been formed in different historical periods and various environments (as shown in Table 1).

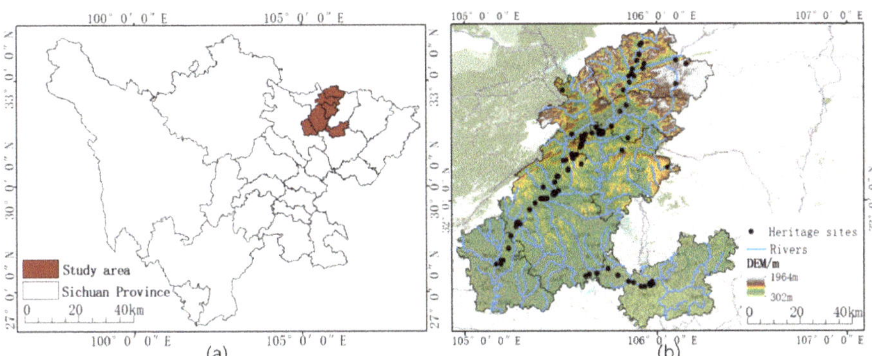

Figure 1. Study area. (**a**) Location of the study area in Sichuan Province. (**b**) Overview of the study area.

Table 1. Heritage resources and resource dating of the Shu Road.

Types	Historic Sites
Road remains	Qingfeng Gorge, Mingyue Gorge1, Mingyue Gorge (Pre Qin Dynasty); Datan Section1, Datan Section2, Jiange Section (The Spring and Autumn Period and the Warring States Period); SantanGou, Longdongbei1, Longdongbei2, Jie Cypress Crossing, Temple of Qiqu mountain Section, Xiaoshikou Section, Houzi Station Section, Songlintang Section1, Songlintang Section2, Mulintang Section, Xiangjiatang Section1, Xiangjiatang Section2 (data deficiencies)
Plants	Huang Cypress, Jiange Cypress, King of Cypress (Qin Dynasty); Hanzhuan Cypress, Adou Cypress, Zhuangyuan Cypress, Tangchang Cypress, Huaitai Cypress, Guest-Greeting Pine, Fuqi Cypress, Zhuxiang Cypress, Jin Cypress (Three Kingdoms)
Buildings	Qiaogouli Bridge, Xiaohe Bridge, Shidonggou Beacon Tower, Shiban Street, Wuhou Bridge, Guangji Bridge (data deficiencies); Jiameng Pass (Qin and Han Dynasties); Xiaojian City Ruins, Jianhua Beacon Tower (Han Dynasty); Yingpan Ruins, Wugong Bridge (Three Kingdoms); Temple of Qiqu Mountain (Jin Dynasty); Jueyuan Temple, Huaguang Tower, Huangze Temple (Tang Dynasty); Kuzhu Village Ruins (Song Dynasty); Tianxiong Pass, Chaotian Pass (Yuan Dynasty); Xinmin Station Bridge, Jianzhou Confucius Temple, Shitaya Bridge, Jianmen Pass, Well of Eight Diagrams, South Gate and Arrow Tower, Bell Tower and Drum Tower, Ancient City Wall, Zhangheng Ancestral Temple, Dazhao Station (Ming Dynasty); Shuanglong Bridge, Tieshuanzi Bridge, Guafu Bridge, Songning Bridge, Jianxi Bridge, Qingliang Bridge, Shuigouwan Bridge, Zhahua Ancient Town Gate, Longmen Academy, Yihetang, Yixin Garden, Zhaohua Kaopeng, Jia Courtyard, Gu Courtyard, Zhang Courtyard, Jianshan Academy, Erxian Ancestral temple, Langzhong Confucius Temple, Baba Temple, Examination Hall, Guanyin Temple, Mosque, Jianzhou White Tower (Qing Dynasty); Zhaohua Yue Tower (the Republic of China)
Ancient ruins	Choubi Stage, Chaoshou Station, Shangting Station, Yanwu Station, Horse blocking wall, Ganchangya, Xin Station, Qipan Pass, Tangfangwan, Zhuyazi, Jiaxiandian, Houzi Station, Guaner Station, Zhaohua Pavilion Ruins, Wulian Stage, Tandu Pass, Baiyang Plank path, Horsepond (data deficiencies); Baiyanba Western Zhou City Ruins (Shang Dynasty); Shangxin Station, Gaomiao Station (Qin and Han Dynasties); Baosanniang Tomb, Jiangwei Tomb, Feiyi Tomb, Jiangewei (Three Kingdoms); Qianfo Ya (Thousand Buddha Cliff) (Wei Dynasty); Guanyin Rock, Heming Mountain Taoism Stone Carving, Baiwei Mountain, Wangyun Stage (Tang Dynasty); Liangshan Station, Shangxin Station, Chuiquan Station (Song Dynasty); Jiaochangba Ancient Architectural Complex, Songxian Pavilion, Jianzhou Pavilion Ruins, Jingu Stage (Qing Dynasty)

2.2. Data Sources

The research data mainly included: (1) information on the ruins of the Shu Road, obtained from the archival records of the preliminary declaration of the Shu Road World Cultural Heritage List and the Sichuan Provincial Culture and Tourism Resources Cloud, which identified a total of 124 heritage resources of various types along the Shu Road; (2) a DEM with a spatial resolution of 30 m, attained from the geospatial data cloud platform of the Chinese Academy of Sciences (https://www.gscloud.cn/home, accessed on 26 November 2022); (3) land use classification data in 2022 with a spatial resolution of 10 m, obtained from Google Earth Engine; (4) data on administrative divisions, rivers, roads, etc., obtained from the National Catalogue Service for Geographic Information (webmap.cn, accessed on 14 December 2022); (5) POI data on restaurants, hotels, scenic spots, etc., obtained from Baidu Maps, containing a total of 4252 POI of public service facilities within the study area.

2.3. Data Processing

In the selection of heritage sites, this paper first collected literature related to the Shu Road, pre-determination of the heritage site, and then the former ancient road area of research, in order to complete the further screening work. DEM data were used for the elevation and slope analysis in ArcGIS.

In terms of land use types, these were based on the Google Earth Engine cloud platform combined with Sentinel-2 data to construct a classification feature set. We used the random forest algorithm to classify the land use types in 2022 into building land, grassland, forest land, cultivated land, and bare land by combining the spectral, textural, and topographic features and evaluating the accuracy of the classification results. To evaluate the results of the classification, a confusion matrix of land use types in the study area was calculated in conjunction with the validation sample (as shown in Table 2). The classification of forested and other land was good, with high producer accuracy. The categorization of construction land and cultivated land was effective, with a small number of cases of the misclassification of construction land as cultivated land and some misclassification between cultivated land and forest land. The poor classification of grass and bare ground and the low accuracy of the producers were due to the relative scarcity of grass and bare ground, while there was confusion between grass and forested land, and it was more difficult to distinguish between bare ground and built-up land. There was an overall accuracy of 85.83% with a kappa coefficient of 82.70%.

Table 2. Confusion matrix for the classification of land use types.

Land Use Type	Number of Pixels							Producer's Accuracy/%
	Grassland	Forest Land	Cultivated Land	Construction Land	Bare Land	Other Land	Total	
Grassland	7	2	3	0	0	0	12	70.00
Forest land	1	30	0	0	0	0	31	93.75
Cultivated land	1	0	22	3	1	0	27	81.48
Construction land	0	0	0	22	3	0	25	78.57
Bare land	1	0	1	3	14	0	19	77.78
Other land	0	0	1	0	0	26	27	100
Total	10	32	27	28	18	26	140	
Producer's accuracy/%	58.33	96.77	81.48	88.00	73.68	96.30		

The data on administrative divisions, rivers and roads as well as restaurants, hotels, and scenic areas were downloaded and then manually screened to eliminate duplicates and irrelevant data. The data were imported into ArcGIS and then analyzed for Euclidean distance for its river road data and kernel density for the POI data (as shown in Figure 2).

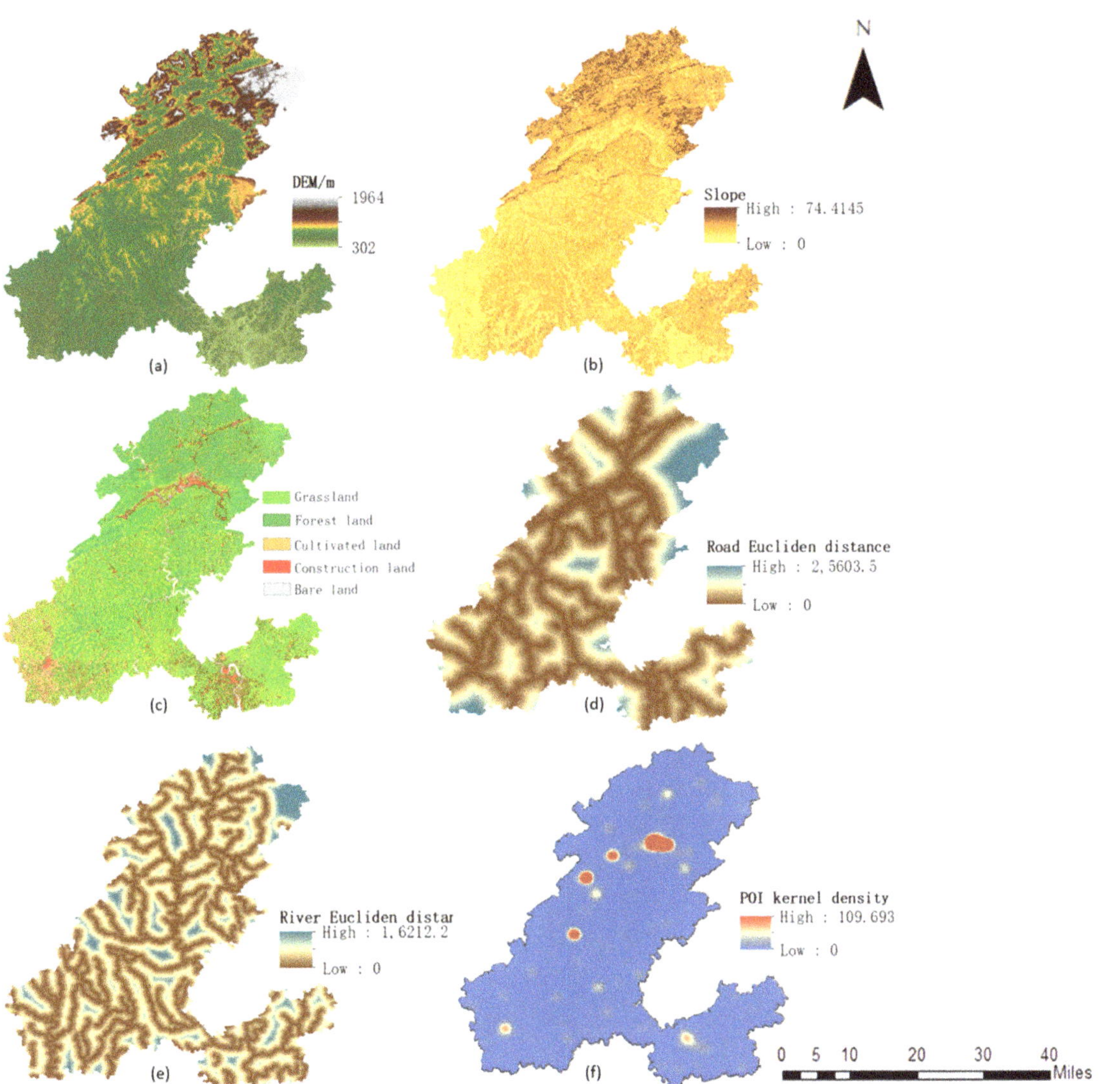

Figure 2. Processing of data. (**a**) Elevation analysis of the study area. (**b**) Slope analysis of the study area. (**c**) Analysis of land use types in the study area. (**d**) Euclidean distance analysis of rivers in the study area. (**e**) Euclidean distance analysis of roads in the study area. (**f**) Kernel density analysis of POI in the study area.

2.4. Research Methods

In this paper, the construction of heritage corridors was mainly carried out in three steps: first, the evaluation index system was established, and the least cumulative resistance model was used to generate heritage corridors; second, the spatial syntax analysis was applied, and the connectivity, choice, and integration were selected for the grading of heritage corridors; lastly, heritage corridors were classified by combining with the kernel density analysis method. The research framework of this paper is shown in Figure 3.

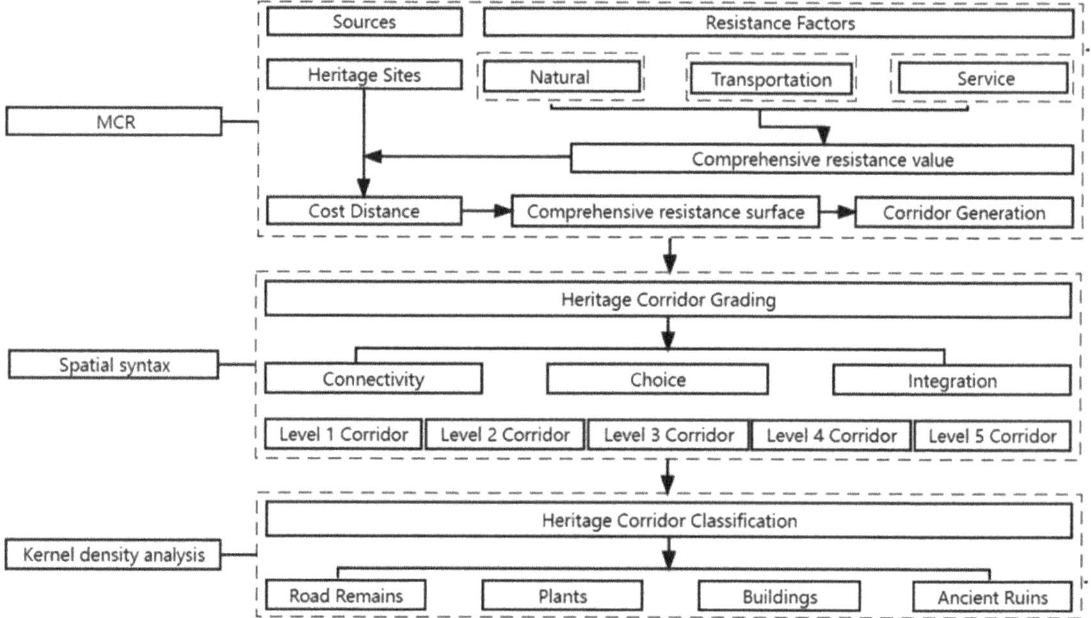

Figure 3. Research framework.

2.4.1. Heritage Corridor Generation

The generation of heritage corridors is mainly based on the establishment of a suitability evaluation index system and the use of the minimum cumulative resistance model to simulate the generation of potential heritage corridors.

In this paper, on the basis of reference to existing studies and considering the actual situation of the Shu Road, 12 types of resistance factors constituting a comprehensive resistance surface were determined from three aspects including the natural environment, traffic network, and public services, and the reasons for selecting the above resistance factors are shown in Table 3.

Table 3. Indicator system for evaluating the suitability of heritage corridors.

Resistance Factor	Resistance Factor Selection Basis	Access Time
Elevation and Slope	Elevation and slope are the basic geomorphological indicators, and the mountains and valleys are the natural skeleton of the heritage corridor. At the same time, the undulating changes of the land surface make the construction of the heritage corridor, the linking of the heritage resource sites, and the accessibility of the corridor more difficult. In the study area, the topography varies greatly, so the elevation and slope have an important influence on the suitability of the construction of heritage corridors.	Calculated from a digital elevation model (DEM) with a spatial resolution of 30 m.
Land use type	The land use type in part affects the orientation and ease of construction of heritage corridors. Areas that are more accessible to human activities such as construction land are more suitable for heritage corridors, while types of land such as cultivated land, forested land, and bare land can be more resistant to the construction of heritage corridors.	Land use data from 2022 Sentinel-2 imagery from the Google Earth Engine (GEE) platform with a product level of L2A.

Table 3. *Cont.*

Resistance Factor	Resistance Factor Selection Basis	Access Time
Rivers	The most important role of rivers in the construction of heritage corridors is to serve as natural environmental elements linking heritage resource sites, enhancing the excellent landscape value of the heritage corridors, and improving the accessibility of the heritage corridors. The study area has rivers such as the Jialing River, and the ancient Shu Road built the Mingyue Gorge trestle as well as the Orange Cypress Ferry and so on based on the rivers during the construction period. Therefore, this paper selected the distance from the river as an indicator factor to measure whether the river system has a significant connecting effect on the heritage corridor.	River data derived from the National Geographic Information Resources Catalog Service System 2021 1:1 million basic geographic information data.
Railroads and Highways	Roads are an important part of the settlement, an important skeleton connecting heritage resource sites and affecting the distribution of heritage resource sites, while road traffic also reflects the accessibility of heritage corridors. Thus, in this paper, railroads, highways, and first, second, and third class national roads were selected as indicators to reflect the influence of road transportation networks on the construction of heritage corridors.	Road data derived from the National Geographic Information Resources Catalog Service System 2021 1:1 million basic geographic information data.
Dining spots, Hotels and Scenic spots	The degree of infrastructure will affect the tourism value of the heritage corridor and its attractiveness to tourists, so this paper screened three indicators: restaurants, hotels, and scenic areas. Dining spots and hotels provide convenience for tourists to visit, eat, and stay in the area, and the number of scenic spots reflects to some extent the level of tourism development in the study area.	Dining spot, hotels scenic spot data obtained from Baidu Maps Platform in 2022.

In order to determine the resistance value of each resistance factor, this paper invited experts in the fields of cultural heritage, tourism, geography, ecology, landscape architecture, and other professional fields to score the weights of the resistance factors of the heritage corridor based on the characteristics of the heritage of the Shu Road and the current situation of the ecological environment, determine the weight of each resistance factor after repeated discussion, and to finally construct a system of evaluation indices, as shown in Table 4.

Table 4. Heritage corridor resistance factors and their weights and relative resistance values.

Resistance Factor	Weight	Resistance Value				
		1	2	3	4	5
Elevation/m	0.1	0~500	500~700	700~900	900~1200	>1200
Slope/°	0.15	<5	5~10	10~15	15~25	>25
Land use type	0.23	Construction land	Grassland	Forest land	Cultivated land	Bare land
Distance from the river/km	0.15	0~0.5	0.5~1	1~2	2~3	>3
Distance from the railroad/km	0.08	0~0.5	0.5~1	1~3	3~5	>5
Distance from the highway/km	0.07	0~0.5	0.5~1	1~3	3~5	>5
Distance from the primary road/km	0.05	0~0.5	0.5~1	1~3	3~5	>5
Distance from the secondary road/km	0.03	0~0.5	0.5~1	1~3	3~5	>5
Distance from the tertiary road/km	0.01	0~0.5	0.5~1	1~3	3~5	>5
Dining locations	0.05	>30	10~30	3~10	0.1~3	0~0.1
Hotels	0.05	>10	5~10	2~5	0.1~2	0~0.1
Scenic spots	0.03	>8	5~8	2~5	0.1~2	0~0.1

Based on the determination of the suitability evaluation index system, this study adopted the minimum cumulative resistance model for the generation of heritage corridors. Minimum cumulative resistance modeling focuses on simulating the minimum work undertaken or the minimum cumulative cost of passing from a "source" through various types of landscapes with different resistance values. The model was first proposed by the Dutch ecologist Knappen and has been mainly used for the study of species dispersal processes [40]. Yu K et al. introduced the minimum cumulative resistance model into the field of heritage corridors for the first time when discussing a new approach to heritage corridor suitability analysis [41]. The minimum cumulative resistance model simulates the resistance posed by different resistance factors to the process of heritage preservation and recreational experience during the generation of heritage corridors, and heritage corridor suitability is inversely proportional to the magnitude of resistance values. The formula is as follows:

$$\text{MCR} = \int {}_{min} \sum_{j=n}^{i=m} \left(D_{ij} \times R_i \right) \tag{1}$$

where MCR is the minimum cumulative resistance value; D_{ij} is the spatial distance of the person experiencing from environmental element i to heritage source j; R_i is the resistance value of environmental element i to the spatial movement process of the experience.

Using the minimum cumulative resistance model and ArcGIS spatial analysis method, we constructed a single-factor resistance surface, weighted, and superimposed it to obtain the comprehensive resistance distribution map, then used the cost distance tool in ArcGIS10.7 to calculate and acquire the suitability evaluation of the Shu Road heritage corridor, and divided the study area into five suitability zones: the high suitability zone, middle-high suitability zone, middle suitability zone, middle-low suitability zone, and low suitability zone by using the natural break-point method.

2.4.2. Heritage Corridor Grading

In order to better study heritage corridors, this paper graded heritage corridors, where objective and logical grading is essential for heritage corridors. In this study, spatial syntax analysis was used and connectivity, choice, and integration were selected to classify the heritage corridors.

Space Syntax was proposed by Prof. Bill Hillier to take the spatial organization and human social relations as the research object, and geometric topology as the theoretical basis for analyzing the complex urban network relationship [42]. Connectivity refers to the number of nodes in the space that directly connect to other nodes [43]. The higher the connectivity of a node, the more influence it has on surrounding nodes. Choice is the probability or frequency of the shortest path from a node to other nodes in a spatial system [44]. Nodes with high choice are more important in the spatial network and are more likely to be passed by the crowd. The formula is as follows:

$$\text{ACH}(x) = \frac{\sum_{i=1}^{n} \sum_{j=1}^{n} \sigma(i,x,j)}{(n-1)(n-2)} (i \neq x \neq j) \tag{2}$$

where n indicates the total number of heritage points within the search radius, and ACH indicates the degree of the angular selection of X.

Integration, which refers to the degree of aggregation and dispersion between a node and other nodes in the space, also represents the associativity of the space [45] A region with a high level of integration exhibits better convenience and accessibility. The formula is as follows:

$$\text{Integration} = \frac{n * n}{\sum_{i=1}^{n} d\theta(x,i)} \tag{3}$$

where n indicates the total number of heritage points within the search radius, and dθ(x,i) indicates the angular topological distance between space x and space i.

Based on the above research, this paper categorized the generated heritage corridors into five classes. The higher the grade of the heritage corridor, the greater its connectivity, choice, and integration, and the more accessible, attractive the heritage corridor will be.

2.4.3. Heritage Corridor Classification

In order to analyze the heritage corridors more comprehensively, this study used kernel density analysis to classify the types of heritage corridors. Kernel density analysis was used to calculate the degree of aggregation of spatial elements throughout the study area in order to visualize the degree of aggregation and the dispersion of the spatial distribution of heritage sites. A greater kernel density value $f(x)$ indicates a higher degree of aggregation [46]. The formula is as follows:

$$f(x) = \frac{1}{nh} \sum_{i=1}^{n} k\left(\frac{x - x_i}{h}\right) \tag{4}$$

where $kx - xih$ is the kernel density formula; h is the search range and $h \neq 0$; n is the number of heritage points in the search range; $x - x_i$ is the distance from the valuation point x to the measurement point x_i, and the density distribution has the highest value at the center of each x_i point, decreases continuously outward, and when the distance reaches a certain threshold value h at the center, the density decreases to zero.

In this paper, we selected the appropriate analysis radius for the kernel density analysis according to the four types of road remains, plants, structures, and ancient sites, in order to characterize the degree of aggregation and dispersion of heritage resource sites in the spatial distribution and classify heritage corridors into a total of eight types.

3. Results

3.1. Heritage Corridor Generation

Based on the established heritage corridor generation methodology described above, the results of the potential heritage corridors generated are shown in Figure 4. The medium-high suitability zone shows a distinct linear distribution characteristic, with a total corridor length of 405.625 km, and the corridor is roughly southwest–northeast oriented, showing a non-closed-ring radial shape.

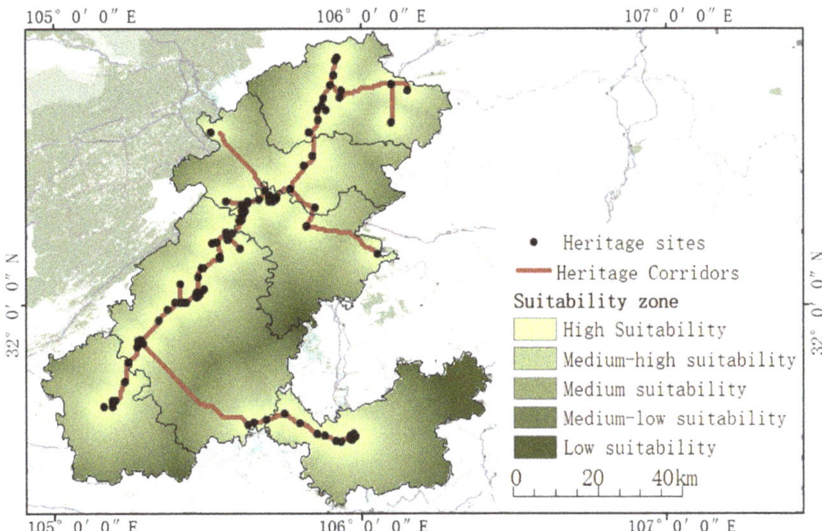

Figure 4. Partition of the suitability of the heritage corridors.

The northern corridors were mainly composed of three branches: the first from Zhahua ancient city to the Daitan at the northern end of the corridors, which was roughly the same direction as the Jialing River due to the undulating terrain; the second started from Mingyue Gorge and passed through the remains of the road to reach Tandu Pass, connecting with the northern Baoxie Road; the third showed an east–west direction, connecting Shuanglong Bridge with the remains of the Xiaoshikou of the road. The central corridors were mainly located in Jiange including the Jianmen Pass military system, the ancient city of Jiange, the Daoist stone statues in Heming Mountain, and numerous bridges. There were two main branches in the south, the first started from the Wulian stage and connected to the Temple of Qiqu Mountain in the south, through which it could enter Deyang to Chengdu. The second one connected Jiange with Langzhong, and this branch roughly followed the same path as the Baixihao River.

3.2. Heritage Corridor Grading

As can be seen from Figure 5a, the degree of connectivity was generally low, and most of the heritage resource sites showed a decentralized distribution, while the areas with a higher degree of connectivity were mostly tourism scenic spots with a large number of heritage resource sites that are concentrated and more maturely developed including the Jianmen Pass, Langzhong Ancient City, Zhaohua Ancient City, and so on.

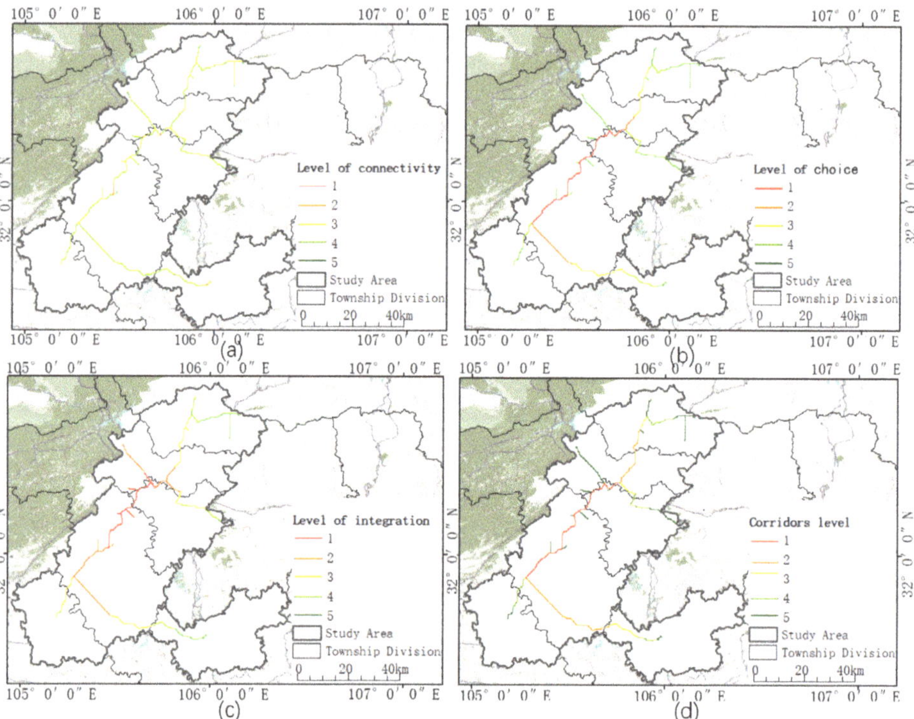

Figure 5. Level analysis of the heritage corridors. (**a**) Connectivity of the heritage corridors; (**b**) choice of the heritage corridors; (**c**) integration of the heritage corridors; (**d**) level of the heritage corridors.

As can be seen from Figure 5b, choice showed an obvious core-edge decreasing distribution pattern, with the highest-value area located in Jiange County and Zhaohua District, where Jiange County and Zhaohua District have a significant advantage in terms of the quantity and quality of the heritage sites, and the area contains the Cuiyun Corridor, Jiange Ancient City, Jianmen Pass, and Zhaohua Ancient City, etc. The second-highest-value area is the corridor connecting Lizhou and Chaotian Districts, and the area is connected to the Mingyue Gorge via Huangze Temple, Thousand Buddha Cliff, and Flying Immortal Pass, etc.

As can be seen in Figure 5c, it can be seen that the corridor integration has formed a more obvious spatial aggregation feature in the middle of the study area, and the integration of the surrounding branches is lower.

As can be seen from Figure 5d, the overall distribution pattern of the heritage corridor grade showed a decreasing pattern from the central high-value area to the surrounding area. The first-level corridor consists of two sections, mainly concentrated in Jiange County, the first section is from Jueyuan Temple to Jianmen Pass, and the second section is from Jianmen Pass to Guanyin Rock, which together with the surrounding branch corridors form a "high-level corridor aggregation area". The second-level and third-level corridors are mainly the corridors connecting Jiange and Langzhong as well as the corridors in the northern part of Lizhou District and Chaotian District. The fourth and fifth level corridors are mainly located in Zitong County and Chaotian District at the edge of the study area.

3.3. Heritage Corridor Classification

As can be seen in Figure 6, a high-density of road remains was located in Langzhong and Chaotian, while the other areas had a lower nuclear density; the high-density areas of plants were mainly located in Jiange and Zitong, with the Cuiyun Corridor as the main representative; the high-density areas of buildings were found in Zhaohua Ancient City, Jiange Ancient City, and Langzhong Ancient City; ancient ruins are scattered throughout the study area, and there were high-density clusters in Zhaohua District.

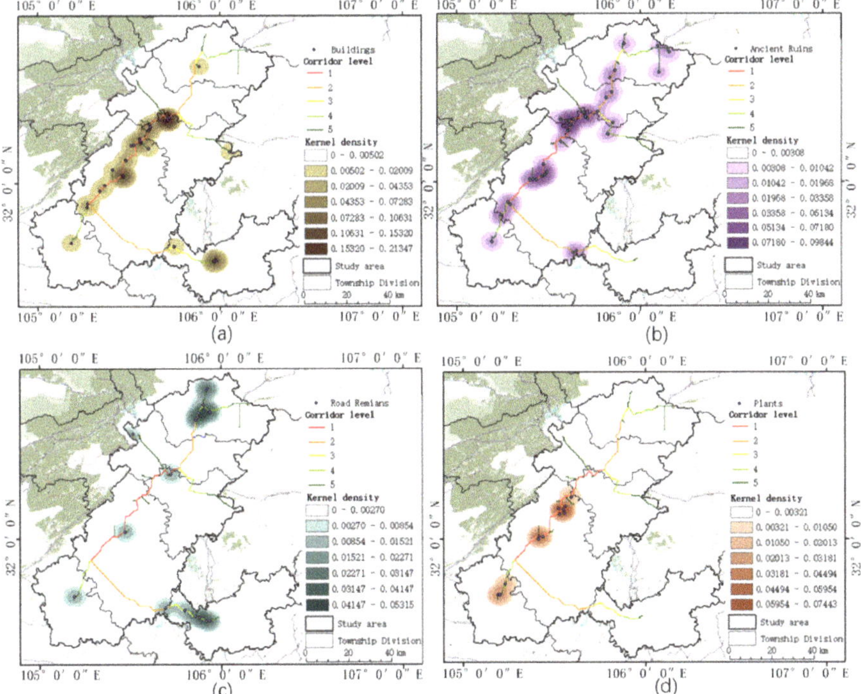

Figure 6. Relationship between heritage kernel density and corridor grade. (**a**) Relationship between the kernel density of buildings and corridor grade; (**b**) relationship between kernel density of ancient ruins and corridor grade; (**c**) relationship between the kernel density of road remains and corridor grade; (**d**) relationship between the kernel density of plants and corridor grade.

3.4. Heritage Corridor Construction

As can be seen from Figure 7 and Table 5, a total of 15 segments of corridors were divided by manual discrimination including two first-level corridors, two second-level corridors, three third-level corridors, four fourth-level corridors, and four fifth-level corridors. On this basis, the 15 sections of the corridor were categorized into eight types: two sections of three composite types, five sections of double composite types, and eight sections of the single type. Corridors located in the central part of the study area are mostly composite high-grade corridors, and the farther away from the center, the lower the grade of the corridor, and mostly single-type corridors such as the first-grade corridors in Jiange County and the first-grade corridors in Zhaohua District are three kinds of composite corridors, and the single-type corridors are concentrated in the fourth- and fifth-grade corridors.

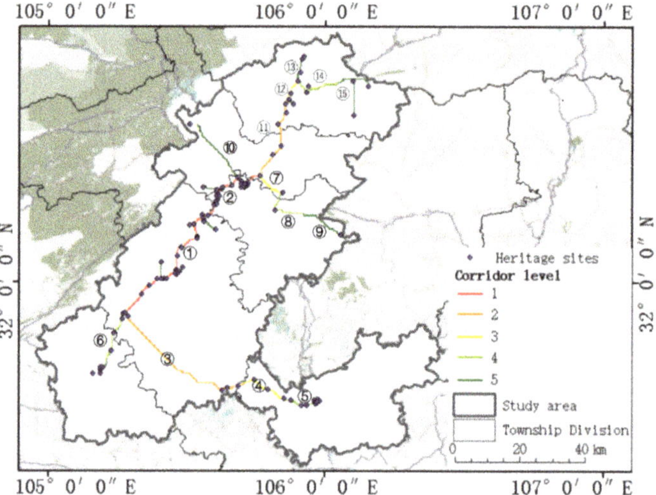

Figure 7. Classification of heritage corridors. Note: The corridor's name corresponding to the corridor's serial number is shown in Table 4. The numbers with circles indicate the serial numbers of the corridor.

Table 5. Grade and type characteristics of the porcelain heritage corridors.

No.	Name of Corridors	Level	Type
1	Wulian Stage—Jiameng Pass	1	Plants—Buildings—Ruins
2	Gaomiao Station—Jie Cypress	1	Roads—Buildings—Ruins
3	Jueyuan Temple—Houzi Station	2	Roads—Ruins
4	Xiaohe Bridge—Songlintang Section	3	Roads—Buildings
5	Songlintang—Guanyin Temple	4	Roads—Buildings
6	Wugong Bridge—Temple of Qiqu Mountain	4	Plants—Ruins
7	Guanyin Rock—Jingu Stage	3	Ruins
8	Jingu Stage—Meiling Pass	4	Ruins
9	Meiling Pass—Shuanglong Bridge	5	Buildings
10	Baosanniang Tomb—Xiaoshikou	5	Roads
11	Guanyin Rock—Mingyue Gorge	2	Roads—Ruins
12	Mingyue Gorge—Qingfeng Gorge	3	Roads
13	Qingfeng Gorge—Datan	5	Roads
14	Qingfeng Gorge—Longdongbei	4	Roads
15	Qipan Pass—Baiyang Plank path	5	Ruins

4. Discussion

4.1. Discussion on Research Methodology

The existing research approaches to heritage corridor construction have mostly focused on the generation of heritage corridors, and few studies have focused on the grading and classification of heritage corridors [47–49].

This study not only analyzed the generation of heritage corridors, but also focused on the grading and classification of heritage corridors, discussed the construction of the heritage corridor system in depth, and proposed a more systematic and complete construction system of heritage corridors. In the construction of an evaluation index system for heritage corridor generation, existing studies have tended to focus on natural factors [50–52] such as elevation, terrain undulation, slope, land use type, and other indicators, while human factors such as road density, public services, and other human activities closely related to the corridor also affect the structural characteristics of the corridor to a certain extent, so this study not only considered the natural environment elements in the selection of resistance factors, but also considered the socio-economic elements to make the evaluation index system more objective. In terms of heritage corridor grading, this paper chose spatial syntax for heritage corridor grading research, as the introduction of spatial syntax can comprehensively analyze the characteristics of the heritage corridor, enabling the grading characteristics of the corridor to achieve visualization, corridor grading protection and development as well as the grading of the configuration of tourism facilities to provide a scientific basis. In terms of heritage corridor classification, this study used kernel density analysis to observe the spatial visualization pattern of elements of heritage sites to obtain the spatial layout and degree of agglomeration and disaggregation of heritage sites in the study area, which increases the scientific process of corridor classification and has a certain value of reference for the conservation strategy, thematic planning, and thematic area delineation of heritage corridors.

Although the results of the heritage corridor construction method obtained based on the minimum cumulative resistance model and spatial syntax and kernel density analysis are reliable, this study still inevitably had some shortcomings. First of all, there was inconsistency in the accuracy between the data of the suitability evaluation index system adopted in the study such as the higher accuracy of the index data obtained based on the interpretation of high spatial resolution remote sensing data and the lower accuracy of the index data obtained based on the National Geographic Information Resources Catalog Service System, which to a certain extent affects the accuracy of the generation of heritage corridors. Secondly, the factors affecting the construction of the Shu Road heritage corridor are complex and variable, and some indicators are difficult to quantify and were therefore not included in the evaluation index system and heritage corridor grading indicators. For example, this study did not consider the impact of the degree of conservation and development of heritage resource sites on the generation and hierarchical classification of heritage corridors. In addition, the application for World Heritage listing of the Shu Road is constantly advancing, and information on the number, location, and chronology of heritage resource sites is being updated and supplemented, so we were not able to comprehensively include heritage resource sites in our study.

The heritage corridor construction system established in this study is an effective method for constructing large-scale cross-regional heritage corridors, so it is applicable to the construction of large-scale cross-regional heritage corridors. Second, this methodology is applicable to other small- and medium-sized heritage corridors. However, it should be noted that the natural environment and socio-economic conditions specific to different regions should be fully taken into account when constructing the resistance surface evaluation index system.

4.2. Strategies for the Protection and Development of the Shu Road Heritage Corridor

The generation of heritage corridors as well as hierarchical classification studies provide a scientific basis for the protection and development of the Shu Road heritage corridors. The generation of heritage corridors connects heritage resource sites and heritage areas, forming a multi-scale heritage protection and development structure of "heritage resource sites–heritage corridors–heritage areas". The hierarchical classification study of heritage corridors explored the refined spatial structure of heritage corridors with multiple levels and types, and delicately portrays the spatial hierarchy and functionality of corridors, which is of reference value for the protection of heritage corridors, the planning of tourism across the whole region, the organization of tourism routes, and the utilization of heritage as a whole. Based on the above research results as a reference, combined with the current situation of the protection and development of the Shu Road, this study proposes the following protection and development strategies:

The Level 1 corridor is divided into two sections, with the Wulian Station–Jianmen Pass section located in Jiange County. This section, as the area with the highest grade and the richest type of heritage resource points, has advantages in terms of protection and development, with maturely developed scenic spots such as Jianmen Pass, Jiange Ancient City, and Cuiyun Corridor in the area, which includes plants, buildings, ruins, and other types of heritage resource points. In terms of protection, as there are a large number of national and provincial cultural protection units in the region, in the core area of the heritage resource sites following the principle of "protection-oriented", the protection of cultural relics, display, and archaeological works have been carried out to ensure that the heritage resource sites are protected in a comprehensive and static manner. In terms of development, this section has the national 5A level tourist attraction Jianmen Pass Scenic Spot and the national 4A level tourist attraction Cuiyun Corridor. Relying on the development of mature scenic spots with the theme of heritage sightseeing, it creates a leisure and sightseeing resort area integrating the experience of distinctive transportation, the experience of outdoor extreme sports, and the experience of the ancient Shu Road and the Three Kingdoms culture. The section of Gaomiao Station–Orange Cypress is located in Zhaohua District, and the area is rich in heritage resources, with the ruins of the Zhaohua Ancient City and Stage as the main ones. In terms of protection, Zhaohua Ancient City has a number of provincial-level cultural relic protection units such as Longmen Academy, Yixinyuan, Yihetang, Zhaohua Examination Sheds, and the Bell and Drum Tower Ancient Architecture Area, where these heritage resource points, in accordance with relevant laws and regulations for protection, constantly update the corresponding protection and development charter to improve the construction of the protection system. In terms of development, this section of the heritage corridor should be centered on the Zhaohua Ancient City, with the Shu Road culture as the core and the Three Kingdoms culture as the supplement, in order to build up the theme of cultural experience and leisure in the ancient town, and to combine Tianxiong Pass, Dazhao Station, and other heritage resource points to create the Shu Road theme tourism and leisure area.

The level 2 and 3 corridors are mainly double-composite corridors, connecting a larger number of heritage resource sites and possessing higher conservation and development values. The Jueyuan Temple–Houzi Station section mainly consists of the remains of the road connecting Jiange and Langzhong and the ruins of the post store, of which Jueyuan Temple is a national key cultural relic protection unit. This section of heritage resource sites is mainly implemented in a single-point protection mode, based on the existing corridor, which relies on the beautiful natural landscape of the village combined with the surrounding villages and towns to develop tourism in the Shu Road original township. The Guangyin Rock–Mingyue Gorge section has heritage resource sites such as the Guanyin Rock Carvings, Huangze Temple, Thousand Buddha Cliff, and the Mingyue Gorge. With the section of the Mingyue Gorge as the core, and the ancient palisades and canyon landscape as the characteristics of the development of the Mingyue Gorge scenic area while at the same time relying on the Thousand Buddha Cliff, Huangze Temple, and other development of the Queen's hometown regarding the humanities scenic area, this section will be created for the Shu Road cultural enlightenment routes and recreational and vacation routes.

Level 4 and 5 corridors are mainly low-grade corridors of a single type. Under the principle of "protection-oriented and rational utilization", important heritage resource sites on the corridors will be protected and repaired, and under the condition of satisfying the protection of the heritage resource sites, the construction of tourism infrastructures will be improved, and leisure and tourism projects related to their themes will be developed. Road-based heritage corridors should be protected and repaired on the basis of their remaining roads, setting up hiking, cycling, and other experiences, restoring the important post stations and pavilions along the roads as well as the commercial and postal service system of the Shu Road, and reproducing the prosperity of the ancient Shu Road. For corridors that are dominated by ruins, the main strategy is to implement monolithic protection, restore important ruins, thoroughly excavate the historical stories of the ruins, improve the interpretation system of the ruins, avoid excessive development and utilization, and carry out regular inspections and repairs.

5. Conclusions

This study mainly focused on the heritage corridor construction system for the Shu Road with the support of ArcGIS and used the minimum cumulative resistance model, spatial syntax, and kernel density analysis to construct the heritage corridor. Through the whole study, we can obtain the following conclusions:

Heritage corridors show highly concentrated non-closed-loop radial distribution characteristics; heritage corridors have multi-center features that show a core-edge attenuation distribution pattern. Based on the natural discontinuity method, the corridor can be divided into five levels, where the heritage corridor grade shows the center of the high, surrounded by the low distribution characteristics. Heritage corridors can be divided into eight types; the corridors located in the center of the study area are multi-complex high-grade corridors, whereas farther away from the center of the corridors, the corridor grade is lower, and most of them are mono-functional corridors. It can be seen that this paper used the minimum cumulative resistance model to generate heritage corridors, adopted the spatial sentence method to assess the corridor centrality characteristics for corridor classification, and combined the kernel density analysis method to form the corridor classification scheme, which together constitute a systematic and comprehensive heritage corridor construction system. The study confirmed that the minimum cumulative resistance model, from a regional perspective, combined with spatial syntax and kernel density analysis, provides a new research idea for the conservation and sustainable development of this type of trans-regional linear cultural heritage, and guides the conservation and development work related to heritage corridors. In the next step of the study, we will supplement the information on the quantity, location, and date of the heritage resource points according to the inscription of the Shu Road, and at the same time, we will consider how to incorporate the information on the protection status of the heritage resource points into the evaluation index system to further improve the methodology of constructing the heritage corridor system proposed in this paper.

Author Contributions: Conceptualization, X.L.; Methodology, F.Y.; Software, F.Y.; Data curation, D.L.; Writing—original draft preparation, F.Y.; Writing—review and editing, X.L.; Visualization, Q.H.; Supervision, X.L. All authors have read and agreed to the published version of the manuscript.

Funding: This work funded by the Chengdu University of Technology "Double First-Class" initiative Construction Philosophy and Social Sciences Key Construction Project (ZDJS202217).

Data Availability Statement: The data presented in this study are available on request from the corresponding author.

Acknowledgments: We thank the reviewers for scrutinizing the manuscript with their insight. We are grateful to the experts who provided us with advice and input in determining the resistance factor.

Conflicts of Interest: The authors declare no conflict of interest.

References

1. Chen, Y.; Dang, A.; Peng, X. Building a cultural heritage corridors based on geodesign theory and methodology. *J. Urban Manag.* **2014**, *3*, 97–112. [CrossRef]
2. Ji, X.; Shao, L. The Application of Landscape Infrastructure Approaches in the Planning of Heritage Corridors Supporting System. *Procedia Eng.* **2017**, *198*, 1123–1127. [CrossRef]
3. Searns, R.M. The evolution of greenways as an adaptive urban landscape form. *Landsc. Urban Plan.* **1995**, *33*, 65–80. [CrossRef]
4. Ahern, J. Greenways as a planning strategy. *Landsc. Urban Plan.* **1995**, *33*, 131–155. [CrossRef]
5. Zou, T.; Hao, Y.; Jiang, L.; Xue, B. Identifying Brand Genes in Tourism Branding Strategy: A Case Study of "Chang'an-Tianshan" Heritage Corridors. In Proceedings of the International Conference Proceedings: Heritage Tourism & Hospitality 2015, Amsterdam, The Netherlands, 26–27 November 2015; pp. 211–222.
6. Li, H.; Jing, J.; Fan, H.; Li, Y.; Liu, Y.; Ren, J. Identifying cultural heritage corridors for preservation through multidimensional network connectivity analysis—A case study of the ancient Tea-Horse Road in Simao, China. *Landsc. Res.* **2021**, *46*, 96–115. [CrossRef]
7. Zhang, S.; Liu, J.; Pei, T.; Chan, C.-S.; Wang, M.; Meng, B. Tourism value assessment of linear cultural heritage: The case of the Beijing–Hangzhou Grand Canal in China. *Curr. Issues Tour.* **2023**, *26*, 47–69. [CrossRef]
8. Shishmanova, M.V. Cultural tourism in cultural corridors, itineraries, areas and cores networked. *Procedia-Soc. Behav. Sci.* **2015**, *188*, 246–254. [CrossRef]
9. Zhang, D.Q.; Shao, W.W.; Feng, T.Q. Discussion on Construction of the Weihe River System Heritage Corridors in Xi'an Metropolitan Area. *Appl. Mech. Mater.* **2014**, *641*, 531–536. [CrossRef]
10. Kashid, M.; Ghosh, S.; Narkhede, P. A Conceptual Model for Heritage Tourism Corridors in the Marathwada Region. In Proceedings of the National Online Conference on Planning, Design and Management, Pune, India, 6–7 May 2022.
11. Hui, C.; Dong, C.; Yuan, Z.; Sicheng, M. Construction of corridors of architectural heritage along the line of ZiJiang River in Hunan Province in the background of the Tea Road Ceremony. In *IOP Conference Series: Materials Science and Engineering*; IOP Publishing: Bristol, UK, 2019; Volume 471, p. 082024.
12. Li, H.; Duan, Q.; Zeng, Z.; Tan, X.; Li, G. Value Evaluation and Analysis of Space Characteristics on Linear Cultural Heritage Corridors Ancient Puer Tea Horse Road. In Proceedings of the Geo-Informatics in Resource Management and Sustainable Ecosystem: Third International Conference, GRMSE 2015, Wuhan, China, 16–18 October 2015; Revised Selected Papers 3; Springer: Berlin/Heidelberg, Germany, 2016; pp. 733–740.
13. Morandi, D.T.; de Jesus França, L.C.; Menezes, E.S.; Machado, E.L.M.; da Silva, M.D.; Mucida, D.P. Delimitation of ecological corridors between conservation units in the Brazilian Cerrado using a GIS and AHP approach. *Ecol. Indic.* **2020**, *115*, 106440. [CrossRef]
14. Lin, F.; Zhang, X.; Ma, Z.; Zhang, Y. Spatial Structure and Corridors Construction of Intangible Cultural Heritage: A Case Study of the Ming Great Wall. *Land* **2022**, *11*, 1478. [CrossRef]
15. Li, Y.; Wang, X.; Dong, X. Delineating an integrated ecological and cultural corridors network: A case study in Beijing, China. *Sustainability* **2021**, *13*, 412. [CrossRef]
16. Zhang, L.; Peng, J.; Liu, Y.; Wu, J. Coupling ecosystem services supply and human ecological demand to identify landscape ecological security pattern: A case study in Beijing–Tianjin–Hebei region, China. *Urban Ecosyst.* **2017**, *20*, 701–714. [CrossRef]
17. Ye, H.; Yang, Z.; Xu, X. Ecological corridors analysis based on MSPA and MCR model—A case study of the Tomur World Natural Heritage Region. *Sustainability* **2020**, *12*, 959. [CrossRef]
18. Li, F.; Ye, Y.; Song, B.; Wang, R. Evaluation of urban suitable ecological land based on the minimum cumulative resistance model: A case study from Changzhou, China. *Ecol. Model.* **2015**, *318*, 194–203. [CrossRef]
19. Jiang, W.; Cai, Y.; Tian, J. The application of minimum cumulative resistance model in the evaluation of urban ecological land use efficiency. *Arab. J. Geosci.* **2019**, *12*, 714. [CrossRef]
20. Huang, L.; Wang, D.; He, C. Ecological security assessment and ecological pattern optimization for Lhasa city (Tibet) based on the minimum cumulative resistance model. *Environ. Sci. Pollut. Res.* **2022**, *29*, 83437–83451. [CrossRef]
21. Tang, F.; Zhang, P.T.; Zhang, G.J.; Zhao, L.; Zheng, Y.; Wei, M.H.; Jian, Q. Construction of ecological corridors in Changli County based on ecological sensitivity and ecosystem service values. *J. Appl. Ecol.* **2018**, *29*, 2675–2684.
22. Li, P.; Cao, H.; Sun, W.; Chen, X. Quantitative evaluation of the rebuilding costs of ecological corridors in a highly urbanized city: The perspective of land use adjustment. *Ecol. Indic.* **2022**, *141*, 109130. [CrossRef]
23. Wei, Z.; Xu, Z.; Dong, B.; Xu, H.; Lu, Z.; Liu, X. Habitat suitability evaluation and ecological corridors construction of wintering cranes in Poyang Lake. *Ecol. Eng.* **2023**, *189*, 106894. [CrossRef]
24. Peng, J.; Zhao, H.; Liu, Y. Urban ecological corridors construction: A review. *Acta Ecol. Sin.* **2017**, *37*, 23–30. [CrossRef]
25. Rodríguez-Espinosa, V.M.; Aguilera-Benavente, F.; Gómez-Delgado, M. Green infrastructure design using GIS and spatial analysis: A proposal for the Henares Corridors (Madrid-Guadalajara, Spain). *Landsc. Res.* **2020**, *45*, 26–43. [CrossRef]
26. Pascual-Hortal, L.; Saura, S. Comparison and development of new graph-based landscape connectivity indices: Towards the priorization of habitat patches and corridors for conservation. *Landsc. Ecol.* **2006**, *21*, 959–967. [CrossRef]
27. Guo, X.; Zhang, X.; Du, S.; Li, C.; Siu, Y.L.; Rong, Y.; Yang, H. The impact of onshore wind power projects on ecological corridors and landscape connectivity in Shanxi, China. *J. Clean. Prod.* **2020**, *254*, 120075. [CrossRef]

28. Xiao, S.; Wu, W.; Guo, J.; Ou, M.; Pueppke, S.G.; Ou, W.; Tao, Y. An evaluation framework for designing ecological security patterns and prioritizing ecological corridors: Application in Jiangsu Province, China. *Landsc. Ecol.* **2020**, *35*, 2517–2534. [CrossRef]
29. Xiao, Y.; Xiao, Y. Space Syntax Methodology Review. In *Urban Morphology and Housing Market*; Springer: Berlin/Heidelberg, Germany, 2017; pp. 41–61.
30. Jiang, B.; Claramunt, C.; Klarqvist, B. Integration of space syntax into GIS for modelling urban spaces. *Int. J. Appl. Earth Obs. Geoinf.* **2000**, *2*, 161–171. [CrossRef]
31. Alkamali, N.; Alhadhrami, N.; Alalouch, C. Muscat City expansion and Accessibility to the historical core: Space syntax analysis. *Energy Procedia* **2017**, *115*, 480–486. [CrossRef]
32. Karimi, K. A configurational approach to analytical urban design: 'Space syntax' methodology. *Urban Des. Int.* **2012**, *17*, 297–318. [CrossRef]
33. Atakara, C.; Allahmoradi, M. Investigating the urban spatial growth by using space syntax and GIS—A case study of Famagusta city. *ISPRS Int. J. Geo-Inf.* **2021**, *10*, 638. [CrossRef]
34. Dursun, P. Space syntax in architectural design. In Proceedings of the 6th International Space Syntax Symposium, Istanbul, Turkey, 12–15 June 2007; pp. 1–56.
35. Alitajer, S.; Nojoumi, G.M. Privacy at home: Analysis of behavioral patterns in the spatial configuration of traditional and modern houses in the city of Hamedan based on the notion of space syntax. *Front. Archit. Res.* **2016**, *5*, 341–352. [CrossRef]
36. Li, W.; Chen, M.; Yao, N.; Luo, Z.; Jiao, Y. Spatial-temporal evolution of roadway layout system from a space syntax perspective. *Tunn. Undergr. Space Technol.* **2023**, *135*, 105038. [CrossRef]
37. Law, S.; Chiaradia, A.; Schwander, C. Towards a multimodal space syntax analysis: A case study of the London street and underground network. In Proceedings of the 8th International Space Syntax Symposium, Santiago de Chile, Chile, 3–6 January 2012; p. 36.
38. Yu, W.; Ai, T.; Shao, S. The analysis and delimitation of Central Business District using network kernel density estimation. *J. Transp. Geogr.* **2015**, *45*, 32–47. [CrossRef]
39. Zhang, J.; Cenci, J.; Becue, V.; Koutra, S. Analysis of spatial structure and influencing factors of the distribution of national industrial heritage sites in China based on mathematical calculations. *Environ. Sci. Pollut. Res.* **2022**, *29*, 27124–27139. [CrossRef] [PubMed]
40. Qiu, S.; Fang, M.; Yu, Q.; Niu, T.; Liu, H.; Wang, F.; Xu, C.; Ai, M.; Zhang, J. Study of spatialtemporal changes in Chinese forest eco-space and optimization strategies for enhancing carbon sequestration capacity through ecological spatial network theory. *Sci. Total Environ.* **2023**, *859*, 160035. (In Chinese) [CrossRef] [PubMed]
41. Yu, K.; Li, W.; Li, D. Suitability analysis of heritage corridor in rapidly urbanizing region: A case study of Taizhou city. *Geogr. Res.* **2005**, *24*, 69–76+162. (In Chinese)
42. Dong, Y.; Jiang, Y.; Quan, D.; Zhu, H. Study on the heritage corridors construction of the western han dynasty mausoleums region. *Curr. Urban Stud.* **2022**, *10*, 55–72. (In Chinese) [CrossRef]
43. Hillier, B.; Leaman, A.; Stansall, P.; Bedford, M. Space syntax. *Environ. Plan. B Plan. Des.* **1976**, *3*, 147–185. [CrossRef]
44. Bafna, S. Space syntax: A brief introduction to its logic and analytical techniques. *Environ. Behav.* **2003**, *35*, 17–29. [CrossRef]
45. Li, Y.; Xiao, L.; Ye, Y.; Xu, W.; Law, A. Understanding tourist space at a historic site through space syntax analysis: The case of Gulangyu, China. *Tour. Manag.* **2016**, *52*, 30–43. (In Chinese) [CrossRef]
46. Hillier, B. Network effects and psychological effects: A theory of urban movement. In Proceedings of the 5th International Space Syntax Symposium, Delft, The Netherlands, 13–17 June 2005.
47. Wang, Y.; Zeng, G. Research on the construction of the Silk Road Cultural Heritage Corridor form the spatial perspective-A case of Gansu section. *World Reg. Stud.* **2022**, *31*, 862–871. (In Chinese)
48. Yang, X.; Chen, J. Construction of linear intangible cultural heritage corridor along the Chinese section of the "Silk Road Economic Belt". *J. Arid Land Resour. Environ.* **2021**, *35*, 202–208. (In Chinese)
49. Li, H.; Wang, Z. Study on the Anti-Japanese Heritage Corridor System based on GIS Spatial Analysis: A case of Main Urban Area in Chongqing. *Urban Dev. Stud.* **2017**, *24*, 86–93. (In Chinese)
50. Ten, Y. Xiaohe Ancient Road Heritage Corridor Construction Based on Minimum Cumulative Resistance Model. *Planners* **2020**, *36*, 66–70. (In Chinese)
51. Yuan, Y.; Xu, J.; Zhang, X. Construction of Heritage Corridor Network Based on Suitability Analysis: A case Study of the Ancient Capital of Luoyang. *Remote Sens. Inf.* **2014**, *29*, 117–124. (In Chinese)
52. Wang, S.; Li, T.; Dong, Y. Spatial structure heritages and establishment of heritage corridor network in Beijing. *J. Arid Land Resour. Environ.* **2010**, *24*, 51–56. (In Chinese)

 *remote sensing*

Article

Spatiotemporal Pattern of Invasive *Pedicularis* in the Bayinbuluke Land, China, during 2019–2021: An Analysis Based on PlanetScope and Sentinel-2 Data

Wuhua Wang [1], Jiakui Tang [1,2,*], Na Zhang [1,2], Yanjiao Wang [1], Xuefeng Xu [1,3] and Anan Zhang [1]

[1] College of Resources and Environment, University of Chinese Academy of Sciences, Beijing 100049, China; wangwuhua20@mails.ucas.ac.cn (W.W.); zhangna@ucas.ac.cn (N.Z.); wangyanjiao21@mails.ucas.ac.cn (Y.W.); xuxuefeng18@mails.ucas.ac.cn (X.X.); zhanganan19@mails.ucas.ac.cn (A.Z.)

[2] Yanshan Earth Key Zone and Surface Flux Observation and Research Station, University of Chinese Academy of Sciences, Beijing 101408, China

[3] Precision Agriculture Lab, School of Life Sciences, Technical University of Munich, 85354 Freising, Germany

* Correspondence: jktang@ucas.ac.cn

Abstract: The accurate identification and monitoring of invasive plants are of great significance to sustainable ecological development. The invasive *Pedicularis* poses a severe threat to native biodiversity, ecological security, socioeconomic development, and human health in the Bayinbuluke Grassland, China. It is imperative and useful to obtain a precise distribution map of *Pedicularis* for controlling its spread. This study used the positive and unlabeled learning (PUL) method to extract *Pedicularis* from the Bayinbuluke Grassland based on multi-period Sentinel-2 and PlanetScope remote sensing images. A change rate model for a single land cover type and a dynamic transfer matrix were constructed under GIS to reflect the spatiotemporal distribution of *Pedicularis*. The results reveal that (1) the PUL method accurately identifies *Pedicularis* in satellite images, achieving F1-scores above 0.70 and up to 0.94 across all three datasets: PlanetScope data (seven features), Sentinel-2 data (seven features), and Sentinel-2 data (thirteen features). (2) When comparing the three datasets, the number of features is more important than the spatial resolution in terms of use in the PUL method of *Pedicularis* extraction. Nevertheless, when compared with PlanetScope data, Sentinel-2 data demonstrated a higher level of accuracy in predicting the distribution of *Pedicularis*. (3) During the 2019–2021 growing season, the distribution area of *Pedicularis* decreased, and the distribution was mainly concentrated in the northeast and southeast of Bayinbuluke Swan Lake. The acquired spatiotemporal pattern of invasive *Pedicularis* could potentially be used to aid in controlling *Pedicularis* spread or elimination, and the methods proposed in this study could be adopted by the government as a low-cost strategy to identify priority areas in which to concentrate efforts to control and continue monitoring *Pedicularis* invasion.

Keywords: *Pedicularis*; positive and unlabeled learning; dynamic variation; change detection; PlanetScope

Citation: Wang, W.; Tang, J.; Zhang, N.; Wang, Y.; Xu, X.; Zhang, A. Spatiotemporal Pattern of Invasive *Pedicularis* in the Bayinbuluke Land, China, during 2019–2021: An Analysis Based on PlanetScope and Sentinel-2 Data. *Remote Sens.* **2023**, *15*, 4383. https://doi.org/10.3390/rs15184383

Academic Editor: Yoshio Inoue

Received: 30 June 2023
Revised: 29 August 2023
Accepted: 1 September 2023
Published: 6 September 2023

1. Introduction

The Bayinbuluke Grassland, which is the second largest grassland in China, boasts a diverse array of grassland species and maintains a relatively intact ecosystem, providing a favorable habitat for various animals [1]. In recent years, the invasion of *Pedicularis* has significantly impacted animal husbandry and the ecological environment of the Bayinbuluke Grassland. *Pedicularis*, with its gorgeous appearance, is a poisonous grass and a semi-parasitic plant that has rapidly spread in alpine grasslands throughout western China [2]. The invasion of non-native grasses such as *Pedicularis* can cause a drastic alteration in ecosystems, leading to significant socioeconomic costs [3]. Therefore, local authorities must urgently investigate the spatial and temporal patterns of *Pedicularis* invasion and develop effective management strategies to control this toxic plant species.

Previous studies have reported that *Pedicularis* had affected an area of 2.33×10^4 hm^2, which expanded to 3.30×10^3 hm^2, from 2000 to 2008 in the Bayinbuluke Grassland [4]. In 2018, Hejing County, Xinjiang Province, strived for CNY 7 million of funding for the *Pedicularis* control project to manually eradicate 2093.33 hm^2 of pastureland *Pedicularis*. The local government invested a significant amount of financial and labor resources into limiting the invasion of *Pedicularis* through physical, chemical, and biological means [5]. The traditional method of field surveying grassland species requires considerable human, material, and financial resources, and fieldwork makes ensuring quality and efficiency challenging [6,7]. Currently, remote sensing technology has significant advantages for dynamically monitoring and analyzing vast grassland resources and their ecological environments [8]. However, only one related study has been conducted on the identification of *Pedicularis* in the Bayinbuluke Grassland using satellite images, and the applied method did not perform well. Gao. S [9] studied the distribution of *Pedicularis* in 2016 using a maximum likelihood algorithm based on GaoFen-1 WFV with a spatial resolution of 16 m; they achieved a low precision of 80.91%, and a large number of samples were required for labeling. Therefore, here, we demonstrate an innovative application of high-resolution remote sensing imagery to discern the invasive species *Pedicularis* by employing the PUL method. Furthermore, we conduct a comprehensive assessment aimed at precisely outlining the advantages and limitations associated with the utilization of various remote sensing images for Pedicularis extraction.

Supervised classification algorithms, such as maximum likelihood, decision trees, support vector machine (SVM), random forest, and deep neural networks, have been widely used for land use and cover classification. Their performance has been validated in previous studies [10–12]. However, these algorithms require labeling all land cover types, which can be a tremendous drain on resources when only one specific land cover type is of interest [13]. Therefore, there is a growing need to develop one-class classifiers that can extract specific land cover types using only feature data of the target of interest. Several one-class classifiers have been proposed, including one-class SVM, isolation forest, and naive Bayes classifier [14–16]. In addition to labeled samples, unlabeled samples can provide useful information for constructing classifiers. Positive and unlabeled learning (PUL), a special one-class classification approach, has been increasingly improved upon and has demonstrated improved land cover classification accuracy in recent years [17]. Previous studies have mainly applied PUL and one-class classification to identify features such as urban buildings, large land targets, and rivers [18–20]. A UAV-based study found that the PUL approach was more appropriate for accurate *Pedicularis* extraction, suggesting its potential as a promising approach for single-species extraction [21]. Owing to the sporadic distribution pattern of *Pedicularis*, the reflectance characteristics of the pixel bands in remote sensing images with lower resolutions showed reduced spectral purity, thereby posing challenges to accurate identification endeavors. Compared with other measurements, both PlanetScope and Sentinel-2, as multi-spectral instruments, exhibit superior spatial resolution and better temporal revisit capabilities. This attribute renders them a judicious choice for Pedicularis identification.

Remote sensing image change monitoring is a technique that quantitatively analyzes and determines the process and characteristics of feature changes based on remote sensing images of the same area over different periods [22]. It is widely used in disaster assessment, urban development, and land use/cover [23]. Common methods for change monitoring include the image difference method, the image ratio method, the principal component transform method, the vegetation index method, and post-classification comparison [24]. To improve the temporal transfer of the algorithm, we selected unlabeled samples from multiple periods. The change-detection-based sample transfer approach is efficient, simple, and robust and has the potential to be used in large-scale ground cover classification [25–27]. This study takes advantage of PUL's ability to conserve negative sample information, reducing the statistical distribution differences between images of the target area by identifying areas of invariance between multiple temporal images [28], and combining the change-detection and post-classification approaches to solve the image classification problem of the target area [29]. A post-classification comparison is the most direct method of change monitoring. The advantage of this method is that it avoids the image sequence consistency conditions required for the direct comparison method, as well as avoids image radiation correction and matching problems. However, this method requires the development of uniform classification criteria enforced via image classification [30,31]. This method is extremely dependent on the accuracy of the classification algorithm, but this method performs well when analyzing the variation in a single species [32]. Therefore, this research uses a post-classification change detection approach for the analysis [33].

The aim of this study was to employ remote sensing techniques to obtain the temporal and spatial distribution of *Pedicularis* in Swan Lake and the buffer zone of the Bayinbuluke Grassland. Remote sensing data were collected in August for three consecutive years (2019–2021) to facilitate a comparison of the changes over time. Geospatial information holds valuable and significant data useful for the strategic planning of eradication efforts targeting *Pedicularis* infestations. Our objectives are as follows:

(1)   Assessing the accuracy of PUL on predictions of the poisonous species *Pedicularis*;
(2)   Comparing the efficacy of Sentinel-2 and PlanetScope satellite imagery in the identification of *Pedicularis*;
(3)   Generating precise distribution maps of *Pedicularis* with time-series data for subsequent spatiotemporal analysis to support the conservation of the Bayinbuluke Grassland ecosystem through time-series remote sensing dynamic monitoring.

This paper is structured as follows. Section 2 describes the materials and methods, including the dataset used for the study, the principles of the PUL method, and the scheme for extending PUL for land cover classification and change detection. Section 3 presents the experimental results, including a comparison of the two types of data sources. Section 4 discusses the issue of change detection and the obtained results. Finally, Section 5 draws some conclusions.

2. Materials and Methods

All calculations and analyses of the research were performed in Python (v3.6.12) and the geographic information system ArcGIS (v10.6, ESRI). The experiment was run on Windows 10 on a machine with two 36-core Intel Xeon 3.10 GHz processors and 128 GB RAM. The technology workflow of the study has been organized in the following sections (Figure 1).

2.1. Study Area

The Bayinbuluke Grassland ($42°18'$~$43°34'$N, $82°27'$~$86°17'$E), in the southern hinterland of the central part of Mountain Tianshan, is located in the northwest region of Hejing county in Xinjiang Province [34]. Furthermore, it is the second largest grassland in China and the most extensive subalpine, alpine meadow grassland in the desert region of China, with a total area of 3523.94 km^2 [35,36], and the main vegetation types are alpine grassland and swampy alpine meadows [37]. The Bayinbuluke Grassland is part of the Kaidu River basin, where water comes mainly from alpine snow melting and the recharge of natural precipitation. Its altitude is 2400~4400 m; the average annual temperature is $-4.7\,°C$, with an extreme high of 28.3 °C and an extreme low of $-48.1\,°C$; the annual precipitation is 216.8~361.8 mm; it experiences about 150~180 d of snow; and the annual dry grass period is seven months. Due to the cold weather, the soil in this area is frozen and there are almost no large trees. There are six main land use types in this grassland: high-cover pasture, low-cover pasture, marshland, timberland, water, and wild land. The water areas include rivers and lakes in the grassland and mountain snowfields [34]. The study area is shown in Figure 2.

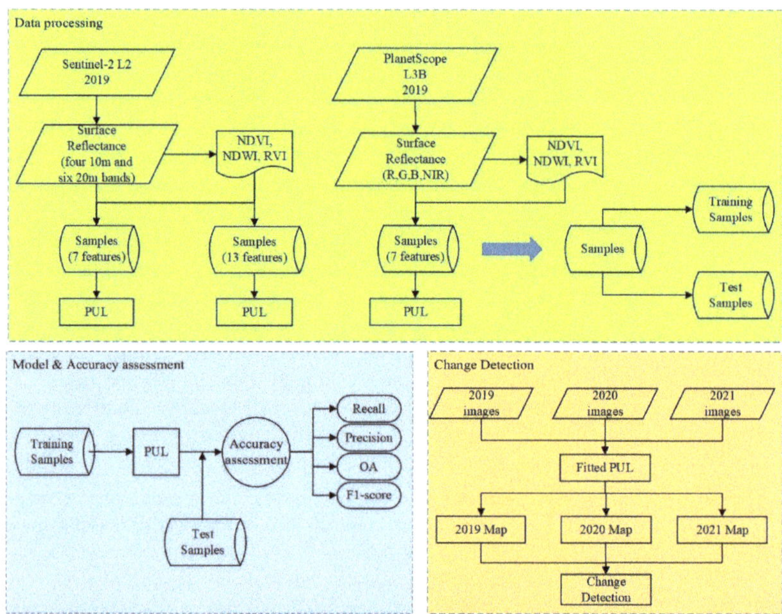

Figure 1. The technology workflow of the study (Blue arrows refer to the components of Samples).

2.2. Data Sources

2.2.1. UAV RGB Imagery

Selecting samples through a visual interpretation of remote sensing images with a low spatial resolution (relative to UAV images) presents a challenge due to the varying degrees of sparsity in the distribution of *Pedicularis* across different regions. Therefore, this study selected *Pedicularis* samples on satellite images via visual interpretation with the aid of UAV RGB (380~760 nm) imagery, which was obtained using a SONY RX1RII. The UAV data selected for the study area were taken on 7 August (f), 8 August (c), and 9 August (d, e), 2019 (Figure 2). The UAV was a DJI M600 (DJI, Shenzhen, China), and the flight information was planned in DJI GS Pro. The flight height of the UAV was 230 m, and the forward and side overlaps were 80% and 70%, respectively. The spatial resolution of all images acquired was 3 cm. The areas photographed by the UAV and where it was located are shown in Figure 2.

2.2.2. Sentinel-2 Imagery

The Sentinel-2 L2-level product used in the study was a Sentinel-2 Multispectral Instrument (MSI) from the European Space Agency (ESA). Sentinel-2 images cover 13 spectral bands in the visible, near-infrared (NIR), and short-wave infrared (SWIR) wavelengths, with four bands at 10 m, six bands at 20 m, and three bands at 60 m spatial resolution. The characteristics of the bands are listed in Table 1. For Sentinel-2 data, four 10 m and six 20 m bands can be used for land cover/land use (LCLU) mapping and change detection. The Sentinel-2 L2-level product provides the surface reflectance of images [38]. Regarding the matching of imaging data, imaging quality, and imaging time, through the Google Earth Engine platform, this study screened remote sensing images from 1 August to 31 August 2019. In order to obtain a better surface reflectance product, the Sentinel-2 L2-level product needs to be filtered and pre-processed. This experiment selected images with less than 10% cloudiness and filled the filtered images via temporal interpolation. Next, the filtered images were mosaicked and clipped through the region of interest (ROI) and all bands were resampled to 10 m. The sensor configuration of Sentinel-2 is as follows [39].

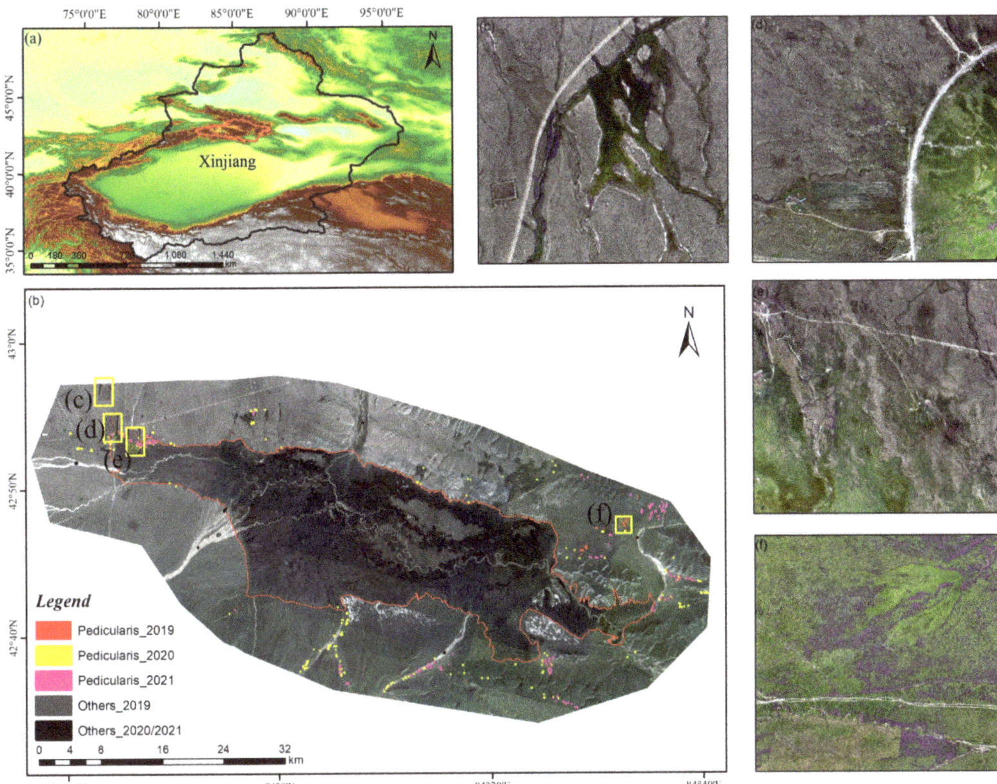

Figure 2. Location of the study area. (**a**) The study area is located in Hejing County, Xinjiang Province, China. (**b**) The Swan Lake (red boundary) and its buffer zone in the Bayinbuluke Grassland (background: an RGB remote sensing image acquired from PlanetScope in August 2019). (**c–f**) Ortho-mosaic image taken via an RGB UAV in August 2019 (the purple pixels are *Pedicularis*).

Table 1. Configuration information for the Sentinel-2 satellite.

Band Name	Sentinel-2A/Sentinel-2B Central Wavelength (nm)	Resolution (Meters)
Band 1—Coastal aerosol	443.9/442.2	60
Band 2—Blue	496.6/492.1	10
Band 3—Green	560.0/559.0	10
Band 4—Red	664.5/664.9	10
Band 5—Vegetation red edge	703.9/703.8	20
Band 6—Vegetation red edge	740.2/739.1	20
Band 7—Vegetation red edge	782.5/779.7	20
Band 8—NIR	835.1/832.9	10
Band 8A—Narrow NIR	864.8/864.0	20
Band 9—Water Vapor	945.0/943.2	60
Band 10—SWIR–Cirrus	1373.5/1376.9	60
Band 11—SWIR	1613.7/1610.4	20
Band 12—SWIR	2202.4/2185.7	20

The 60 m spatial resolution bands were not used in the experiment.

2.2.3. PlanetScope Imagery

The PlanetScope images with a 3 m spatial resolution were from the Planet Labs. It had four reflectance bands—blue, green, red, and near-infrared—with near-daily global coverage. The PlanetScope images can be accessed for free by researchers through a research and education license (https://developers.planet.com/ (accessed on 1 May 2022)). This study used a Level-3B surface reflectance product, which was geometrically corrected, radiometrically corrected, and atmospherically corrected. We also processed it with ENVI (v5.3, ESRI) for mosaic, registration, and reflectance calculations (divided by 10,000). The daily acquisition of PlanetScope data allowed us to select images with optimal imaging quality for our analysis. For our experiment, we selected PlanetScope images from 7 August and 12 August 2019. Similarly, for the year 2020, we selected images acquired on 5 August, 24 August, and 30 August. For the year 2021, we selected images acquired on 11 August, 18 August, and 23 August. The configuration information for the PlanetScope satellite is as Table 2 [40].

Table 2. Configuration information for the PlanetScope satellite.

Band Name	Spatial Resolution (m)	Spectral Wavelength (nm)
Blue		464–517
Green		547–585
Red	3.0	650–682
NIR		846–888

2.3. Datasets and Data Analysis

2.3.1. Generation of Additional Features

Identifying *Pedicularis* on satellite images is a challenging operation. It is difficult to identify *Pedicularis* accurately through object-oriented methods. Therefore, it is necessary to use pixel-based classification to extract *Pedicularis*, and feature engineering is a particularly useful method. Some work in the literature has indicated that spectral features and combined vegetation indices are more essential than textural features and principal components in target identification and classification [41,42].

In order to better extract *Pedicularis* from the land cover, we compared the spectral characteristics of *Pedicularis* and familiar grasses. These spectral curves were obtained from the results in a previous study [43]. The normalized difference vegetation index (NDVI), the normalized ratio vegetation index (RVI), and the difference water index (NDWI) were obtained using Equations (1)–(3), respectively. The NDVI is one of the most influential parameters for characterizing changes in vegetation greenness, and it is often employed in studies of land cover [44]. RVI is widely used to estimate and monitor the biomass of green plants [45]. *Pedicularis* is a water-loving plant, often growing next to rivers [4,46]. Therefore, this experiment calculated the NDWI to extract *Pedicularis* [47]. Numerous studies have verified that the three indices are efficient for land use/land cover and target extraction [44,45,48–51].

$$NDVI = \frac{\rho_{nir} - \rho_{red}}{\rho_{nir} + \rho_{red}}, \tag{1}$$

$$RVI = \frac{\rho_{nir}}{\rho_{red}}, \tag{2}$$

$$NDWI = \frac{\rho_{green} - \rho_{nir}}{\rho_{green} + \rho_{nir}}. \tag{3}$$

2.3.2. Construction of the Datasets

This study was based on UAV RGB images from the PlanetScope and Sentinel-2 satellites, with visual interpretation using ROIs plotted on the UAV RGB images to select the *Pedicularis* samples. These data were obtained in August for each of the years 2019~2021, and the details are presented in Section 2.2. Subsequently, we selected a sample of non-*Pedicularis* species for 2019 to 2021, ensuring that the chosen locations exhibited a consistent distribution of non-*Pedicularis* species in 2020 and 2021. Finally, the samples used as input data for the construction of the model were of three types: PlanetScope data (7 features),

Sentinel-2 data (7 features), and Sentinel-2 data (13 features). The sample information is shown in Table 3. The location of the sample selection is shown in Figure 2b.

Table 3. Ground sample information for the study regions.

		Training Samples (Pixels)		Test Samples (Pixels)	
		Pedicularis	Others	*Pedicularis*	Others
2019	Sentinel-2 data (7/13 features)	7690	7690	1923	5598
	PlanetScope data (7 features)	62,063	62,063	15,515	455,615
2020	Sentinel-2 data (7/13 features)	1943	6477	486	1900
	PlanetScope data (7 features)	15,568	51,893	15,515	15,434
2021	Sentinel-2 data (7/13 features)	2395	7983	599	1900
	PlanetScope data (7 features)	19,520	65,066	4880	15,434

Others: unlabeled samples. Seven features: four 10 m bands and three vegetation indices. Thirteen features: four 10 m, six 20 m bands, and three vegetation indices.

The constructed models were applied to three different datasets for change detection in multi-period remotely sensed images. The classification results were evaluated and analyzed separately, and the samples were partitioned into training and test sets at an 8:2 ratio. To ensure balance within the training set, we performed downsampling such that the proportion of positive class samples to unlabeled samples was 3:7 in 2020/2021. Since the sample size in 2019 was sufficient, 1:1 was chosen for modelling to ensure the model's generalizability. The details of the selected samples are shown in Table 3.

2.4. Methodology

2.4.1. Classification Method

This study used the PUL method, which was demonstrated to be feasible for identifying *Pedicularis* in another study [21]. This method is based on the bootstrap aggregation (bagging) technique: the algorithm iteratively trains many binary classifiers to distinguish known positive (P) examples from random subsamples of the unlabeled (U) dataset and to average their predictions. PUL is a semi-supervised learning algorithm based on a positive and unlabeled sample [52]. This study chose decision trees as the base classifiers and described PUL's classification results in detail. Similar to all machine learning approaches, the PUL methodology encounters challenges in addressing the intricacies of spatiotemporal migration within the recognition process. However, this algorithm is a few-shot learning approach, which is well-suited for addressing the challenge of identifying *Pedicularis* in the presence of limited samples, particularly within a vast spatial extent.

The steps are as follows: (1) determine a set of reliable negation (RN) examples, which has a small number of positive samples, from U and transform the problem into a binary classification problem; (2) train binary classifiers based on P and RN by iteratively applying existing classification algorithms; and (3) iterate over the previous two steps, with the number of bootstrap samples T also being a user-defined parameter. Finally, the probability of each unlabeled sample being judged as a positive sample is calculated.

In the PUL classifier, positive and negative samples are selected from the training dataset to train the model. In an example conducted in 2019, Sentinel-2 had 7690 positive and 7690 negative samples. To improve the training efficiency of PUL, we trained 'n' decision trees to fit the training dataset. Firstly, we selected 10% of the positive class samples as positive samples (y = 1); secondly, the remaining 90% of positive class samples and all negative class samples were labeled as unlabeled samples (y = 0); then, the same proportion of unlabeled samples was randomly selected as negative samples (y = −1) for training; and, finally, 'k' iterations were performed in this manner. The probability of each sample being positive was calculated to obtain the classification result of this classifier. The parameter 'k' was adjusted from 100 to 2000 with a step size of 100. The 'n' ranged from 100 to 1000 with a step size of 100. PlanetScope datasets were modelled using the same methodology and range of parameters.

2.4.2. Accuracy Assessment

Using PUL, we obtained a confusion matrix and calculated the evaluation metrics on the test dataset, including recall, precision, overall accuracy (OA), F1-score, and AUC (area under the ROC curve), which can be calculated using the ROC curve (receiver operating characteristic curve). These metrics were utilized to assess the model's performance. Precision represents predictions for a positive class in the truly labeled dataset and can be obtained using Equation (4). Recall represents the evaluation of samples predicted to be in the positive class and can be obtained using Equation (5). Overall accuracy (OA) is the sum of the true positives plus true negatives divided by the total number of tested individuals, as shown in Equation (7). F1-score is a metric that combines the strengths of precision and recall, is well suited for evaluating models in situations where there is an imbalance between categories, and can be calculated using Equation (8). AUC can be viewed as the probability of randomly selecting a pair of positive and negative samples from a sample, which is the area under the ROC curve drawn using the FPR and TPR (Equations (5) and (6)). The AUC is less sensitive to class imbalance than OA and reflects the model's performance under sample imbalance more accurately [53].

$$precision = \frac{TP}{TP + FP},\tag{4}$$

$$TPR = recall = \frac{TP}{TP + FN},\tag{5}$$

$$FPR = \frac{FP}{TN + FP}.\tag{6}$$

$$OA = \frac{TP + TN}{TP + TN + FP + FN},\tag{7}$$

$$F1 - score = \frac{2 \times precision \times recall}{precision + recall}\tag{8}$$

TP, FP, FN, and TN are the classifications true positive, false positive, false negative, and true negative, respectively.

2.4.3. Change Detection

This study took the distribution area and proportion, and the change rate of *Pedicularis* from the classification results for 2019–2021 [54]. To further reflect the direction, amount, and rate of *Pedicularis* change, this study generated a dynamic transfer matrix from the change detection results in 2019–2021, which can reflect the flow of species change [55]. Based on these results, we analyzed the change in the dynamics of *Pedicularis*. The change rate of the area of *Pedicularis* was calculated using Equation (8).

$$C_i = \frac{W_{bi} - W_{ai}}{W_{ai}} \times \frac{1}{t} \times 100\% .\tag{9}$$

where C_i is the change rate of *Pedicularis* during the study period; W_{ai} and W_{bi} are the distribution areas of *Pedicularis* at the beginning and end of the study period, respectively; t is the study period, measured by year; and the calculation results indicate the annual change rate of *Pedicularis* [56].

A transfer matrix model can depict a species' evolutionary patterns, capturing both species distribution changes and their migration directions. Using the classification outcomes of the study area in 2019, 2020, and 2021, we derived the transfer areas for each stage and constructed a transfer matrix.

$$S_{ij} = \begin{bmatrix} S_{11} & \cdots & S_{1n} \\ \vdots & \ddots & \vdots \\ S_{n1} & \cdots & S_{nn} \end{bmatrix}. \tag{10}$$

where S is the species area; i and j are the species types at the beginning and end of the study, respectively; and n is the number of species types.

3. Results

3.1. Comparison of Classification Accuracy on Sentinel-2 and PlanetScope

Three classifiers were developed for the positive and unlabeled learning (PUL) method using three different types of datasets. Based on the results presented in Table 4, the highest extraction accuracy for *Pedicularis* was observed when using Sentinel-2 data (13 features), yielding an F1-score of 0.9405. On the other hand, the lowest classification accuracy was obtained using PlanetScope data, resulting in an F1-score of 0.7049. Notably, the classification accuracy for *Pedicularis* using Sentinel-2 data (7 features) was significantly higher compared with that obtained using PlanetScope data and slightly lower than the accuracy observed for Sentinel-2 data (13 features).

Table 4. Assessment metrics of models during the 2019–2021 period.

Year	Datasets	Types	Recall	Precision	Accuracy	F1-Score
	Sentinel-2 data (7 features)	*Pedicularis*	0.9212	0.9286	0.9617	0.9248
		Others	0.9757	0.9730		
2019	Sentinel-2 data (13 features)	*Pedicularis*	0.9278	0.9536	0.9700	0.9405
		Others	0.9845	0.9754		
	PlanetScope data (7 features)	*Pedicularis*	0.8458	0.6042	0.8678	0.7049
		Others	0.8728	0.9610		
	Sentinel-2 data (7 features)	*Pedicularis*	0.8861	0.7307	0.9169	0.8009
		Others	0.9241	0.9721		
2020	Sentinel-2 data (13 features)	*Pedicularis*	0.8710	0.9045	0.9583	0.8874
		Others	0.9786	0.9702		
	PlanetScope data (7 features)	*Pedicularis*	0.8340	0.6132	0.8708	0.7067
		Others	0.8793	0.9584		
	Sentinel-2 data (7 features)	*Pedicularis*	0.8971	0.8629	0.9454	0.8796
		Others	0.9591	0.9703		
2021	Sentinel-2 data (13 features)	*Pedicularis*	0.8864	0.9277	0.9593	0.9065
		Others	0.9802	0.9678		
	PlanetScope data (7 features)	*Pedicularis*	0.8985	0.7399	0.9313	0.8115
		Others	0.9377	0.9791		

Our findings indicate that the highest classification accuracy for *Pedicularis* was observed in 2019. While recall and precision were generally comparable across most models, recall was found to be higher than precision for the models built using PlanetScope data. Moreover, our results demonstrate that the difference between accuracy and F1-score is minimal when the model's accuracy is high but becomes more pronounced when the accuracy is low.

To evaluate the temporal transferability of the model, we tested the model developed in 2019 on data from 2020 and 2021. As illustrated in Figure 3, the model's performance on this test was unsatisfactory, with the F1-scores for both years being less than 0.62. Specifically, the F1-score for the PlanetScope dataset in 2020 was insufficient, with a value of only 0.3300. Subsequently, upon training the model with the corresponding year's training set, a notable improvement in accuracy was observed. These results demonstrate the potential of positive and unlabeled learning (PUL) for small-sample learning.

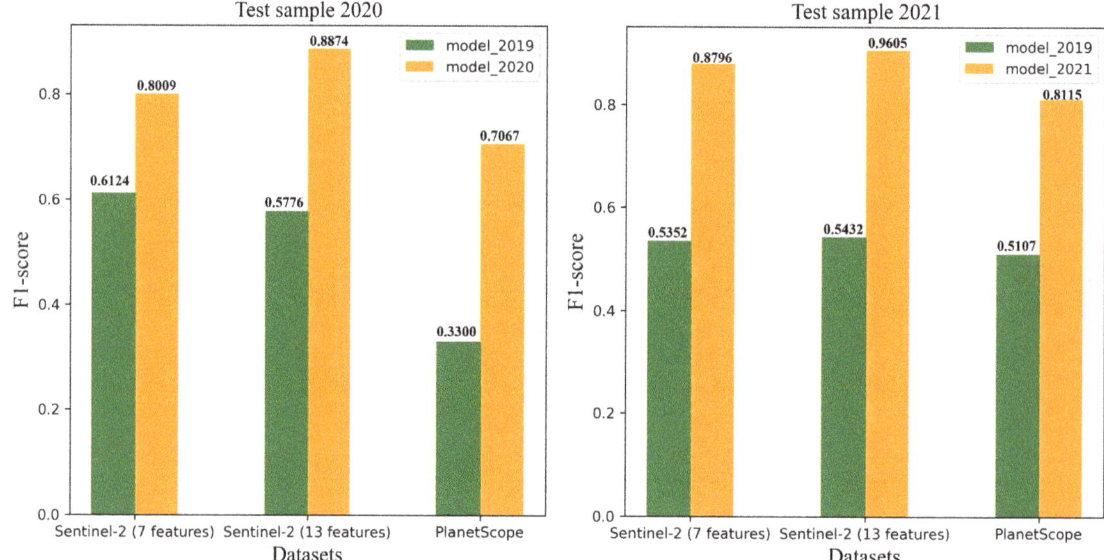

Figure 3. Comparison of accuracy between the 2019 model and model built in the same year as the test set.

As evidenced by the ROC and AUC curves depicted in Figure 4, changes in precision and recall exhibit an inverse relationship while the AUC values remain stable. In order to maximize the AUC, we could appropriately adjust the threshold to manipulate the precision and recall, ultimately achieving a balance between the two for accurate classification of *Pedicularis*. However, in scenarios with a large F1-score and accuracy gap, the AUC of the model remains low compared with that of other models.

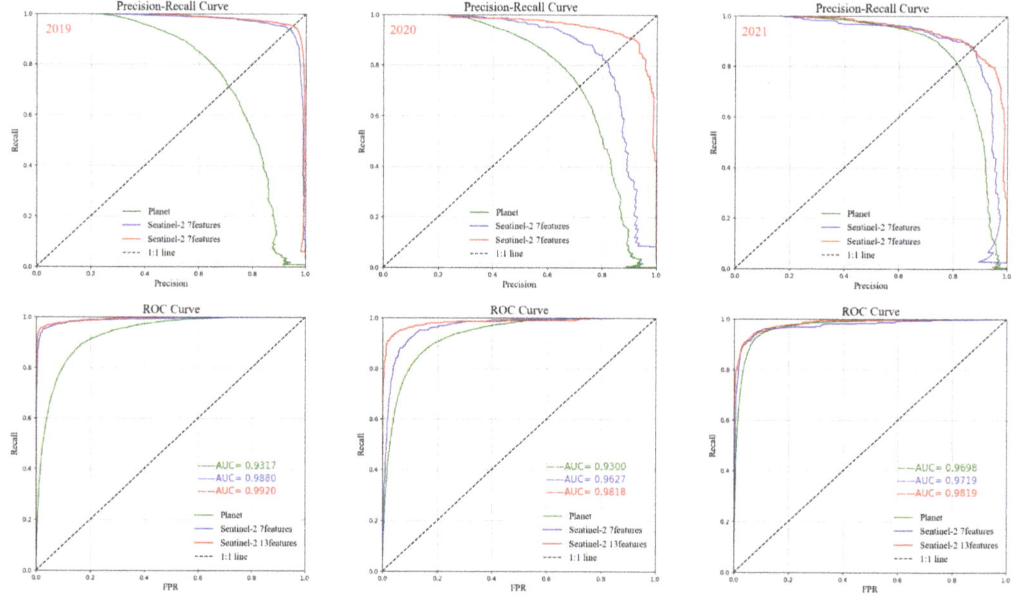

Figure 4. ROC and AUC curves for the models developed from 2019 to 2021.

3.2. Changes in Dynamics of Pedicularis

The classification results obtained using Sentinel-2 data (13 features) were the most accurate among all models. Accordingly, changes in the distribution area of *Pedicularis* from 2019 to 2021 were monitored based on this classification result. The analysis revealed that *Pedicularis* had the largest area, 195.7803 km^2, in 2019, accounting for 5.55% of the total region, while its size decreased to 3.54% in 2020.

Figure 5 illustrates the spatiotemporal distribution of *Pedicularis* from 2019 to 2021 using Sentinel-2 data (13 features). The results showed a gradual decrease in its distribution in the northwest and an increase in the south over the past three years. Additionally, the total distribution area of *Pedicularis* showed a decreasing and then increasing trend over this period.

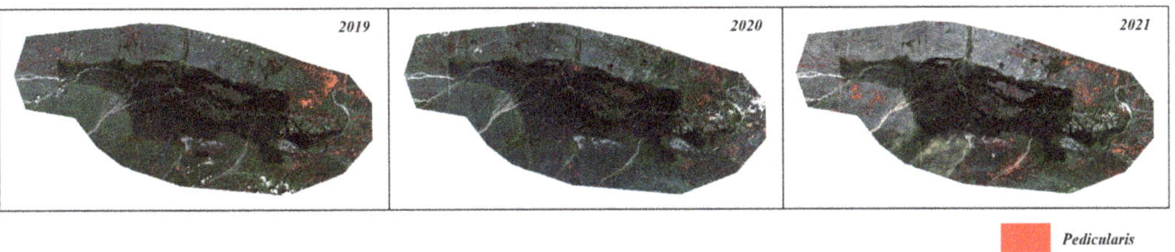

Figure 5. Spatiotemporal pattern of *Pedicularis* in 2019–2021 using Sentinel-2 data (13 feature).

The change in the distribution area of *Pedicularis* in the Bayinbuluke Grassland was analyzed from 2019 to 2021. Comparing the changes in *Pedicularis* in 2019 and 2020, it was observed that 173.89 km^2 of the *Pedicularis* distribution area disappeared, while 103.0704 km^2 was converted from distribution areas of other species to that of *Pedicularis*, indicating an overall improvement. From 2020 to 2021, there was a decrease of 106.0108 km^2 in *Pedicularis*, followed by an increase of 138.2587 km^2. However, the total amount of *Pedicularis* from 2019 to 2021 still decreased, with a decrease in distribution area of 38.5740 km^2.

4. Discussion

4.1. Influencing Factors of Classification Accuracy

We can identify *Pedicularis* clearly from the high-resolution drone images. However, the low spatial resolution of the images and the interplay of spectral information from different land covers produce mixed pixels. These phenomena result in the color and geometric textural information of *Pedicularis* not being easily acquired on satellite images. The PUL method used in this study can effectively improve the classification accuracy of *Pedicularis* based on satellite imagery; relevant studies on UAV imagery can support this method [21].

During the mosaicking of data, an uneven color balance across multiple images can have an impact on the outcome of image classification [57]. Sentinel-2 only requires two images to be sufficient to cover the study area; PlanetScope has a higher temporal resolution and a smaller coverage area per image. It requires 41 images to cover the study area, making the classifier more likely to learn some noise.

The accuracy of *Pedicularis* extraction was significantly enhanced by increasing the number of features from 7 to 13 in the Sentinel-2 classification results. This finding aligns with other pixel-based classification methods [58]. Moreover, the classification results of the model constructed using Sentinel-2 data (13 features) provide a more realistic distribution of *Pedicularis*.

Comparing the identification results of PlanetScope data (seven features) and Sentinel-2 data (seven features), the identification result for the 10 m resolution is higher than that for the 3 m resolution, which is unexpected. Our analysis shows that this is mainly caused by the significant difference in spectral reflectance between the images due to PlanetScope requiring more images [59,60]. Additionally, the band characteristics of the pixels are contingent on the complexity of the species within each pixel: the more intricate the object within the pixel, the greater the divergence between its spectral features and the pure pixel features, resulting in the reduced recognition accuracy of scattered patterns. Conversely, during sample selection, we noticed that Pedicularis was easier to identify in Sentinel-2 images compared to PlanetScope images, based solely on color considerations.

It has been found that the spectral signature of the same object using the same sensor may vary at different times, and this phenomenon also affects the classification results. Consequently, identifying features resistant to temporal and spectral variations is crucial for pixel-based classification methods, and can also help to improve the reliability of monitoring changes after classification.

4.2. Spatiotemporal Pattern of Pedicularis

Figure 5, Tables 5 and 6 collectively elucidate the dynamic trajectory of the spatial distribution of *Pedicularis* and characterize its temporal pattern, marked by an initial decrease, followed by a subsequent increase from 2019 to 2021. While the distribution of *Pedicularis* sharply decreased in the northwestern region between 2019 and 2020, other regions showed no significant migration trend. Furthermore, in 2021, the distribution area of *Pedicularis* in the southern region showed an increasing trend. As depicted in Figure 5, the distribution of *Pedicularis* was predominantly along rivers, suggesting that *Pedicularis* is a water-loving plant, which is consistent with local investigation findings [43]. Following a comprehensive survey, it has been observed that the precipitation levels in the region have exhibited a decline from 2019 to 2020. This observation explains the decrease also seen in the distribution of Artemisia marcescens [61]. This case also contributes to the investigation into the driving mechanisms underlying *Pedicularis* distribution. By utilizing GIS methods to analyze the spatial distribution of *Pedicularis*, a further understanding of its invasion routes and drivers can be attained.

Table 5. Statistics of distribution area of *Pedicularis* in different periods using Sentinel-2 data (13 features).

Year	Area (km^2)	Area Ratio (%)
2019	195.7803	5.55%
2020	124.9584	3.54%
2021	157.2063	4.46%

Table 6. Transition matrix of land cover change for 2019~2021.

	2019–2020		2020–2021		2019–2021	
	Pedicularis	**Others**	*Pedicularis*	**Others**	*Pedicularis*	**Others**
Pedicularis	21.8880	173.8923	18.9476	106.0108	33.6437	162.1330
Others	103.0704	3225.0944	138.2587	3260.7280	123.5590	3204.6058

This article discusses the feasibility of a spatiotemporal analysis based on PlanetScope and Sentinel-2 imagery. Both satellites have a high temporal resolution and meet the needs of long-time-series monitoring. However, post-classification change monitoring can avoid influencing the classification results due to different spatiotemporal and data sources. The accuracy of the classification results dramatically affects the detection of changes in the distribution of *Pedicularis*. It further shows that investigating a classifier or feature factor that can resist spatial and temporal variation and developing a better transfer learning method are essential.

4.3. A Case of Pedicularis Eradication

We learned that the government conducted a local campaign for the removal of *Pedicularis* in July 2019. To verify the process of eliminating *Pedicularis*, we looked up remote sensing images from PlanetScope, which has a high spatial resolution, for July 2019.

These images correspond to the position in Figure 2f. As seen from Figure 6, *Pedicularis* was physically removed by local people on 17 July, 23 July, and 29 July 2019. It is, therefore, essential to identify the location of *Pedicularis* and its dynamics. This contributes to the local government's efforts to control the invasion of *Pedicularis* and to protect the ecological environment [1,7].

Figure 6. Local variations in *Pedicularis* in PlanetScope images from July 2019 within the red box. (**a–c**) refer to PlanetScope images taken on 17 July, 23 July, and 29 July 2019. (the purple pixels are *Pedicularis*).

5. Conclusions

The main contribution of this work is the use of a new PUL method to extract *Pedicularis*. This extracted *Pedicularis* distribution is subsequently employed for the dynamic detection of Pedicularis and for conducting a spatiotemporal analysis of the prediction results. With the use of the spatial distribution of *Pedicularis* from 2019 to 2021, we can achieve real-time monitoring and the effective eradication of *Pedicularis*. In contrast to the one-class classifier, which only uses positive-class samples, PUL makes full use of numerous unlabeled samples to improve the accuracy of the classification results, contributing to curbing the expansion of the distribution area of poisonous *Pedicularis*. The conclusions are as follows:

(1) The proliferation of *Pedicularis* in the Bayinbuluke Grassland has resulted in significant ecological damage, necessitating the substantial expenditure of resources and efforts by the provincial government for rehabilitation efforts. In addressing this issue, change-detection methods utilizing remote sensing technology offer a practical approach for informed management and mitigation. Sentinel-2 images have the advantages of a large width, easily acquirable data, and high accuracy in extracting *Pedicularis*. The resolution of PlanetScope is higher than that of Sentinel-2, which is more advantageous when removing *Pedicularis* from small areas. The results of the study show that the PUL method is able to achieve a high recognition accuracy across different images.

(2) Within the confines of the same sensor platform, the influence of feature count on improvements to the identification accuracy becomes obvious with an ample sample size, as evidenced by an increasing feature count coinciding with increased recognition accuracy. However, within an equivalent feature framework, the correlation between resolution elevation and accuracy enhancement does not invariably hold, implying that the resultant classification outcome is dependent on the inherent data quality obtained using the sensor apparatus.

(3) The post-classification comparison algorithm avoids spectral differences in remote sensing images, especially long-time-series images from different sensors. It enables the rapid monitoring of regional variations in the distribution of different land types. However, it is highly dependent on the stability of the model, and a transferred, high-accuracy classification model needs to be further developed. The distribution of *Pedicularis* is concentrated in the northwestern and southwestern parts of Bayinbuluke

Swan Lake. From 2019 to 2021, the distribution area of *Pedicularis* exhibited a fluctuating trend, initially increasing and then subsequently decreasing, with the 2021 area measuring 157.2063 km^2. Despite better eradication efforts in the northeast region, the distribution area of *Pedicularis* did not exhibit significant changes, indicating that grassland managers may not have done enough to control the growth of *Pedicularis*.

Author Contributions: Methodology: W.W. and J.T.; validation: W.W., J.T. and N.Z.; formal analysis: W.W.; investigation: N.Z., Y.W. and J.T.; writing—original draft preparation: W.W.; writing—review and editing: W.W., J.T., X.X. and A.Z.; N.Z. contributed the same as the corresponding authors. All authors have read and agreed to the published version of the manuscript.

Funding: This research was funded by the Strategic Priority Research Program of the Chinese Academy of Sciences (XDA20050103) and the National Key Research and Development Program of China (No. 2020YFC1807102).

Data Availability Statement: Not applicable.

Conflicts of Interest: The authors declare no conflict of interest.

References

1. Fu, P.; Yao, J.; Hu, J.; Guo, X. Capital Endowments, Policy Perceptions and Herdsmen's Willingness to Reduce Livestock:A Case Study from the World Natural Heritage Site of Bayinbuluke. *Acta Agrestia Sin.* **2021**, *29*, 780–787.
2. Hameed, A.; Zafar, M.; Ahmad, M.; Sultana, S.; Bahadur, S.; Anjum, F.; Shuaib, M.; Taj, S.; Irm, M.; Altaf, M.A. Chemo-taxonomic and biological potential of highly therapeutic plant Pedicularis groenlandica Retz. using multiple microscopic techniques. *Microsc. Res. Tech.* **2021**, *84*, 2890–2905. [CrossRef] [PubMed]
3. Elkind, K.; Sankey, T.T.; Munson, S.M.; Aslan, C.E. Invasive buffelgrass detection using high-resolution satellite and UAV imagery on Google Earth Engine. *Remote Sens. Ecol. Conserv.* **2019**, *5*, 318–331. [CrossRef]
4. Sui, X.; Li, A.; Guan, K. Impacts of climatic changes as well as seed germination characteristics on the population expansion of Pedicularis verticillata. *Ecol. Environ. Sci.* **2013**, *22*, 1099–1104.
5. Yanyan, L.I.U.; Yukun, H.U.; Jianmei, Y.U.; Kaihui, L.I.; Guogang, G.A.O.; Xin, W. Study on Harmfulness of Pedicularis myriophylla and Its Control Measures. *Arid Zone Res.* **2008**, *25*, 778–782.
6. Hongtao, J.I.A.; Pingan, J.; Luming, C.; Chengyi, Z.; Yukun, H.U. Estimation of Organic Carbon Storage of Bayinbuluke Alpine Grassland Ecosystem. *Xinjiang Agric. Sci.* **2006**, *43*, 480–483.
7. Pingan, J.; Hui, L.I.; Hongtao, J.I.A.; Zihong, D. Impacts of fencing on soil animals diversity beneath mountainous lawn vegetation in Bayinbuluke. *J. Northwest Sci-Tech Univ. Agric. For.* **2007**, *35*, 69–74.
8. Choudhary, K.; Boori, M.S.; Kupriyanov, A. Landscape Analysis through Remote Sensing and GIS Techniques: A Case Study of Astrakhan, Russia. In Proceedings of the 8th International Conference on Graphic and Image Processing (ICGIP), Tokyo, Japan, 29–31 October 2016.
9. Gao, S.; Zheng, J.; Ma, T.; Wu, J.; Nasongcaoketu; Maidi, K. Research on the Applicability of Remote Sensing Monitoring of Inedible Grass Pedicularis sp. by GF-1 WFV Satellite in Bayanbulak Grassland. *Xinjiang Agric. Sci.* **2017**, *54*, 1949–1956.
10. Huber, N.; Ginzler, C.; Pazur, R.; Descombes, P.; Baltensweiler, A.; Ecker, K.; Meier, E.; Price, B. Countrywide classification of permanent grassland habitats at high spatial resolution. *Remote Sens. Ecol. Conserv.* **2023**, *9*, 133–151. [CrossRef]
11. Lopatin, J.; Dolos, K.; Kattenborn, T.; Fassnacht, F.E. How canopy shadow affects invasive plant species classification in high spatial resolution remote sensing. *Remote Sens. Ecol. Conserv.* **2019**, *5*, 302–317. [CrossRef]
12. Kakembo, V.; Smith, J.; Kerley, G. A Temporal Analysis of Elephant-Induced Thicket Degradation in Addo Elephant National Park, Eastern Cape, South Africa. *Rangel. Ecol. Manag.* **2015**, *68*, 461–469. [CrossRef]
13. Singh, P.S.; Singh, V.P.; Pandey, M.K.; Karthikeyan, S.; IEEE. One-class Classifier Ensemble based Enhanced Semisupervised Classification of Hyperspectral Remote Sensing Images. In Proceedings of the 2nd IEEE International Conference on Emerging Smart Computing and Informatics (ESCI), All India Shri Shivaji Memorial Soc, Inst Informat Technol, Pune, India, 12–14 March 2020; pp. 22–27.
14. Hossain, M.A.; Jia, X.; Benediktsson, J.A. One-Class Oriented Feature Selection and Classification of Heterogeneous Remote Sensing Images. *IEEE J. Sel. Top. Appl. Earth Obs. Remote Sens.* **2016**, *9*, 1606–1612. [CrossRef]
15. Li, W.; Guo, Q.; Elkan, C. A Positive and Unlabeled Learning Algorithm for One-Class Classification of Remote-Sensing Data. *IEEE Trans. Geosci. Remote Sens.* **2011**, *49*, 717–725. [CrossRef]
16. Mack, B.; Roscher, R.; Waske, B. Can I Trust My One-Class Classification? *Remote Sens.* **2014**, *6*, 8779–8802. [CrossRef]
17. Li, W.K.; Guo, Q.H. A maximum entropy approach to one-class classification of remote sensing imagery. *Int. J. Remote Sens.* **2010**, *31*, 2227–2235. [CrossRef]
18. Braun, A.C. Evaluation of One-Class Svm for Pixel-Based and Segment-Based Classification in Remote Sensing. In Proceedings of the ISPRS-Technical-Commission III Symposium on Photogrammetric Computer Vision and Image Analysis (PCV), Saint Mande, France, 1–3 September 2010; pp. 160–165.

19. Chang, S.; Du, B.; Zhang, L. A Subspace Selection-Based Discriminative Forest Method for Hyperspectral Anomaly Detection. *IEEE Trans. Geosci. Remote Sens.* **2020**, *58*, 4033–4046. [CrossRef]
20. Dambros, C.S.; Morais, J.W.; Azevedo, R.A.; Gotelli, N.J. Isolation by distance, not rivers, control the distribution of termite species in the Amazonian rain forest. *Ecography* **2017**, *40*, 1242–1250. [CrossRef]
21. Wang, W.; Tang, J.; Zhang, N.; Xu, X.; Zhang, A.; Wang, Y. Automated Detection Method to Extract Pedicularis Based on UAV Images. *Drones* **2022**, *6*, 399. [CrossRef]
22. Morshed, N.; Yorke, C.; Zhang, Q. Urban Expansion Pattern and Land Use Dynamics in Dhaka, 1989–2014. *Prof. Geogr.* **2017**, *69*, 396–411. [CrossRef]
23. Franklin, S.E.; Ahmed, O.S.; Wulder, M.A.; White, J.C.; Hermosilla, T.; Coops, N.C. Large Area Mapping of Annual Land Cover Dynamics Using Multitemporal Change Detection and Classification of Landsat Time Series Data. *Can. J. Remote Sens.* **2015**, *41*, 293–314. [CrossRef]
24. Angulo, D.; Angulo, F.; Olivar, G. Dynamics and Forecast in a Simple Model of Sustainable Development for Rural Populations. *Bull. Math. Biol.* **2015**, *77*, 368–389. [CrossRef] [PubMed]
25. Wu, T.J.; Luo, J.C.; Zhou, Y.N.; Wang, C.P.; Xi, J.B.; Fang, J.W. Geo-Object-Based Land Cover Map Update for High-Spatial-Resolution Remote Sensing Images via Change Detection and Label Transfer. *Remote Sens.* **2020**, *12*, 174. [CrossRef]
26. Yu, W.J.; Zhou, W.Q.; Qian, Y.G.; Yan, J.L. A new approach for land cover classification and change analysis: Integrating backdating and an object-based method. *Remote Sens. Environ.* **2016**, *177*, 37–47. [CrossRef]
27. Chen, X.H.; Chen, J.; Shi, Y.S.; Yamaguchi, Y. An automated approach for updating land cover maps based on integrated change detection and classification methods. *ISPRS-J. Photogramm. Remote Sens.* **2012**, *71*, 86–95. [CrossRef]
28. Qian, Y.G.; Zhou, W.Q.; Yu, W.J.; Han, L.J.; Li, W.F.; Zhao, W.H. Integrating Backdating and Transfer Learning in an Object-Based Framework for High Resolution Image Classification and Change Analysis. *Remote Sens.* **2020**, *12*, 4094. [CrossRef]
29. Xu, Y.D.; Yu, L.; Zhao, F.R.; Cai, X.L.; Zhao, J.Y.; Lu, H.; Gong, P. Tracking annual cropland changes from 1984 to 2016 using time-series Landsat images with a change-detection and post-classification approach: Experiments from three sites in Africa. *Remote Sens. Environ.* **2018**, *218*, 13–31. [CrossRef]
30. Feng, X.; Li, P.; Cheng, T. Detection of Urban Built-Up Area Change From Sentinel-2 Images Using Multiband Temporal Texture and One-Class Random Forest. *IEEE J. Sel. Top. Appl. Earth Obs. Remote Sens.* **2021**, *14*, 6974–6986. [CrossRef]
31. Kempeneers, P.; Sedano, F.; Strobl, P.; McInerney, D.O.; San-Miguel-Ayanz, J. Increasing Robustness of Postclassification Change Detection Using Time Series of Land Cover Maps. *IEEE Trans. Geosci. Remote Sens.* **2012**, *50*, 3327–3339. [CrossRef]
32. Crowson, M.; Hagensieker, R.; Waske, B. Mapping land cover change in northern Brazil with limited training data. *Int. J. Appl. Earth Obs. Geoinf.* **2019**, *78*, 202–214. [CrossRef]
33. Colditz, R.R.; Acosta-Velazquez, J.; Diaz Gallegos, J.R.; Vazquez Lule, A.D.; Teresa Rodriguez-Zuniga, M.; Maeda, P.; Cruz Lopez, M.I.; Ressl, R. Potential effects in multi-resolution post-classification change detection. *Int. J. Remote Sens.* **2012**, *33*, 6426–6445. [CrossRef]
34. Bao, A.; Cao, X.; Chen, X.; Xia, Y. Study on Models for Monitoring of Aboveground Biomass about Bayinbuluke grassland Assisted by Remote Sensing. In Proceedings of the Conference on Remote Sensing and Modeling of Ecosystems for Sustainability, San Diego, CA, USA, 13 August 2008.
35. Chen, X.; Yang, Z.; Wang, T.; Han, F. Landscape Ecological Risk and Ecological Security Pattern Construction in World Natural Heritage Sites: A Case Study of Bayinbuluke, Xinjiang, China. *ISPRS Int. J. Geo-Inf.* **2022**, *11*, 328. [CrossRef]
36. Xu, X.; Wang, X.; Zhu, X.; Jia, H.; Han, D. Landscape Pattern Changes in AlpineWetland of Bayanbulak Swan Lake during 1996–2015. *J. Nat. Resour.* **2018**, *33*, 1897–1911.
37. Liu, Q.; Yang, Z.P.; Han, F.; Shi, H.; Wang, Z.; Chen, X.D. Ecological Environment Assessment in World Natural Heritage Site Based on Remote-Sensing Data. A Case Study from the Bayinbuluke. *Sustainability* **2019**, *11*, 6385. [CrossRef]
38. Grabska, E.; Frantz, D.; Ostapowicz, K. Evaluation of machine learning algorithms for forest stand species mapping using Sentinel-2 imagery and environmental data in the Polish Carpathians. *Remote Sens. Environ.* **2020**, *251*, 112103. [CrossRef]
39. De Vroey, M.; de Vendictis, L.; Zavagli, M.; Bontemps, S.; Heymans, D.; Radoux, J.; Koetz, B.; Defourny, P. Mowing detection using Sentinel-1 and Sentinel-2 time series for large scale grassland monitoring. *Remote Sens. Environ.* **2022**, *280*, 113145. [CrossRef]
40. Kimm, H.; Guan, K.Y.; Jiang, C.Y.; Peng, B.; Gentry, L.F.; Wilkin, S.C.; Wang, S.B.; Cai, Y.P.; Bernacchi, C.J.; Peng, J.; et al. Deriving high-spatiotemporal-resolution leaf area index for agroecosystems in the US Corn Belt using Planet Labs CubeSat and STAIR fusion data. *Remote Sens. Environ.* **2020**, *239*, 111615. [CrossRef]
41. Holloway-Brown, J.; Helmstedt, K.J.; Mengersen, K.L. Interpolating missing land cover data using stochastic spatial random forests for improved change detection. *Remote Sens. Ecol. Conserv.* **2021**, *7*, 649–665. [CrossRef]
42. Saber, A.; El-Sayed, I.; Rabah, M.; Selim, M. Evaluating change detection techniques using remote sensing data: Case study New Administrative Capital Egypt. *Egypt. J. Remote Sens. Space Sci.* **2021**, *24*, 635–648. [CrossRef]
43. Gao, S.; Lin, J.; Ma, T.; Wu, J.; Zheng, J. Extraction and Analysis of Hyperspectral Data and Characteristics fromPedicularis on Bayanbulak Grassland in Xinjiang. *Remote Sens. Technol. Appl.* **2018**, *33*, 908–914.
44. Carlson, T.N.; Ripley, D.A. On the relation between NDVI, fractional vegetation cover, and leaf area index. *Remote Sens. Environ.* **1997**, *62*, 241–252. [CrossRef]
45. Gnyp, M.L.; Miao, Y.X.; Yuan, F.; Ustin, S.L.; Yu, K.; Yao, Y.K.; Huang, S.Y.; Bareth, G. Hyperspectral canopy sensing of paddy rice aboveground biomass at different growth stages. *Field Crops Res.* **2014**, *155*, 42–55. [CrossRef]

46. Wang, D.; Cui, B.C.; Duan, S.S.; Chen, J.J.; Fan, H.; Lu, B.B.; Zheng, J.H. Moving north in China: The habitat of Pedicularis kansuensis in the context of climate change. *Sci. Total Environ.* **2019**, *697*, 133979. [CrossRef] [PubMed]
47. Chai, L.; Jiang, H.; Crow, W.T.; Liu, S.; Zhao, S.; Liu, J.; Yang, S. Estimating Corn Canopy Water Content From Normalized Difference Water Index (NDWI): An Optimized NDWI-Based Scheme and Its Feasibility for Retrieving Corn VWC. *IEEE Trans. Geosci. Remote Sens.* **2021**, *59*, 8168–8181. [CrossRef]
48. Francini, S.; McRoberts, R.E.; Giannetti, F.; Mencucci, M.; Marchetti, M.; Mugnozza, G.S.; Chirici, G. Near-real time forest change detection using PlanetScope imagery. *Eur. J. Remote Sens.* **2020**, *53*, 233–244. [CrossRef]
49. Sui, Y.; Shao, F.; Wang, C.; Sun, R.; Ji, J. Complex network modeling of spectral remotely sensed imagery: A case study of massive green algae blooms detection based on MODIS data. *Phys. A-Stat. Mech. Its Appl.* **2016**, *464*, 138–148. [CrossRef]
50. Zhao, B.; Yang, F.; Zhang, R.; Shen, J.; Pilz, J.; Zhang, D. Application of unsupervised learning of finite mixture models in ASTER VNIR data-driven land use classification. *J. Spat. Sci.* **2021**, *66*, 89–112. [CrossRef]
51. Zhou, X.-X.; Li, Y.-Y.; Luo, Y.-K.; Sun, Y.-W.; Su, Y.-J.; Tan, C.-W.; Liu, Y.-J. Research on remote sensing classification of fruit trees based on Sentinel-2 multi-temporal imageries. *Sci. Rep.* **2022**, *12*, 11549. [CrossRef]
52. Mordelet, F.; Vert, J.P. A bagging SVM to learn from positive and unlabeled examples. *Pattern Recognit. Lett.* **2014**, *37*, 201–209. [CrossRef]
53. Halligan, S.; Altman, D.G.; Mallett, S. Disadvantages of using the area under the receiver operating characteristic curve to assess imaging tests: A discussion and proposal for an alternative approach. *Eur. Radiol.* **2015**, *25*, 932–939. [CrossRef] [PubMed]
54. Wang, L.; Zheng, S.; Wang, X. The Spatiotemporal Changes and the Impacts of Climate Factors on Grassland in the Northern Songnen Plain (China). *Sustainability* **2021**, *13*, 6568. [CrossRef]
55. Zhang, Z.; Yang, X.; Xie, F. Macro analysis of spatiotemporal variations in ecosystems from 1996 to 2016 in Xishuangbanna in Southwest China. *Environ. Sci. Pollut. Res.* **2021**, *28*, 40192–40202. [CrossRef] [PubMed]
56. Cao, Y.; Kong, L.; Ouyang, Z. Characteristics and Driving Mechanism of Regional Ecosystem Assets Change in the Process of Rapid Urbanization—A Case Study of the Beijing–Tianjin–Hebei Urban Agglomeration. *Remote Sens.* **2022**, *14*, 5747. [CrossRef]
57. Fan, C.; Chen, X.; Zhong, L.; Zhou, M.; Shi, Y.; Duan, Y. Improved Wallis Dodging Algorithm for Large-Scale Super-Resolution Reconstruction Remote Sensing Images. *Sensors* **2017**, *17*, 623. [CrossRef] [PubMed]
58. Zhao, L.; Li, Q.; Zhang, Y.; Wang, H.; Du, X. Normalized NDVI valley area index (NNVAI)-based framework for quantitative and timely monitoring of winter wheat frost damage on the Huang-Huai-Hai Plain, China. *Agric. Ecosyst. Environ.* **2020**, *292*, 106793. [CrossRef]
59. Zhao, Y.; Liu, D. A robust and adaptive spatial-spectral fusion model for PlanetScope and Sentinel-2 imagery. *Gisci. Remote Sens.* **2022**, *59*, 520–546. [CrossRef]
60. Ye, N.; Morgenroth, J.; Xu, C.; Chen, N. Indigenous forest classification in New Zealand-A comparison of classifiers and sensors. *Int. J. Appl. Earth Obs. Geoinf.* **2021**, *102*, 102395. [CrossRef]
61. Wang, Q.; Zhai, P.-M.; Qin, D.-H. New perspectives on 'warming–wetting' trend in Xinjiang, China. *Adv. Clim. Chang. Res.* **2020**, *11*, 252–260. [CrossRef]

MDPI
St. Alban-Anlage 66
4052 Basel
Switzerland
www.mdpi.com

Remote Sensing Editorial Office
E-mail: remotesensing@mdpi.com
www.mdpi.com/journal/remotesensing